PLANT METABOLITES UNDER ENVIRONMENTAL STRESS

Mechanisms, Responses, and Adaptation Strategies

PLANT METABOLITES UNDER ENVIRONMENTAL STRESS

Mechanisms, Responses, and Adaptation Strategies

Edited by

Nivas M. Desai, PhD, FAPSR

Manasi Patil, PhD

Umesh R. Pawar, PhD

APPLE ACADEMIC PRESS

First edition published 2023

Apple Academic Press Inc.
1265 Goldenrod Circle, NE,
Palm Bay, FL 32905 USA

760 Laurentian Drive, Unit 19,
Burlington, ON L7N 0A4, CANADA

CRC Press
6000 Broken Sound Parkway NW,
Suite 300, Boca Raton, FL 33487-2742 USA

4 Park Square, Milton Park,
Abingdon, Oxon, OX14 4RN UK

Library and Archives Canada Cataloguing in Publication

Title: Plant metabolites under environmental stress : mechanisms, responses, and adaptation strategies / edited by Nivas M. Desai, PhD, Manasi Patil, PhD, Umesh R. Pawar, PhD.
Names: Desai, Nivas M., editor. | Patil, Manasi, editor. | Pawar, Umesh R., editor.
Description: First edition. | Includes bibliographical references and index.
Identifiers: Canadiana (print) 20220422605 | Canadiana (ebook) 20220422613 | ISBN 9781774910948 (hardcover) | ISBN 9781774910955 (softcover) | ISBN 9781003304869 (ebook)
Subjects: LCSH: Plant metabolites. | LCSH: Plants—Metabolism. | LCSH: Plants—Effect of stress on.
Classification: LCC QK881 .P56 2023 | DDC 572/.42—dc23

Library of Congress Cataloging-in-Publication Data

Names: Desai, Nivas M., editor. | Patil, Manasi, editor. | Pawar, Umesh R., editor.

Title: Plant metabolites under environmental stress : mechanisms, responses, and adaptation strategies / edited by Nivas M. Desai, Manasi Patil, Umesh R. Pawar.

Description: First edition. | Palm Bay, FL, USA : Apple Academic Press, 2023. | Includes bibliographical references and index. | Summary: "Plant Metabolites under Environmental Stress: Mechanisms, Responses, and Adaptation Strategies provides a timely update on the knowledge of plant responses to various environmental stresses such as salinity, waterlogging, drought, pollution, heat, temperature, oxidative stress, and mineral deficiencies. Chapters focus on physiological and biochemical mechanisms identified in plants that are crucial for them to adapt to biotic and abiotic stress and the methods for improving plant tolerance mechanisms. The book also throws light on plant secondary metabolites such as phenolic compounds and plant growth regulators in ameliorating the stressful conditions in plants. The scientific knowledge and expertise presented in this volume will be valuable to agronomists, plant physiologists, horticulturists, research scholars of botany and agriculture science, academicians of plant sciences as well students in these areas"-- Provided by publisher.

Identifiers: LCCN 2022042195 (print) | LCCN 2022042196 (ebook) | ISBN 9781774910948 (hardback) | ISBN 9781774910955 (paperback) | ISBN 9781003304869 (ebook)

Subjects: LCSH: Plant metabolites. | Plants--Metabolism. | Plants--Effect of stress on.

Classification: LCC QK881 .P5353 2023 (print) | LCC QK881 (ebook) | DDC 572/.42--dc23/eng/20220916

LC record available at https://lccn.loc.gov/2022042195

LC ebook record available at https://lccn.loc.gov/2022042196

ISBN: 978-1-77491-094-8 (hbk)
ISBN: 978-1-77491-095-5 (pbk)
ISBN: 978-1-00330-486-9 (ebk)

About the Editors

Nivas M. Desai, PhD, FAPSR
Department of Botany and Plant Protection,
Sadguru Gadage Maharaj College, Karad,
Satara District, Maharashtra, India

Nivas M. Desai, PhD, is a Fellow of the Association of Plant Science Research (FAPSR). With over a decade in the field of plant science research, Dr. Desai has a unique identity that shines through his more than 32 international and national publications. His research areas are marine botany, plant physiology, functional foods, and photochemistry. He is a recipient of a DST Fastrack Young Scientist Fellowship and has successfully completed a project on developing biofuel from marine cyanobacteria. He holds two Australian and two Indian patents. He has also attended more than 30 international and national conferences. He is also the recipient of several Young Scientist Awards and one Research Excellence Award. He is the author of many book chapters and has edited six books.

Manasi Patil, PhD
Plant Physiologist and Assistant Professor,
Department of Botany and Plant Protection,
Sadguru Gadage Maharaj College, Karad,
Satara District, Maharashtra, India

Manasi Patil, PhD, is a Plant Physiologist and an Assistant Professor at Sadguru Gadage Maharaj College, Karad, India. She has received a PhD from the Department of Botany, Shivaji University Kolhapur, India. Her research interests are stress physiology, oil seed plants, and the application of plant growth regulators. She has many research publications to her credit and has attended several conferences. She holds one Australian and two Indian patents. She has also received major and minor projects from the University Grants Commission and Shivaji University Kolhapur, India.

Umesh R. Pawar, PhD
Assistant Professor, PG Department of Botany, SPK College Sawantwadi, Maharashtra, India

Umesh R. Pawar, PhD, is a hardcore botanist and an Assistant Professor in the PG Department of Botany, SPK College Sawantwadi, India. His research area is in the field of molecular characterization of mangroves. He has received his PhD from Annamalai University, Tamil Nadu, India. He has several publications in his credits and attended many conferences and has one international and three national patents to his credit.

Contents

Contributors ... *ix*

Abbreviations ... *xi*

Foreword ... *xv*

Preface ... *xvii*

Introduction ... *xix*

1. **Endophytic Fungi: A Tool for Managing Abiotic Stress in Plants ... 1**
Shakuntala Ghorai

2. **Comparative Study of Physiological and Biochemical Response of *Tribulus terrestris* L. and *Pedalium murex* L. to Salt Stress ... 25**
Varsha Sanjay Khude

3. **Adaptation Strategies of *Rhododendron* Under Environmental Stress ... 45**
Shreyasi Biswas and Mahuya Mukhopadhyay

4. **Waterlogging Stress: ROS Generation and Its Effects on Plant ... 61**
Swarupa Agnihotri and Vijaya Nikam

5. **Remediation of Organic Textile Dyes and Effluent by Freshwater Filamentous Cyanobacteria: A Review ... 73**
Madhulika Gupta

6. **Dust Pollution and Plants: A Case Study of Mumbai, Maharashtra, India ... 85**
Nitesh Joshi

7. **Air Pollution Induced Stress in Avenue Trees: A Case Study of Mysore City, Karnataka, India ... 117**
C. Hemavathi and J. Shobha

8. **Plant Responses to Drought Stress: Morphological, Physiological, Molecular Approaches, and Drought Resistance ... 149**
Pratap Vyankatrao Naikwade

9. Devastating Effect of Amphan Cyclone on the Biodiversity of the Sundarbans Mangroves and the Need of Its Rehabilitation 185
Madhumita Majumder

10. Phenolic Compounds Under Stress 203
Babita Rana and Kavita Chahal

11. Stress and Its Influence on the Generation of Reactive Oxygen Species and Oxidative Damage in Plants 219
J. Naveen, G. Hithamani, and H. G. Pushpalatha

12. Phenolic Compounds: Stress Markers of Plants and Their Protective Role.. 249
G. Hithamani, J. Naveen, and H. G. Pushpalatha

13. Nitrogen and Organic Acid Metabolism in *Portulaca oleracea* L. Under Drought Stress 277
Vishal V. Naik, Amol V. Patil, and Baburao A. Karadge

Index 293

Contributors

Swarupa Agnihotri
Department of Botany, VIVA College, Virar (W), University of Mumbai, Maharashtra, India,
E-mail: Swarupaagnihotri13@gmail.com

Shreyasi Biswas
Department of Microbiology, Lady Brabourne College, Kolkata–700017, West Bengal, India

Kavita Chahal
Department of Botany, Government College, Bichhua, Chhindwara, Madhya Pradesh, India

Shakuntala Ghorai
Assistant Professor, Department of Microbiology, Raidighi College, Raidighi, West Bengal–743383,
India, E-mail: shakuntalaghorai@raidighicollege.org.in

Madhulika Gupta
Assistant Professor, Department of Microbiology, Lady Brabourne College, P1/2 Suhrawardy Avenue,
Kolkata–700017, West Bengal, India, E-mail: madhulikajhum@gmail.com

C. Hemavathi
Government First Grade College for Women, Vijayanagar, Mysore–570017, Karnataka, India

G. Hithamani
Department of Biochemistry, CSIR–Central Food Technological Research Institute, Mysore–570020,
Karnataka, India

Nitesh Joshi
Rizvi College of Arts Science and Commerce, Department of Botany, Mumbai, Maharashtra, India,
E-mail: niteshcjoshi@gmail.com

Baburao A. Karadge
Department of Botany, Shivaji University, Kolhapur, Maharashtra, India

Varsha Sanjay Khude
Associate Professor, Botany Department, Devchand College, Arjunnagar, Maharashtra, India,
E-mail: drvarshakhude81@gmail.com

Madhumita Majumder
Department of Botany, Raidighi College, Raidighi, P.O.+P.S.–Raidighi, South 24 Parganas,
West Bengal–743383, India, E-mails: madhumita2363@rediffmail.com;
madhumitamajumder1981@gmail.com

Mahuya Mukhopadhyay
Department of Microbiology, Lady Brabourne College, Kolkata–700017, West Bengal, India,
E-mails: mahuya.mukhopadhyay@ladybrabourne.com; moumahuya1@yahoo.com;
hornbillgorumara@gmail.com

Vishal V. Naik
Department of Botany, Bharati Vidyapeeth's M. B. S. K. K. M., Kadegaon, Sangli, Maharashtra, India,
E-mail: purslanevvn@gmail.com

Pratap Vyankatrao Naikwade
Department of Botany, ASP College, Devrukh, Maharashtra, India,
E-mail: naikwade.pratap@gmail.com

J. Naveen
Department of Biochemistry, CSIR–Central Food Technological Research Institute, Mysore–570020, Karnataka, India

Vijaya Nikam
Department of Botany and Plant Protection, S.G. M. College, Karad, Maharashtra, India

Amol V. Patil
Department of Botany, Dr. Bapuji Salunkhe College, Miraj, Maharashtra, India

H. G. Pushpalatha
Assistant Professor, Maharani's Science College for Women, Mysore–570001, Karnataka, India,
E-mail: pushpa_2001@rediffmail.com

Babita Rana
Department of Botany, G. N. Khalsa College, Matunga–400019, Mumbai, Maharashtra, India,
E-mail: babitarana2009@yahoo.com

J. Shobha
Department of Studies in Botany, University of Mysore, Manasagangotri, Mysore–570006, Karnataka, India

Abbreviations

$^1Chl^*$	singlet chlorophyll
$^1O_2$	singlet oxygen
$^3Chl^*$	triplet chlorophyll
AA	ascorbic acid
AAQMS	ambient air quality monitoring stations
ABA	abscisic acid
AF	Arizona fescue
AMV	alfalfa mosaic virus
AOX	alternative oxidase
APTI	air pollution tolerance index
APX	ascorbate peroxidase
AQI	air quality index
AREB	abscisic acid responsive element-binding proteins
ASC-GSH	ascorbate-glutathione
BYDV-PAV	barley yellow dwarf virus-PAV
Cl	chlorine
CLCuBuV	cotton leaf curl Burewala virus
$CO_3^{\bullet-}$	carbonate
CSSVD	cocoa swollen shoot virus disease
DEGs	differentially expressed genes
DMPD	N,N-dimethyl phenylenediamine
DREB	dehydration-responsive element-binding proteins
DRI	dust retention indices
ER	endoplasmic reticulum
ERM	ericoid mycorrhizal
ESTs	expressed sequence tags
ETC	electron transport chain
GAL	γ-lactone dehydrogenase
$H_2O^{\bullet}$	hydroperoxyl radical
H_2O_2	hydrogen peroxide
HOBr	hypobromous acid
HOI	hypoiodous acid
HR	hypersensitive reaction
HSPs	heat-shock proteins

IAA	indole-3-acetic acid
IMD	Indian Meteorological Departments
LEA	late embryogenesis abundant
LOX	lipoxygenase
LPO	lipid peroxidation
MCGM	municipal corporation of Greater Mumbai
MPCB	Maharashtra Pollution Control Board
mtETC	mitochondrial electron transport chain
Na	sodium
Ni	nickel
NiR	nitrite reductase
NR	nitrate reductase
O_2^-	superoxide
$O_2^{\bullet-}$	superoxide anion
OH	hydroxyl
$OH^{\bullet}$	hydroxyl radical
PAL	phenylalanine ammonia-lyase
PCD	programmed cell death
PEG	polyethylene glycol
PGPR	plant growth-promoting rhizobacteria
PM	particulate matter
P_N	NET photosynthetic rate
POD	peroxidase activities
POX	peroxidase
PSI	photosystem I
PUFA	polyunsaturated fatty acids
RBOHD	respiratory burst oxidase homolog D
RO*	excited carbonyl
$RO^{\bullet}$	alkoxy radical
ROI	reactive oxygen intermediates
$ROO^{\bullet}$	peroxy radical
ROOH	hydroperoxides
ROS	reactive oxygen species
RT-PCR	reverse transcription-polymerase chain reaction
SG	sleepy grass
SMCD	Saskatchewan microbial collection database
SO_3^-	sulfonic
SOD	superoxide dismutase
SPM	suspended particulate matter

$SQ^{\bullet -}$	semiquinone
TAN	titratable acid number
TBARS	thiobarbituric acid reactive substances
TDW	total dry weight
TFs	transcription factors
USEPA	United States Environmental Protection Agency
VOC	volatile organic compounds
XOD	xanthine oxidase

Foreword

Plants are the basic component of the Earth, essentially important for the survival and existence of human life as a source of food, fuel, medicine, and all other requirements. However, the plants are increasingly under environmental stress. As a result, plant productivity is being diminished with climate change, ultimately posing a potential threat of hunger and poverty. Hence, it is crucial to understand the mechanisms of plant responses to environmental stresses for effective conservation and sustainable utilization of plant resources.

Considering the significance and sensitivity of plants to environmental stresses, Dr. Nivas M. Desai has brought out the present book *Plant Metabolites under Environmental Stress: Mechanisms, Responses, and Adaptation Strategies.* The book deals with the adaptive responses of plants to drought, salinity, cyclone, flood, air, and industrial pollutants; the physiological and biochemical mechanisms in terms of phenolics, nitrogen, organic acids, and secondary metabolites to overcome the stresses; the role of endophytic fungi in managing abiotic stresses in plants; and alleviation of effluent pollution using cyanobacteria.

My hearty congratulations and best wishes are due to Dr. Nivas M. Desai for his great effort in providing scientific knowledge about plant metabolisms under various environmental stresses. This book would be of immense help to researchers, academicians, and students.

—Prof. Dr. K. Kathiresan, MSc, PhD, DSc, FNIE, FMSI, FNABS
Annamalai University
Centre of Advanced Study in Marine Biology,
Faculty of Marine Sciences, Parangipettai, Tamil Nadu, India
E-mails: kathiresan57@gmail.com; kathirsum@rediffmail.com

Preface

The global increase in human population at a faster rate is one of the big concerns in all the aspects of living, right from food security to better livelihood. The increase in the human population with such a faster rate is expected to increase up to 10 billion by the end of 2050. In addition to this, agriculture produce is not increasing at the expected rate that is needed to feed every individual. The burning issue of climate change is causing negative impacts, including environmental stresses. These stresses may be biotic or abiotic such as heat, cold, drought, waterlogging, salinity, temperature, heavy metal toxicity, nutrient imbalance, etc. To cope with increasing food requirements, improvement in crop productivity with minimum loss is a big concern for all nations.

The present book, Plant Metabolites under Environmental Stress: Mechanisms, Responses, and Adaptation Strategies, provides scientific knowledge and expertise of the mechanism of plant responses to various environmental stresses. Chapter 1 gives an overview of endophytic fungi and their role in managing abiotic stress in plants. Chapter 2 focuses on the physiological and biochemical response of *Tribulus terrestris* L. and *Pedalium murex* L. to salinity stress. Chapter 3 provides an overview of environmental adaptations and stress tolerance of plants. Chapter 4 gives an idea about waterlogging stress, ROS generation, and its effect on the plant. In Chapter 5, the author discusses the aspects of "Remediation of Organic Textile Dyes and Effluent by Freshwater Filamentous Cyanobacteria." Chapter 6 covers the dust pollution mechanism in plants. Chapter 7 deals with the "Air Pollution and Stress in Avenue Trees." Chapter 8 discusses plant responses to drought stress. Chapter 9 comprehensively deals with the devastated effect of the Amphan cyclone on the biodiversity of the Sundarbans mangroves. Chapter 10 presents the role of phenolic compounds during stress. Chapter 11 deals with stress and its influence on the generation of reactive oxygen species (ROS) and oxidative damage to plants. Chapter 12 deals with the role of phenolics and secondary metabolites in response to various abiotic stress factors. Chapter 13 covers nitrogen and organic acid metabolism in *Portulaca oleracea* L. under drought stress.

We express our thanks to all the contributors. We would like to thank Prof (Dr.) K. Kathiresan for writing the foreword. Finally, it is a profound pleasure to thank AAP for taking up the publication of this book. We hope that this book will provide current knowledge on environmental stress in plants and will lead to new discussions and efforts to deal with various abiotic and biotic stress factors.

Introduction

The plant response to various environmental stresses is an important phenomenon in plant life. The knowledge of such plant metabolism is crucial to understand their mechanisms as well as to develop some new crop varieties that can cope with abiotic and biotic stresses and can be suited for the surrounding environment where they grow better.

Stress in a plant refers to the external conditions that can adversely affect the growth and development of a plant, which ultimately decreases productivity. Stresses cause a variety of plant responses either by altering gene expressions, cellular metabolism, alterations in growth performance, crop yield, etc. Plant stress reflects some morphological changes due to sudden changes in environmental conditions. The stress-tolerant plant can acclimate to stress conditions in a time-time dependent manner. Plant stress is generally categorized into two classes: abiotic stress and biotic stress. Abiotic stress may be due to physical or chemical factors, whereas biotic stress caused due to exposure of the plant to biological agents like diseases and insect pests. The plant can be recovered if the stress is mild or if the plant is exposed for a shorter period, as the effect is temporary, but on the other hand, if the effect is long-term or severe, the inhibition of flowering, limitation in seed formation, and induced senescence may lead to the death of the crop plant. Such plants are considered stress-susceptible plants. But several plants in nature can escape the stressful conditions.

Biotic stress in plants can be caused by living organisms either by viruses, bacteria, fungi, nematodes, insects, weeds, etc., with the agents causing biotic stress to deprive the host of its nutrients that can lead to the death of plants. Due to pre- and post-harvest losses, biotic stress is considered a major problem. Many biotic stresses affect photosynthesis due to reduced leaf area. Abiotic stress such as water stress (drought), waterlogging stress (excess water), extreme temperature (heat, cold, freezing), salinity, and even mineral toxicity negatively impact the growth and development as well as seed quality of a plant. The plant can develop some stress tolerance mechanisms by altering some metabolic pathways through the production of secondary metabolites or osmolytes to survive in adverse conditions.

Plant Metabolites under Environmental Stress: Mechanisms, Responses, and Adaptation Strategies, provides readers a timely update on the knowledge of plant responses to various environmental stresses such as salinity, water-logging, drought, pollution, heat, temperature, oxidative stress, and mineral deficiencies. Chapters focus on physiological and biochemical mechanisms identified in plants that are crucial to adapt biotic and abiotic stress and the methods of improving plant tolerance mechanisms. The book also throws light on secondary plant metabolites such as phenolic compounds and plant growth regulators in ameliorating the stressful conditions in plants. The authors have provided detailed references for extended reading to support the information in the book.

CHAPTER 1

Endophytic Fungi: A Tool for Managing Abiotic Stress in Plants

SHAKUNTALA GHORAI

Assistant Professor, Department of Microbiology, Raidighi College, Raidighi, West Bengal–743383, India,
E-mail: shakuntalaghorai@raidighicollege.org.in

ABSTRACT

Fungal endophytes are residing inside the plant tissues since the early days of plant colonization in soil. These invaders are mostly harmless and come in handy in every stressful situation faced by the host in their lifetime. Biotic and abiotic stress alleviation by the fungal endophytes may be attributed to a wide array of metabolites produced by those insiders like alkaloids, terpenoids, quinones, steroids, furandiones, xanthones, isoprenoids, glycosides, polyketides, depsipeptides, isocoumarines, phenols, shikimates, and perylene derivatives, etc. Many of these metabolites are potent antiviral, immune-suppressive, antidiabetic, hypolipidemic, antiarthritic, and anti-inflammatory agents. Abiotic stresses of the host plants may be managed via induction of stress response in the invaders and then production of stress-relieving metabolites by these endophytes. Several studies indicated that mitigation of stress response in the plant is dependent on the relative abundance of the colonizing endophytes. This review discussed the role of various fungal endophytes in ameliorating daunting effects of several abiotic stress factors like draught, heat, salt and metal etc. in plants. This pragmatic association between plant and fungi may be exploited further through genomic, proteomic and metabolomic approaches to fine tune their applicability in sustainable agriculture in the long run.

Plant Metabolites under Environmental Stress: Mechanisms, Responses, and Adaptation Strategies.
Nivas M. Desai, Manasi Patil, Umesh R. Pawar (Eds.)

1.1 INTRODUCTION

Life on the Earth is dependent on the primary producer organisms. Productivity and health of the primary producers like plants are naturally prime concerns for the sustainability of our ecosystem. With simultaneous advancement in industrialization and expansion of the human population, arable lands are gradually depleting with time. In addition to those factors, the productivity of the crop plants is affected by several biotic and abiotic stresses. Irregular levels of available water, heat, salinity, and heavy metals are the abiotic stresses that are responsible for the loss of productivity of crop plants. It is estimated that the human population will reach around 9.7 billion by 2050 (www.ourworldindata.org), and by that time, we will be needing approximately 3 billion tons of cereals annually (www.fao.org). Several strategies are employed to improve crop production in a sustainable way, like improved technical strategies, breeding, precision farming, eco-adapted production, etc. (Schroder et al., 2019). Endophytic fungi can play a critical role in eco-adapted, and climate-smart crop production practices and technologies as they are natural inhabitants of the plant species, and they are already famous for their part in alleviating abiotic stress conditions as well as improving biomass and yield of plants. The present chapter elaborates on the multifaceted character those endophytes play in ameliorating the above-mentioned abiotic stress factors.

1.2 FUNGAL ENDOPHYTES

Plants harbor a wide array of microorganisms on their aerial surfaces, inside the internal tissues, and throughout the rhizosphere. They may be commensal, parasitic, or mutualistic in nature. Among those invaders, microfungi or endophytic fungi are the symptomless colonizers of apparently healthy plant tissues. Hirsch and Braun (1992) defined endophytic fungi as the group of fungi that colonize living plant tissue without causing any immediate, overt negative effects. Endophytic microfungi are ubiquitous as they are found in plants from tropical, temperate, and boreal forest; herbaceous plants from extreme artic, alpine, and xeric climate; from mesic temperate and tropical forests. They also colonize in mosses and hepatics, ferns, and fern allies, angiosperms, gymnosperms, estuarine plants, herbaceous annuals, deciduous, and evergreen perennials (Stone et al., 2004). Microorganisms invading internal tissues (endophytes) are of special interest owing to their complex interactions with the host plants and enabling the host in a consorted and synergistic way to alleviate stressful conditions faced by the host in their lifetime. Those stress situations may be attributed by biotic or abiotic sources. Abiotic sources include water scarcity or drought conditions, water excess availability or flood, heat, salt, heavy metals, etc. Diverse metabolites released from the endophytes have significant roles in plant growth stimulation, crop protection, and phytoremediation. Some of them demonstrated anti-proliferative activity against cancer cells; entomopathogenic activity against pests, bugs, aphids; and antibiotic activity (Sudha et al., 2016).

1.3 ABIOTIC STRESSES

1.3.1 DROUGHT

According to Farroq et al. (2009) draft affects the plant system in morphological, physiological, and molecular level. Morpho-anatomical changes associated to water unavailability are stomatal closure, reduced leaf development and leaf abscission, etc. Physiological changes include dehydrated leaf mesophyll cells, reduced rate of photosynthesis due to photosynthetic enzyme degradation, reduced flow of carbon-di-oxide, leaf transpiration, root, and shoot growth. In order to maintain the cell volume, cells accumulate different molecules like proline glycine, betaine, etc. (Fathi and Tari, 2016). Reactive oxygen species (ROS) produced in response to stress often lead to lipid peroxidation (LPO) and degradation of proteins, lipids, and nucleic acids. Germination and vegetative both stages are severely affected by this stress (Figure 1.1).

Although the extent of cooperation depends on the type of host and endosymbiont, fungal endophytes generally improve plant's total biomass, chlorophyll content and stomatal conductance. Colonization favors the host in many ways, including increase in shoot biomass, tiller density, plant height, Fv/Fm, photosynthetic activity, relative water content, ascorbate peroxidase (APX), polyphenol oxidase, superoxide dismutase (SOD), glutathione, and phenolic compounds under stress condition than well-watered condition (Dastogeer, 2018). When germination was tested in durum wheat cultivar AC Avonlea in the presence of six endophytic Ascomycota mitosporic fungal isolates deposited in the Saskatchewan microbial collection database (SMCD), namely SMCD 2204, SMCD 2206, SMCD 2208, SMCD 2210, SMCD 2214, and SMCD 2215, myco-symbionts showed enhanced stress tolerance against drought compared to the non-symbionts (Hubbard et al., 2012). Under drought stress, cucumber plants assimilated nutrients like magnesium, potassium, calcium, growth parameters and plant biomass, etc., were significantly increased when it was inoculated with *Penicillium sp.* (Waqas et al., 2012). Proportions of protein components like glutenins, gliadins, albumin-globulins, etc., varied significantly during grain filling in wheat varieties under drought stress (Balla et al., 2011). Moreover, polymeric protein fraction (unextractable) and glutenin to gliadin ration were also decreased which indicated the lowering of the grain quality as a whole. Under drought stress, a significant negative correlation was found between the starch granule size and relative protein content of all the five wheat varieties. It also influenced the moisture content and water absorption ability of the plants.

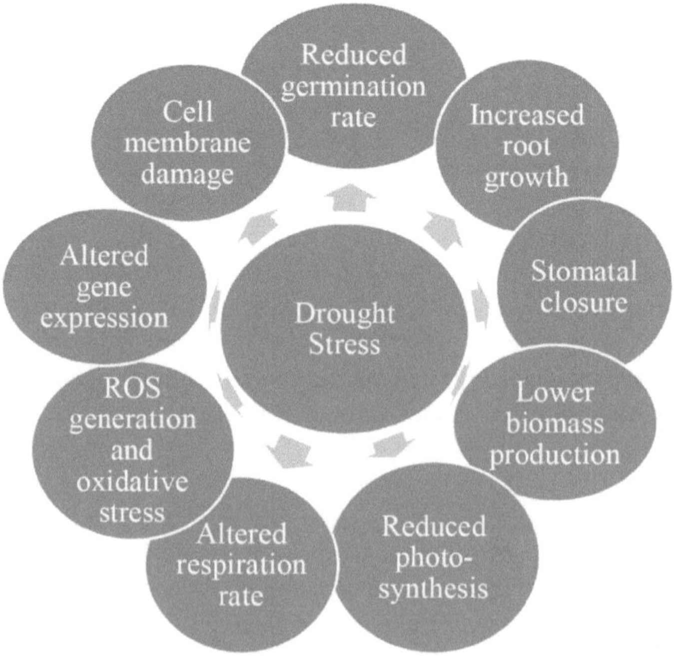

FIGURE 1.1 Drought stress responses in plants (Ghatak et al., 2017).

An increase in biomass was observed in drought-challenged rice when they were inoculated with fungal endophytes *Fusarium culmorum* and *Curvularia protuberata* (Redman et al., 2011). The presence of *Chaetomium globosum* and *Penicillium resedanum* were responsible for the increase in shoot length, biomass, and photosynthetic activity in *Capsicum annuum* when subjected to drought stress (Khan et al., 2012, 2014). Similar observations were noted when drought-affected tomato plants were inoculated with *Alternaria sp.* and *Trichoderma harzianium* (Azad and Kaminskij, 2016). In drought-challenged tall fescue, *Neotyphodium coenophialum* mediated upregulation of free glucose, fructose, trehalose, sugar alcohols, proline, and glutamic acid etc. was observed in shoots and roots (Nagabhyru et al., 2013). Under water stress, proline, soluble sugar, and catalase levels were improved in the presence of endophytic fungus *Chaetomium globosum* in wheat (Cong et al., 2015). Under similar stressed conditions, tomato plants showed less ROS production in the presence of endophytes compared to the control plants (Azad and Kaminskij, 2016).

Under severe water stress, *the Nicotiana benthamiana* plant showed significantly higher valine content in the presence of E-284 (unidentified ascomycetous endophytic fungus), whereas GABA and leucine accumulation were higher in the presence of E-162 (one endophyte resembling *Cladosporium cladosporioides*) (Dastogeera et al., 2017). They also regulated higher accumulation of organic acids like heptadecanoate and succinate. Arabinose, galactopyranose, fucose, and rhamnose were accumulated in higher amounts in endophytic plants at moderate stress conditions compared to the non-endophytic plants. Among different sugar alcohols accumulated, mannitol, and xylitol were of significantly higher amounts. Glutathione reductase activity was much higher in drought-stressed sleepy grass (SG) and Arizona fescue (AF) (perennial bunchgrasses native to elevations of >2,000 m in the southwestern USA) in the presence of *Neotyphodium* endophytes in comparison to non-endophytic grasses (Hamilton and Bauerle, 2012). Li et al. (2012) showed through the interaction between the grass *Achnatherum sibiricum* (L.) and its fungal endophyte *Neotyphodium* sp. that the benefits of endophyte infection like increase in aboveground biomass and total biomass was dependent on both nitrogen and phosphorous availability.

When durum wheat *Triticum turgidum* L. was inoculated with six different endophytic fungi from SMCD namely SMCD 2204 (Class *Dothideomycetes*), SMCD 2206, 2208, 2210, 2214, and 2215 (Class *Incertae sedis*), those endophytes improved the plant's adaptation to heat and drought (Hubbard et al., 2013). SMCD 2204, 2206, 2208, 2210, 2214, and 2215 increased the photochemical efficiency of photosystem II, Fv/Fm in heat- and/or drought-challenged wheat in pre-reproductive phase. In the case of mature plants, SMCD 2206, 2210, and 2215 improved average seed weight and grain yield or total seed weight under heat and drought stress. *T. harzianum* T22 colonized tomato seeds were germinated in water-deficient state and in the presence of methyl viologen (Mastouri et al., 2012). Those colonizing endophytes upregulated chloroplastic and cytosolic monodehydroascorbate reductase, dehydroascorbate reductase, glutathione reductase as well as cytosolic APX in cytosol of shoots of plants. This indicated the possible upregulation of ascorbate-glutathione (ASC-GSH) cycle.

Lactuca sativa colonized with fungal endophytes from the roots of *Colobanthus quitensis (Caryophyllaceae)* and *Deschampsia antarctica (Poaceae)* showed greater accumulation of proline and lower concentration of thiobarbituric acid reactive substances (TBARS) under 75% irrigation (Molina-Montenegro et al., 2016). Those Antarctic root endophytes (>75% of total morphotypes were *Penicillium chrysogenum* and *Penicillium brevicompactum*) significantly improved nitrogen, phosphorous, and calcium content of the plant.

Epichloe sinica contributed enhanced seed germination potential and rate in *Roegneria kamoji.* Those endophytes also alleviated PEG-6000 mediated drought stress by lowing concentration of ROS (Bu et al., 2019). Under water stress, *Penicillium minioluteum* colonized *Chenopodium quinoa* showed 40% improvement in root formation in comparison to non-colonized plants (Gonzales-Tuber et al., 2018). Root colonization of *Trichoderma atroviride* in *Zea mays L.* almost alleviated the effects of drought stress. They lowered LPO, hydrogen peroxide (H_2O_2) content while increasing antioxidant enzyme activity, pigment content, and photosynthetic activity of the maize plants (Guler et al, 2016). Fungal endophytes *Penicillium citrinum*, *Aurobassium pullunts*, and *Dothideomycetes* sp. inoculated in *Citrus reticulata* significantly lowered drought-induced responses like reduced growth, chlorophyll, carotenoid content, chlorophyll fluorescence (Fv/Fm), catalase, monodehydroascorbate reductase and dehydroascorbate reductase activity with increased levels of H_2O_2, superoxide, LPO, APX, peroxidase (POX), SOD, ascorbate, and glutathione, etc. (Sadeghi et al., 2019). *Talaromyces omanensis* significantly increased drought tolerance in tomato as evident from higher shoot dry weight, root length, flower numbers, and fruit weight (Halo et al., 2020). Colonization also resulted in improved reproductive characteristics, and chlorophyll fluorescence with increased phloem and cortex width. This may be attributed to a higher concentration of gibberellic acid in the colonized plants. *Piriformospora indica* mitigated drought and mechanical stress responses like greater leaf proline concentration, root diameter, catalase activity, and APX activity; lower leaf water potential, root volume, leaf area, chlorophyll content, root, and shoot fresh weight, etc., in wheat plants (Hosseini et al., 2017).

Colonization of endophytic fungus *Epichloë coenophiala* in tall fescue enabled the plants to be more energy efficient by spending less energy to take up water from nominal plant available water range (Hosseini et al., 2016). *Piriformospora indica* colonized maize root cells showed improved root fresh and dry weight, leaf area, SPDA values under drought conditions (Xu et al., 2017). Enhance proline and reduced MDA levels were noted in the endophyte inoculated plants. Antioxidant activity was also improved as evident from higher levels of SOD, POX, and catalase. Increased mRNA levels of ZmANAC072, ZmDREB2A, ZmCIPK3, ZmERD1, ZmCBL1, ZmHAT, and ZmPLDδ signified overexpression of drought-related genes in the colonized plants. Meta-analysis of RNA-Seq data of *Arabidopsis thaliana, Solanum lycopersicum, Zea mays, Vitis vinifera, Malus X domestica, Solanum tuberosum, Triticum aestivum* revealed down-regulation of callose synthase 3, cellulose synthase-like 1 and glucan synthase-like 10 involved in cell wall organization (Benny et al., 2019) during drought stress. Genes engaged in fatty acid biosynthesis like 3-ketoacyl CoA 1,3-ketoacyl CoA 2 and 3-ketoacyl CoA 6 were also down-regulated. BTB-POZ and MATH domain 4, sucrose non-fermenting 1-related protein kinase 2, calcineurin B-like protein 1, CBL interacting protein kinase 1 and bush and chlorotic dwarf genes responsive to osmotic stress were up-regulated. Water deprivation-related gene upregulation was mapped in case of homeobox 7, lipid transfer protein, sucrose non-fermenting 1-related protein kinase 2, calcineurin B-like protein 1 and short-chain dehydrogenase reductase 1. Homeobox 7, indole 3-butyric acid response 5, sucrose non-fermenting 1-related protein kinase 2 and calcineurin B-like protein 1 genes of abscisic acid (ABA) activated signaling pathway became up-regulated.

ABA-related homeobox 7, lipid transferrin protein 3, potassium transport 3, dual specificity protein phosphate 1E, shaker family potassium ion channel, sucrose non-fermenting 1-related protein kinase 2, CBL interacting protein kinase 1 and bush-and-chlorotic dwarf were also up-regulated. Up-regulation of pyrophosphorylase 6, salt overly sensitive 1, sucrose non-fermenting 1-related protein kinase 2, glyceraldehyde 3-phosphate dehydrogenase, calcineurin B-like protein 1 and CBL interacting protein kinase 1 were identified in response to salt stress. Genes related to hormone metabolism also showed altered expression in drought-stressed plants. Ethylene-related Feronia, pleiotropic drug resistance 12 and response regulator 2 were up-regulated and salicylic acid-related ATP-binding cassette G40, wall-associated kinase 2 and glutathione s-transferase 25 were down-regulated. Auxin-related cytochrome P-450, indole-3-butyric acid response 5, RUBI conjugating enzyme 1 and phytohormone-associated protein 2; cytokinin-related photosystem 1 subunit 4-2, heat shock protein

93-V and glycolate oxidase 1; ethylene activated signaling pathway-related ethylene response factor 1, S phase kinase-associated protein 1 and DREB subfamily A-4 were found to be up-regulated. Drought-induced upregulated expression was identified in case of transcription factors (TFs) alfin like 1, ureidoglycine aminohydrolase, WRKY 20, zinc ion binding, homeobox 7, homeobox 16, cycling DOF factor 2, set domain group 26, IAA-leucine resistance 3. More numbers of TFs were down-regulated, they included protodermal factor 2, MYC4, TMO5-like 1, TSL-Kinase interacting protein 1, methyltransferase 1, HSI2-like 1, NGA 1, tesmin CSC domain protein, FMN-linked oxidoreductases, DEMETER, SNF7, response regulator 2, calmodulin-binding activator 2 and histone 3.3. Six out of 9 studies indicated the results and genes were *Arabidopsis thaliana* orthologs of each gene in the analyzed plant species. Colonization of *Pirimiforma indica* in drought-stressed rice regulated growth and hyposensitivity response by miR159/miR396 mediated targeting of transcription factor MYB and GRF (Fard et al., 2017).

1.3.2 HEAT

Heat stress-induced formation of different ROS in the host plant. Desert plant-associated endophytic fungi act as a source of growth promoter, natural compounds and secondary metabolites, antibiotic agents, etc., to mediate stress (Ali et al., 2018). *Epichloe typhina* colonized tall fescue, a commercial cultivar 'Millennium,' showed lower electrolyte leakage under heat stress (Xu et al., 2017). Under drought stress, reduction in leaf relative water content was greater in case uninoculated plants. The presence of the endophyte helped in lowering the levels of superoxide radicals, H_2O_2 and MDA during drought and heat challenge. During drought condition, colonized plants showed higher activity of SOD and APX whereas catalase, guaiacol POX and APX activity increased during heat stress. *Aspergillus violaceofuscus* was inoculated to sunflower (*H. annuus*) and soybean (*G. max*) seedlings under heat stress (Ismail et al., 2020). Inoculated *H. annuus* accumulated 5% higher flavonoids, 15% phenolics, and 68% lower proline compared control plants during heat treatment. In the case of *G. max,* these values were 22.9%, 6% and 41%, respectively. Catalase and ascorbic acid (AA) oxidase enzyme levels in leaves were reduced in both the inoculated plants after heat treatment in comparison to control sets. The endophytes were responsible for a significant decrease in ABA levels in both the plants during heat treatment. ABA contents were 42 ng/g and 32 ng/g in colonized

G. max and *H. annuus* in comparison to 262 ng/g and 60 ng/g in uninoculated plants, respectively. Both the plantlets also have much higher protein, soluble sugar, and lipid contents at heat stress.

Aspergillus japonicus EuR-26 inoculated soyabean and sunflower showed distinct increase in levels of growth parameter, chlorophyll content, shoot length, shoot fresh and dry weights under heat treatment (Ismail et al., 2018). Endophytes improved flavonoid and phenolic content in sunflower and soybean seedlings with reduced amount of proline, catalase, AA oxidase and exogenous ABA levels. Nutritive value of soyabean and sunflower were also increased, as evident from total sugar, protein, and lipid content (Table 1.1).

Around 242 sporulating rice endophytic fungi were identified from rice cultivars based on their morphological trait (Su-Han et al., 2019). Predominant species were *Nigrospora oryzae*, *Curvularia lunata*, *Daldinia eschscholtzii*, and *Exserohilum sp.*3. Endophyte distribution, abundance, and community were dependent on rice cultivars, collection areas, collection time and tissue segment type, etc.

At 40°C, *Helianthus annuus* L. (sunflower) and *Glycine max* L. (soybean) aligned with *Rhizopus oryzae* had more chlorophyll content, shoot, and root length, fresh weight shoot, fresh weight root, dry weight shoot and dry weight root in comparison to 25°C (Ismail et al., 2020). Endophyte treated sunflower and soybean had 34 ng/g and 29.9 ng/g of ABA, whereas untreated plants contained 60 ng/g and 262 ng/g, respectively at 40°C. Significantly flavonoids content was reduced in colonized plants, but proline, phenolics content, catalase, and AA oxidase levels were increased. Total soluble sugar and lipid content were increase in both stressed plants in the presence of endophytes with slight increase in protein content in soybean plant.

Gliocladium cibotii increased chlorophyll content (8%), shoot length (20%), root length (9%), fresh shoot (26%) and root weight (31%), shoot (38%) and root dry weight (58%) in thermally stressed *Glycine max* and *Helianthus annuus* seedlings (Ismail et al., 2021). It also mediated an increase in flavonoids and phenolics levels with a decrease in proline and ABA levels. Stressed *Glycine max* and *Helianthus annuus* seedlings accumulated 35% and 79% lower amount of MDA in the presence of endophytes, respectively. At 40°C, antioxidant activity of the colonized *G. max* and *H. annuus* plantlets increased considerably as evident from the increase in concentration of catalase (13% and 43%), POX (39% and 37%), AA oxidase (2% and 28%), SOD (27% and 22%) and glutathione reductase (25% and 36%) respectively. In *G. max* and *H. annuus* plantlets endophytes contributed respectively 31 and 24%, 25 and 18%, 21 and 14% increase in soluble sugar levels, lipid, and total protein levels.

TABLE 1.1 Stress Alleviation by Fungal Endophytes

Endophytic Fungi	Host	Condition	Improvements	References
Aspergillus niger	*Helianthus annus*	40°C	Higher chlorophyl – 5.9%, shoot length – 17%, root length – 35.5%, fresh shoot wight – 2.8%, fresh root weight – 23.5%, shoot dry weight – 30%, root dry weight – 50% Reduction in hydrogen peroxide 45.8%, MDA – 49%, ABA – 17% Increase in proline – 61%, phenolics – 14%, CAT – 34.6%, POD – 64%, AAO – 3%, SOD – 27.8%, GR – 42.8%, lipid – 39%, protein – 35%, soluble sugars – 29%	Ismail et al. (2020)
Aspergillus niger	*Glycine max*	40°C	Higher chlorophyll – 9.9%, shoot length – 32.3%, root length – 33%, fresh shoot weight – 47.6%, fresh root weight – 23%, dry shoot weight – 5%, dry root weight – 16.7% Reduction in hydrogen peroxide level – 37%, MDA – 30%, ABA – 18% Increase in proline – 26%, phenolics – 34%, CAT – 32%, POD – 58%, AAO – 18.9%, SOD – 24.8%, GR – 18.9%, lipid – 48%, protein – 51%, soluble sugars – 11%	Ismail et al. (2020)
Thermomyces lanuginosus	*Cullen plicata*	45°C	Three-fold increase in photosynthesis rate. Significant increase in root length, ratio of dry weight of the roots to the dry weight of the shoots, soil water content, leaves number, fruits number, fruit weight, stomata number, total carbohydrate (310 mg/g DW), flavonoid (4.7 mg/g DW), ascorbic acid content (44.2 mg/g FW), phenylalanine ammonia-lyase (70.6 U/g FW), catalase (80.9 U/g FW), peroxidase (51.8 U/g FW)	Ali et al. (2019)
Thermomyces sp.	Cucumber plant	Heat stress	Efficiency in Photosystem II, photosynthetic rate, water use. Increase in root length, significant accumulation of total sugars, flavonoids, saponins, soluble proteins, and antioxidant enzyme activities	Ali et al. (2018)

TABLE 1.1 *(Continued)*

Endophytic Fungi	Host	Condition	Improvements	References
Thermomyces lanuginosus	*Cullen plicata* Delile	Heat stress	Growth improvement, resistance to heat stress	Ali et al. (2019)
Paecilomyces formosus LHL10 and *Penicillium funiculosum* LHL06	*Glycine max* L.	Heavy metals (Ni, Cd, Al), high temperature, drought stress	Growth promotion, higher photosynthetic activity, glutathione, catalase, superoxide dismutase activity, lowered lipid peroxidation, enhanced micronutrient uptake, reduced metal accumulation and translocation by down-regulation of heavy metal ATPase gene (GmHMA 13, GmHMA14, GmHMA18) expression. Regulation of drought-related GmDREB2, GmDREB1B, GmERD1, GmRD20 and heat shock protein 90 (GmHsp90A2 and GmHsp90A1) lead to reduced abscisic acid and jasmonic acid levels	Bilal et al. (2019)

Paecilomyces formosus LWL1 released phytohormones and organic acids which helped japonica rice plants to grow during heat stress (Waqas et al., 2015). Plant height, fresh weight, dry weight, chlorophyll content and total protein content (18.76–33.22%) were significantly improved along with lowered endogenous levels of stress signaling compounds like ABA (25.71%) and jasmonic acid (34.57%) inthe presence of the endophyte. A number of genes linked to abiotic and biotic stress tolerance are upregulated in response to salinity stress in rice (Sampangi-Ramaiah et al., 2020).

1.3.3 SALT

Salinity causes a reduction in the growth of plant species. Chloride induced stress leads to leaf chlorosis, stunting, and lowering of biomass in plants. *Penicillium minioluteum* LHL09 inoculated *Glycine max.* L showed 13.8% and 22.75% more shoot length in the presence of 70 and 140 mM NaCl. Leaf area, chlorophyll content, shoot fresh and dry weight, etc., were also enhanced significantly (Khan et al., 2011). Nitrogen assimilation was more than 2% and ABA content was reduced 42.53% under saline environment. Daidzein content (138.51 ng/g DW) and genistein content (109.6 ng/g DW) was increased significantly at 140 mM and 70 mM NaCl, respectively. *Cochliobolus lunatus* inoculated in *Abelmoschus esculentus* L. showed no growth inhibition in the presence of 50 and 100 mM of NaCl (Bibi et al., 2019). Under salt stress, leaf number, root, and shoot lengths, fresh, and dry weight were reduced, whereas inoculated plants had more leaves, longer root and shoot length, >30% increase in fresh weight and dry weight. Endophyte colonized plants improved relative water content (24.83%), proline content (29.48%) whereas lowered malondialdehyde content (19.55%) in comparison to the control plants. In the presence of 100 mM salt, plants accumulated more soluble sugars (15%), indole acetic acid (160% in leaves, 107.83% in stem, 81.20% in root), phenolics (29.31% in leaf, 27.68% in stem, 18.71% in root) and flavonoids (20.29% in leaf, 16.95% in stem and 13.84% in root) when aligned with the endophyte.

Fungal endophyte SXSp1, *Meyerozyma Caribbea*n, increased 26.2 μg/ml chlorophyll a, 17.76 μg/ml chlorophyll b, 16.96 μg/ml carotenoid in salt-stressed maize plants on an average (Gul Jan et al., 2019). In the presence of the colonizers, relative water content was 24% higher in the stressed plants. It also improved IAA (indole-3-acetic acid) content by 31.67%, proline content by two-fold, POX activity by 23%, catalase activity by 25% and reduced ABA content by 17.2%. *Porostereum spadiceum* AGH786 inoculated in

Glycine max L. Hwangkeum improved more than 31% in leaf area of seedling under salt stress (Rehman et al., 2017). In the presence of 70 and 140 mM of NaCl, gibberellin 1 content in the soybean seedlings were 41% and 33% higher compared to control. The levels of gibberellin 4 were 27% and 53% higher in the respective stressed plants. The presence of endophytes enhanced the levels of ABA, jasmonic acid, and isoflavone in the soybean plants under salt stress.

During salt treatment, *Oryza sativa* L. plants accumulated higher protein, proline, and relative water content in shoots and roots when aligned with fungal endophyte *Piriformospora indica* (Bagheri et al., 2013). Activities of antioxidant enzymes like SOD, POX, catalase, APX, glutathione reductase were increased at 100 mM NaCl. Endophytes reduced LPO in stressed plants. *Aspergillus aculeatus* enhanced growth of Bermuda grass and reduced LPO, peroxide level during salt treatment (Xie et al., 2017). Improved photosynthetic efficiency, IAA, and glutathione level were also observed. *Yarrowia lipolytica* inoculated maize plants showed significantly higher amounts of chlorophyll a, chlorophyll b, and carotenoid content compared to control plants under salinity (Gul Jan et al., 2019). Those colonized plants showed higher relative water content, shoot, and root length, IAA level whereas reduced electrolyte leakage and ABA level. The presence of endophytes controlled the production of proline, catalase, and POX during stress treatment. *Penicillium brevicompactum* and *P. chrysogenum,* isolated from Antarctic plant roots, accelerated yield under salt stress by 42% and 68% in *Lactuca sativa* L. and *Solanum lycopersicum* L., respectively (Molina-Montenegro et al., 2020). Photosynthetic efficiency was increased by 43% and 37% along with water use efficiency by 54% and 60% in the respective stressed plants. The plants accumulated higher amount of nitrogen and potassium in the presence of endophytes. qRT-PCR data revealed a higher expression level of vacuolar Na^+/H^+ antiporter NHX1 which can be correlated with higher amount of Na^+ in leaves.

Fusarium verticillioides enhanced length and fresh weight of soybean plants under salinity stress condition (Radhakrishnan et al., 2013). It increased protein content, antioxidant enzymes activity and salicylic acid level while lowering LPO and ABA level. *Cochliobolus lunatus* isolated from *Mirabilis jalapa* enhanced shoot length, root length, plant dry weight, chlorophyll, carotenoids, xanthophyll, phenolics, flavonoids, IAA, total soluble sugars, and relative water content but reduced malondialdehyde level in salt-stressed okra plants (Bibi et al., 2019). Stress relief in the host plant may be attributed to the presence of pinocembrin, chlorogenic acids, wogonin, calycosin, and daidzein produced by the endophyte. *Fusarium sp.* induced a significant increase in root and shoot length, biomass, total chlorophyll content, net carbon assimilation rate and cell membrane stability in salt-sensitive *Oryza sativa* L during salt treatment (Sampangi-Ramaiah et al., 2020). Colonization Significantly reduced Na^{+}/K^{+} ratio under stress. The level of malondialdehyde was very high in both control and experimental conditions, whereas proline accumulation decreased upon colonization. No signification alteration in SOD activity was noted during stress. During salt stress, calcium signaling and transport genes were highly upregulated in the presence of the endophytes in roots and shoots as evident from qPCR data. Ca^{2+} and calmodulin activate NaCl inducible genes which in turn upregulate transcription factor R2R3-type MYB2 of salt responsive genes and WRKY transcription factor. Overexpression of PeWRKY1 was found to be associated with salinity tolerance, improved root growth, and photosynthesis in poplar plants under stress. Moreover, colonization induced ubiquitin-mediated protein degradation. This was evident in another experiment with transgenic tobacco where salinity tolerance was linked with overexpression of wheat F-box genes which are part of gene family involved in ubiquitin-mediated degradation of cellular proteins.

1.3.4 METALS

Fungal endophytes employ several mechanisms to eliminate heavy metal contamination in the rhizospheric soil. Metal ions can be tolerated by extracellular chelation or binding with cell wall or they may be intracellularly compartmentalized or bound to detoxicants. Toxic metal tolerance imparted by endophytic fungi can be mediated by biosorption, bioaccumulation, metal transformation or secretion of chelating agents like organic acids, siderophores, and phenolic compounds, etc. (Domka et al., 2019). *P fortini*, dark septate root endophyte, for example, produced black biopolymer melanin from phenolics and that was found to bind with heavy metals. Melanin-bound heavy metals were kept separated from living plant cell via compartmentalization in fungal cell wall. *Clethra barbinervis* Sieb. et Zucc. seedlings were unable to grow without the presence of root fungal endophytes in heavy metal contaminated soil in mining area (Yamaji et al., 2016). *Phialocephala fortinii, Rhizodermea veluwensis, and Rhizoscyphus sp.* w., etc., were symbiotically associated with *C. barbinervis* roots under heavy metal stress conditions. Symbionts mediated metal tolerance by different siderophores released in response to mental stress. Growth improvement and

K uptake were noted during stress mediation. Multiple mechanisms might be involved as metal tolerance induced by *Rhizoscyphus* sp. was not associated with siderophore production.

Curvularia affinis and *Neosartorya stramenia and Neosartorya hiratsukae* isolated from mangrove plant *Avicennia* sp. were capable of removing 25 mg of Cu/g biomass, 24 mg of Zn/g of biomass and 24 mg of Zn/g biomass respectively when treated with heavy metal contaminated wastewater (Onn et al., 2016). Mangrove *Kandelia candel* from South China Coast inoculated with root endophyte *Purpureocillium lilacinum* A5 showed enhanced chlorophyll a, b, and relative water content in the presence of 25, 100, 200, and 400 mg/kg of Cu (Gong et al., 2017). Endophytes increased the concentration of carbonate-bound Cu, Mn-Fe Cu complexes and organic-bound Cu in soil by 3.9, 2.7, and 3.1 times respectively. A5 associated plants reduced Cu ion concentration by 4.8-fold. 237 culturable endophytes were isolated from *Dysphania ambrosioides* which is a hyperaccumulator of heavy metals (Li et al., 2016). 31 isolates exhibited heavy metal tolerance ability. Dominant endophyte genera of the plant from slag heap area were *Plectosphaerella, Cladosporium,* and *Verticillium*, whereas *Phoma, Peyronellaea*, and *Alternaria*, etc., were dominant genera from wasteland plants. Among the isolates *Penicillium* (FT2G59) and *P. columnaris* (FT2G7) were better performers. FT2G59 showed tolerance against Pb, Zn, and Cd simultaneously and minimum inhibitory concentrations were 30–50, >680 and 20–30 mmol/l, respectively. *P. columnaris* (FT2G7) exhibited tolerance against Cd with a minimum inhibitory concentration 30–50 mmol/l.

Arabis hirsuta, Acacia decurrens, Symplocos paniculata, Rabbosia eriocalyx, Arenaria serpyllifolia and *Rosa longicuspis,* grown in the wasteland of Huize Pb-Zn mine of Southwest China, were investigated for their ability to endure metal toxicity (Li et al., 2012). *Phoma, Alternaria, and Peyronellaea* were mostly dominated in the isolates. Isolates L124, S383 and S487 showed enhanced growth in the presence of Pb^{2+} and Zn^{2+} with tolerance index (percent of dry weight of treated mycelia to dry weight of control mycelia) between 107.1 and 8245.10. *Alternaria* and *Peyronellaea* were less sensitive to Zn^{2+} with tolerance index more than 50%. Fungal endophytes *Penicillium janthinellum, Paecilomyces formosus, Exophiala sp.* and *Preussia sp.* were studied for Ni removing ability (Bilal et al., 2017). *Paecilomyces formosus* LHL 10 showed substantial tolerance to Ni toxicity along with significant accumulation of S and K. In presence of Ni, *P. formosus* LHL 10 inoculated soybean plants showed increase in length and fresh/dry weight of root and shoot (11–47%) as well as chlorophyll content (10%). IAA and gibberellin biosynthetic genes were overexpressed and malondialdehyde content was reduced in colonized soybean plants. Under Ni induced stress, inoculated plants showed higher glutathione content, 13–36% higher POX activity, 19.9% higher polyphenol oxidase activity, 50% higher catalase activity along with 32–36% lower ABA and jasmonic acid content compared to uninoculated plants. The down-regulation of endogenous hormones and upregulation of antioxidant enzymes helped the treated plants to alleviate heavy metal-induced stress.

Root endophyte *Serendipita indica* inoculated sweet basil showed higher fresh weight, shoot, and root dry weight, plant height and number of leaves compared to control plants grown in Pb and Cu contaminated soil (Sabra et al., 2018). The colonized plants had a higher tendency to accumulate P along with a lower tendency to accumulate Cu and Pb in shoots. Mucor strain MHR-7 isolated from *Brassica campestris* L., tolerant against multiple heavy metal like Cr^{6+}, Mn^{2+}, Co^{2+} and Zn^{2+}, stimulated growth of the host plant in metal-contaminated broth (Zahoor et al., 2017). *Vachellia farnesiana* (L.) plants, hyperaccumulator of Pb (0.149 g/Kg, Cu (0.0815 g/Kg) and Zn (2.495 g/Kg) was the dominant plant of Huautla mine region, were richly inhabited by endophytic fungi of *Neocosmospora* and *Aspergillus* genera (Salazar-Ramirez et al., 2020). Strain H17 from *Neocosmospora* genus was tolerant against 500 ppm of Cu, 700 ppm of Zn and 700 ppm of Pb with growth inhibition of 55%, 86% and 15%, respectively. This strain also produced purple pigment in the presence of 100–700 ppm of Cu and the intensity of pigmentation increased with increase in Cu concentration. Strain H21 from genus *Aspergillus* tolerated maximum 1,000 ppm of Cu, 350 ppm of Zn and 700 ppm of Pb with growth inhibition of 88%, 97% and 89%, respectively. At 500 and 700 ppm of Pb, this strain produced a clear halo region surrounding the mycelial mat with pH 3. Metals are non-degradable in nature. Moreover, they are cytotoxic, carcinogenic, or genotoxic. They inhibit normal cell functioning like seed germination, elongation of shoot and root, chlorophyll production, seedling development, transpiration, water/protein content and overall plant growth. Pb tolerant *Trametes hirsuta,* isolated from *Chenopodium album* L. grown in Pb contaminated industrial area, when inoculated to *Triticum aestivum* L. showed higher fresh biomass, root biomass, chlorophyll content, and accumulation of metal in root and shoot in the presence of Pb (Malik et al., 2020). Dark septate endophytes improve salt tolerance of native and invasive lineages of *Phragmites australis* (Mateu et al., 2020). Inoculated grass seedlings showed enhanced survival rate in laboratory studies under salt stress, whereas higher aboveground biomass was observed under mesohaline conditions during greenhouse assay.

1.4 FUNGAL ENDOPHYTES: A PROMISING TOOL

Endophytic fungi could be a major source of a wide array of metabolites like alkaloids, terpenoids, quinones, steroids, furandiones, xanthones, isoprenoids, glycosides, polyketides, depsipeptides, isocoumarines, phenols, shikimates, and perylene derivatives, etc. (Rajamanikyam et al., 2017). Many endophytes provide metabolites having antiviral, immune-suppressive, anti-diabetic, hypolipidemic, antiarthritic, and anti-inflammatory activity, etc. Abiotic stresses in plants can be mitigated in two ways broadly 1. Inducing host stress responses immediately after exposure to the stress factor 2. Production of stress-relieving metabolites by endophytes (Lata et al., 2018). Torres et al. (2012) postulated that relief of oxidative stress is the main role of endophytes in the host plant, which extends protection during biotic and abiotic stress treatment or environment. The mechanism of stress management is dependent on the release of ROS and auxin from the colonizers to the host plant which in turn stimulates antioxidant systems of the host issues. Meta-analysis on 92 endophytic fungi and 42 host plants related research data revealed that the presence of endophytes in host improved biomass accumulation under drought stress and nitrogen deficiency and excessive salinity (Rho et al., 2018). Mitigation of salinity, drought, or heat stress responses mediated via fungal endophytes *Penicillium resedanum* LK6 showed comparable effects with exogenous addition of gibberellic acid in *Capsicum annuum* L (Khan et al., 2015).

Alleviation of stress-induced responses is dependent on the relative abundance of the colonizing endophytes (Pandey et al., 2016). A dose-dependent response study of *Trichoderma hazarinum* Th-56 endophytes in drought-affected rice demonstrated the most effective concentration of the endophytes was 30 g/L in mitigating stress levels in the plant. IAA-producing endophytic fungus, *Alternaria alternata* strain LQ1230, from a forage plant *Elymus dahuricus* Turcz grown in an arid area, was investigated for its role on drought stress performance of wheat (*Triticum aestivum* L.) (Qiang et al., 2019). Inoculation of LQ1230 significantly increased N accumulation of 55.6% under moderate and severe drought treatment compared to non-inoculated plants, 45.0% and 39.5% increase in POX activity was estimated in endophyte colonized plants with respect to endophyte-free plants under the conditions mentioned earlier respectively. The soluble sugars and proline accumulation were increased by 43.8% and 36.4%, respectively, in the LQ1230 inoculated plants compared to the controls plants in moderate drought-challenged condition. The endophyte also significantly ($P < 0.01$) improved total dry weight (TDW) of wheat plantlets by 19.8 and 32.0%

under moderate and severe drought treatment compared to non-inoculated plants. In comparison to control, a 32.4% increase in shoot dry weight was observed under severe drought stress and 95.2% increase in root dry weight was noted under medium drought stress in the inoculated plants.

Gaeumannomyces cylindrosporous, a highly Pb tolerant dark septate endophyte isolated from Pb-Zn mine tailings in China, was inoculated in *Zea mays* L. seedlings to transfer resistance against metal toxicity in the plant (Ban et al., 2017). Under different concentrations of Pb, those endophytes increased the height, basal diameter, root length and total biomass of the maize seedlings. The root and shoot biomass were 2.81 and 1.35 times, respectively, with respect to the control at 1,000 μg/g of Pb. Chlorophyll a and chlorophyll (a+b) contents were noted 66.9% and 59.5% higher in the inoculated plants compared to the non-inoculated ones. Inoculated seedlings showed 2.14- and 4.11-times higher net photosynthetic rate than control ones at 500 and 1,000 μg/g of Pb, respectively. At the same conditions, transpiration rates were 76.2% and 114.3%, respectively, in the inoculated seedlings. In the presence of 1,000 μg/g of Pb, total Pb content in root and shoot increased 1.36 times in colonized seedlings and translocation of the heavy metal from root to shoot was also recorded. Endophytic fungi from *Artemisia absinthium* were tested for fungicidal activity on *Botrytis cinerea* to control postharvest decay of table grapes (Noumeur et al., 2016). Majority of the isolates reduced radial fungal growth from 33–50% and highest inhibition was achieved from B11C29. Three of them were *Penicillium sp.* isolates, which were very effective in reducing the percent of infection and lesion diameter and all the isolates were effective during *in vivo* trial at a concentration of 10^{-6} spores/ml.

1.5 FUTURE PROSPECTS AND CONCLUSION

Endophytic fungi mediated stress tolerance in plants, as evident from various studies, could well be a result of specific, consorted, and synergistic signaling cascades generated through altered gene expression of the host. Foreign endophyte colonization is generally inhibited by the native endophytes in a plant tissue. Bottlenecks in fungal endophyte mediated crop improvement are co-existence with native microorganisms, stability factor after colonization, toxicity, and co-lateral reactions in plants and regulatory approval, etc. (Suryanarayanan et al., 2018; Chitnis et al., 2020). Fungal endophytes are coming up as a promising treasure trove for research as their presence induces several orchestrated mechanisms in the host plant. In soybean, endophytes upregulated genes involved in phenylpropanoid and derivative pathway and iron scavenging siderophores (Bajaj et al., 2018). During 1988–2019, 88 fungal endophytes were patented for their applications in abiotic stress tolerance and plant growth promotion. Most utilized genera belong to *Alternaria*, *Aspergillus*, *Chaetomium*, *Fusarium*, *Penicillium*, and *Muscodor* (Ortega et al., 2020). Metabolites from fungal endophytes can compete with standard cancer drugs as evident from *in-silico* interaction studies of fungal diketopiperazines, including hexahydropyrrollo [1,2-a] pyrazine-1,4-dione and its related compounds with cancer protein Hsp90 via Autodock Vina. One compound 4,4'-(1,2-dimethyl-1,2-ethanediyl) bis-2,6-piperazinedione scored better with docking score of – 7.5 Kcal/mol compared to the standard drugs like Geldanamycin, 17-(Allylamino) Geldanamycin and Alvespimycin (17DMAG) (Sharma et al., 2014). With these trends, more avenues are expected to open up in fungal endophyte research and application in the near future.

KEYWORDS

- **abscisic acid**
- ***Arizona fescue***
- **indole-3-acetic acid**
- **Saskatchewan microbial collection database**
- **sleepy grass**
- **total dry weight**

REFERENCES

Ali, A. H., Abdelrahman, M., Radwan, U., El-Zayat, S., & El-Sayed, M. A., (2018). Effect of *Thermomyces* fungal endophyte isolated from extreme hot desert-adapted plant on heat stress tolerance of cucumber. *Appl. Soil Ecol., 124*, 155–162.

Ali, A. H., Radwan, U., El-Zayat, S., & El-Sayed, M. A., (2018). Desert plant-fungal endophytic association: The beneficial aspects to their hosts. *Biol. Forum Int. J., 10*, 138–145.

Ali, A. H., Radwan, U., El-Zayat, S., & El-Sayed, M. A., (2019). The role of the endophytic fungus *Thermomyces lanuginosus* on mitigation of heat stress to its host desert plant *Cullen plicata. Biol. Futura, 70*, 1–7.

Azad, K., & Kaminskyj, S., (2016). A fungal endophyte strategy for mitigating the effect of salt and drought stress on plant growth. *Symbiosis, 68*, 73–78.

Bagheri, A. A., Saadatmand, S., Niknam, V., Nejadsatari, T., & Babaeizad, V., (2013). Effect of endophytic fungus, *Piriformospora indica*, on growth and activity of antioxidant enzymes of rice (*Oryza sativa* L.) under salinity stress. *Int. J. Adv. Biol. Biomed. Res., 1*, 1337–1350.

Bajaj, R., Huang, Y., Gebrechristos, S., Mikolajczyk, B., Brown, H., Prasad, R., Varma, A., & Bushley, K. E., (2018). Transcriptional responses of soybean roots to colonization with the root endophytic fungus *Piriformospora indica* reveals altered phenylpropanoid and secondary metabolism. *Sci. Rep., 8*, 10227.doi: 10.1038/s41598-018-26809-3.

Balla, K., Rakszegi, M., Li, Z., Bekes, F., Bencze, S., & Veisz, O., (2011). Quality of winter wheat in relation to heat and drought shock after anthesis. *Czech. J. Food Sci., 29*, 117–128.

Ban, Y., Xu, Z., Yang, Y., Zhang, H., Chen, H., & Tang, M., (2017). Effect of dark septate endophytic fungus *Gaeumannomyces cylindrosporus* on plant growth, photosynthesis and Pb tolerance of maize (*Zea mays* L.) *Pedosphere, 27*, 283–292.

Benny, J., Pisciotta, A., Caruso, T., & Martinelli, F., (2019). Identification of key genes and their chromosome regions linked to drought responses in leaves across different crops through meta-analysis of RNA-seq data. *BMC Plant Biol., 19*, 194.

Bibi, N., Jan, G., Jan, F. G., Hamayun, M., Iqbal, A., Hussain, A., Rehman, H., et al., (2019). *Cochliobolus sp.* acts as a biochemical modulator to alleviate salinity stress in okra plants. *Plant Physiol. Biochem., 139*, 459–469.

Bilal, S., Khan, A. L., Shahzad, R., Asaf, S., Kang, S. M., & Lee, I. J., (2017). Endophytic *Paecilomyces formosus* LHL10 augments *Glycine max* L. adaptation to Ni-contamination through affecting endogenous phytohormones and oxidative stress. *Front. Plant Sci., 8*, 870.

Bilal, S., Shahzad, R., Imran, M., Jan, R., Kim, K. M., & Lee, I. J., (2020). Synergistic association of endophytic fungi enhances *Glycine max* L. resilience to combined abiotic stresses: Heavy metals, high temperature and drought stress. *Ind. Crops Prod., 143*, 111931.

Bu, Y., Guo, P., Ji, Y., & Zhang, S., (2019). Effects of *Epichloe sinica* on *Roegneria kamoji* seedling physiology under PEG-6000 simulated drought stress. *AGRIS, 77*, 123–132.

Chitnis, V. R., Suryanarayanan, T. S., Nataraja, K. N., Prasad, R., Oelmuller, R., & Shaanker, R. U., (2020). Fungal endophyte-mediated crop improvement: The way ahead. *Front. Plant Sci., 11*, 561007.

Cong, G. Q., Yin, C. L., He, B. L., Li, L., & Gao, K. X., (2015). Effect of the endophytic fungus *Chaetomium globosum* ND35 on the growth and resistance to drought of winter wheat at the seedling stage under water stress. *Acta Ecol. Sin., 35*, 6120–6128.

Dastogeer, K. M. G., (2018). Influence of fungal endophytes on plant physiology is more pronounced under stress than well-watered conditions: A meta-analysis. *Planta, 248*, 1403–1416.

Dastogeer, K. M. G., Li, H., Sivasithamparam, K., Jones, M. G. K., Du, X., Ren, Y., & Wylie, S. J., (2017). Metabolic responses of endophytic *Nicotiana benthamiana* plants experiencing water stress. *Environ. Exp. Bot., 143*, 59–71.

Domka, A. M., Rozpadek, P., & Turnau, K., (2019). Are fungal endophytes merely mycorrhizal copycats? The role of fungal endophytes in the adaptation of plants to metal toxicity. *Front. Microbiol., 10*, 371.

Fard, M., Ghabooli, M., Mehri, N., & Bakhshi, B., (2017). Regulation of miR159 and miR396 mediated by *Piriformospora indica* confer drought tolerance in rice. *J. Plant Mol. Breed., 5*, 10–18.

Farooq, M., Wahid, A., Kobayashi, N., Fujita, D., & Basra, S. M. A., (2009). Plant drought stress: Effects, mechanisms and management. *Agron. Sustain. Dev., 29*, 185–212.

Fathi, A., & Tari, D. B., (2016). Effect of drought stress and its mechanisms in plants. *Int. J. Life Sci., 10*, 1–6.

Ghatak, A., Chaturvedi, P., & Weckwerth, W., (2017). Cereal crop proteomics: Systemic analysis of crop drought stress responses towards marker-assisted selection breeding. *Front. Plant Sci., 8*, 1–25. doi: 10.3389/fpls.2017.00757.

Gong, B., Liu, G., Liao, R., Song, J., & Zhang, H., (2017). Endophytic fungus *Purpureocillium sp.* A5 protect mangrove plant *Kandelia candel* under copper stress. *Braz. J. Microbiol., 48*, 530–536.

Gonzales-Teuber, M., Urzua, A., Plaza, P., & Bascunan-Godoy, L., (2018). Effects of root endophytic fungi on response of *Chenopodium quinoa* to drought stress. *Plant Ecol., 219*, 231–240.

Guler, N. S., Pehlivan, N., Karaoglu, S. A., Guzel, S., & Bozdeveci, A., (2016). *Trichoderma atroviride* ID20G inoculation ameliorates drought stress-induced damages by improving antioxidant defense in maize seedlings. *Acta Physiol. Plant., 38*, 132.

Halo, B. A., Al-Yahyai, R. A., & Al-Sadi, A. M., (2020). An endophytic *Talaromyces omanensis* enhances reproductive, physiological and anatomical characteristics of drought-stresses tomato. *J. Plant Physiol., 249*, 153163.

Hamilton, C. E., & Bauerle, T. L., (2012). A new currency for mutualism? Fungal endophytes alter antioxidant activity in hosts responding to drought. *Fungal Divers., 54*, 39–49.

Hirsch, G., & Braun, U., (1992). Communities of parasitic microfungi. In: Wintehroff, W., (eds.), *Handbook of Vegetation Science* (pp. 225–250). Kluwer Academics Dordrecht.

Hosseini, F., Mosaddeghi, M. R., & Dexter, A. R., (2017). Effect of the fungus *Piriformospora indica* on physiological characteristics and root morphology of wheat under combined drought and mechanical stresses. *Plant Physiol. Biochem., 118*, 107–120.

Hosseini, F., Mosaddeghi, M. R., Hajabbasi, M. A., & Sabzalian, M. R., (2016). Role of fungal endophyte of tall fescue (*Epichloe coenophiala*) on water availability, wilting point and integral energy in texturally-different soils. *Agric. Water Manag., 163*, 197–211.

Hubbard, M., Germida, J. J., & Vujanovic, V., (2014). Fungal endophytes enhance wheat heat and drought tolerance in terms of grain yield and second-generation seed viability. *J. Appl. Microbiol., 116*, 109–122.

Hubbard, M., Germida, J., & Vujanovic, V., (2012). Fungal endophytes improve wheat seed germination under heat and drought stress. *Botany, 90*, 137–149.

Ismail, Hamayun, M., Hussain, A., Iqbal, A., Khan, S. A., & Lee, I. J., (2018). Endophytic fungus *Aspergillus japonicus* mediates host plant growth under normal and heat stress conditions. *BioMed Res. Int., 3*, 1–11.

Ismail, Hamayun, M., Hussain, A., Iqbal, A., Khan, S. A., & Lee, I. J., (2020). *Aspergillus Niger* boosted heat stress tolerance in sunflower and soybean via regulating their metabolic and antioxidant system. *J. Plant Interact., 15*, 223–232.

Ismail, Hamayun, M., Hussain, A., Iqbal, A., Khan, S. A., Khan, M. A., & Lee, I. J., (2021). An endophytic fungus *Gliocladium cibotii* regulates metabolic and antioxidant system of *Glycine max* and *Helianthus annuus* under heat stress. *Pol. J. Environ. Stud., 30*, 1631–1640.

Ismail, Hamayun, M., Hussain, A., Khan, S. A., Iqbal, A., & Lee, I. J., (2020). An endophytic fungus *Aspergillus violaceofuscus* can be used as heat stress adaptive tool for *Glycine max* L. and *Helianthus annuus* L. *J. Appl. Bot. Food Qual., 93*, 112–120.

Ismail, Hussain, A., Mehmood, A., Qadir, M., Husna, Iqbal, A., Hamayun, M., & Khan, N., (2020). Thermal stress alleviating potential of endophytic fungus *Rhizopus oryzae* inoculated to sunflower (*Helianthus annus* L.) and soybean (*Glycine max* L.). *Pak. J. Bot., 52*, 1857–1865.

Jan, F. G., Hamayun, M., Hussain, A., Iqbal, A., Jan, G., Khan, S. A., Khan, H., & Lee, I. J., (2019). A promising growth-promoting *Meyerozyma caribbica* from *Solanum xanthocarpum* alleviated stress in maize plants. *Biosci. Rep., 39*, BSR20190290.

Jan, F. G., Hamayun, M., Hussain, A., Jan, G., Iqbal, A., Khan, A., & Lee, I. J., (2019). An endophytic isolate of the fungus *Yarrowia lipolytica* produces metabolites that ameliorate the negative impact of salt stress on the physiology of maize. *BMC Microbiol., 19*, 3.

Khan, A. L., Shinwari, Z. K., Kim, Y., Waqas, M., Hamayun, M., Kamran, M., & Lee, I. J., (2012). Role of endophyte *Chaetomium globosum* Ik4 in growth of *Capsicum annuum* by production of gibberellins and indole acetic acid. *Pak. J. Bot., 44*, 1601–1607.

Khan, A. L., Waqas, M., & Lee, I. J., (2015). Resilience of *Penicillium resedanum* LK6 and exogenous gibberellin in improving *Capsicum annuum* growth under abiotic stresses. *J. Plant Res., 218*, 259–268.

Khan, L., Latif, A., Hamayun, M., Ahmad, N., Hussain, J., Kang, S. M., Kim, Y. H., et al., (2011). Salinity stress resistance offered by endophytic fungal interaction between *Penicillium minioluteum* LHL09 and *Glycine max* L. *J. Microbiol. Biotechnol., 21*, 893–902.

Kord, H., Fakheri, B., Ghabooli, M., Solouki, M., Emamjomeh, A., & Khatabi, B., (2019). Salinity-associated microRNAs and their potential roles in mediating salt tolerance in rice colonized by the endophytic root fungus *Piriformospora indica. Funct. Integr. Genomics, 19*, 659–672.

Lata, R., Chowdhury, S., Gond, S. K., & White, Jr. J. F., (2018). Induction of abiotic stress tolerance in plants by endophytic microbes. *Lett. Appl. Microbiol., 66*, 268–276.

Li, H. Y., Li, D. W., He, C. M., Zhou, Z. P., Mei, T., & Xu, H. M., (2012). Diversity and heavy metal tolerance of endophytic fungi from six dominant plant species in Pb-Zn mine wasteland in China. *Fungal Ecol., 5*, 309–315.

Li, X., Li, W., Chu, L., White, Jr. J. F., Xiong, Z., & Li, H., (2016). Diversity and heavy metal tolerance of endophytic fungi from *Dysphania ambrosioides*, a hyperaccumulator from Pb-Zn contaminated soils. *J. Plant Interact., 11*, 186–192.

Li, X., Ren, A., Han, R., Yin, L., Wei, M., & Gao, Y., (2012). Endophyte-mediated effects on the growth and physiology of *Achnatherum sibiricum* are conditional on both N and P availability. *Plos One, 7*, e48010.

Malik, A., Butt, T. A., Naqvi, S. T. A., Yousaf, S., Qureshi, M. K., Zafar, M. I., Farooq, G., et al., (2020). Lead tolerant endophyte *Trametes hirsuta* improved the growth and lead accumulation in the vegetative parts of *Triticum aestivum* L. *Heliyon., 6*, e04188.

Mastouri, F., Bjorkman, T., & Harman, G. E., (2012). Trichoderma harzianum enhances antioxidant defense of tomato seedlings and resistance to water deficit. *Mol. Plant-Microbe Interact., 25*, 1264–1271.

Mateu, G., Baldwin, A. H., Maul, J. E., & Yarwood, S. A., (2020). Dark septate endophyte improves salt tolerance of native and invasive lineages of *Phragmites australis. ISME J., 14*, 1943–1954.

Molina-Montenegro, M. A., Acuna-Rodriguez, I. S., Torres-Diaz, C., Gundel, P. E., & Dreyer, I., (2020). Antarctic root endophytes improve physiological performance and yield in crops under salt stress by enhanced energy production and Na^+ sequestration. *Sci. Rep., 10*, 5819.

Molin-Montenegro, M. A., Oses, R., Torres-Diaz, C., Atala, C., Zurita-Silva, A., & Ruiz-Lara, S., (2016). Root-endophytes improve the ecophysiological performance and production of an agricultural species under drought condition. *AoBP, 8*, plw062.

Nagabhyru, P., Dinkins, R. D., Wood, C. L., Bacon, C. W., & Schardl, C. L., (2013). Tall fescue endophyte effects on tolerance to water-deficit stress. *BMC Plant Biol., 13*, 127.

Noumeur, S. R., Mancini, V., & Romanazzi, G., (2016). Activity of endophytic fungi from *Artemisia absinthium* on *Botrytis cinerea*. *Acta Hortic., 1144*, 101–104.

Onn, M. L., Teen, L. P., Mujahid, A., Proksch, P., & Muller, M., (2016). Initial screening of mangrove endophytic fungi for antimicrobial compounds and heavy metal biosorption potential. *Sains Malays., 45*, 1063–1071.

Ortega, H. E., Torres-Mendoza, D., & Cubilla-Rios, L., (2020). Patents on endophytic fungi for agriculture and bio-and phytoremediation applications. *Microorganisms., 8*, 1237.

Pandey, V., Ansari, M. W., Tula, S., Yadav, S., Sahoo, R. K., Shukla, N., Bains, G., et al., (2016). Dose-dependent response of *Trichoderma harzianum* in improving drought tolerance in rice genotypes. *Planta, 243*, 1251–1264.

Qiang, X., Ding, J., Lin, W., Li, Q., Xu, C., Zheng, Q., & Li, Y., (2019). Alleviation of the detrimental effect of water deficit on wheat (*Triticum aestivum* L.). *Plant Soil, 439*, 373–391.

Radhakrishnan, R., Khan, A. L., & Lee, I. J., (2013). Endophytic fungal pre-treatments of seeds alleviate salinity stress effects in soybean plants. *J. Microbiol., 51*, 850–857.

Rajamanikyam, M., Vadlapudi, V., Amanchy, R., & Upadhyayula, S. M., (2017). Endophytic fungi as novel resources of natural therapeutics. *Braz. Arch. Biol. Technol., 60*, e17160542.

Redman, R. S., Kim, Y. O., Woodward, C. J. D. A., Greer, C., Espino, L., Doty, S., & Rodriguez, R. J., (2011). Increased fitness and adaptation of rice plants to cold, drought and salt stress via habitat adapted symbiosis: A strategy for mitigating impacts of climate change. *Plos One, 6*, E14823.

Rehman, G., Jan, S., & Lee, I. J., (2017). Gibberellins producing endophytic fungus *Porostereum spadiceum* AGH786 rescues growth of salt-affected soybean. *Front. Microbiol., 8*. doi: 10.3389/fmicb.2017.00686.

Rho, H., Hseih, M., Kandel, S. L., Cantillo, J., Doty, S. L., & Kim, S. H., (2018). Do endophyte promote growth of host plants under stress? A meta-analysis on plant stress mitigation by endophytes. *Microb. Ecol., 75*, 407–418.

Sabra, M., Aboulnasr, A., Franken, P., Perreca, E., Wright, L. P., & Camehl, I., (2018). Beneficial root endophytic fungi increase growth and quality parameters of sweet basil in heavy metal contaminated soil. *Front. Plant. Sci., 9*, 1726.

Sadeghi, F., Samsampour, D., Sevahooei, M. A., Bagheri, A., & Soltani, J., (2019). Fungal endophytes alleviate drought-induced oxidative stress in mandarin (*Citrus reticulata* L.): Toward regulating the ascorbate-glutathione cycle. *Sci. Hortic.* doi: 10.1016/j.scienta.2019.108991.

Salazar-Ramirez, G., Flores-Vallejo, R. C., Rivera-Leyva, J. C., Tovar-Sanchez, E., Sanchez-Reyes, A., Mena-Portales, J., Sanchez-Carbente, M. R., et al., (2020). Characterization of fungal endophytes isolated from the metal hyperaccumulator plant *Vachellia farnesiana* growing in mine tailings. *Microorganisms, 8*, 2. doi: 10.3990/microorganisms8020226.

Sampangi-Ramaiah, M. H., Jagadheesh, Dey, P., Jambagi, S., Vasantha, K. M. M., Oelmuller, R., Nataraja, K. N., et al., (2020). An endophyte from salt-adapted pokkali rice confers salt tolerance to a salt-sensitive rice variety and targets a unique pattern of genes in its new host. *Sci. Rep., 10*, 3237.

Schroder, P., Sauvetre, A., Gnadinger, F., Pesaresi, P., Chmelikova, L., Dogan, N., Gerl, G., et al., (2019). Sustainable increase of crop production through improved technical strategies, breeding and adapted management–a European perspective. *Sci. Total Environ., 678*, 146–161.

Sharma, R., & Vijaya, K. B. S., (2014). *In silico* interaction studies on inhibitory action of endophytic fungal diketopiperazine and its related compounds on heat-shock protein 90 (Hsp90). *Asian J. Biomed. Pharm. Sci., 4*, 25–29.

Stone, J. K., Polishook, J. D., & White, Jr. J. F., (2004). Endophytic fungi. In: Mueller, et al., (eds.), *Biodiversity of Fungi: Inventory and Monitoring Methods* (pp. 241–270). Burlington: Elsevier.

Sudha, V., Govindaraj, R., Baskar, K., Al-Dhabi, N. A., & Duraipandiyan, V., (2016). Biological properties of endophytic fungi. *Braz. Arch. Biol. Technol., 59*, e16150436.

Su-Han, N. H., Songkuarn, P., Nuankaew, S., Boonyuen, N., & Piasai, O., (2019). Diversity of sporulating rice endophytic fungi associated with Thai rice cultivars (*Oryza sativa* L.) cultivated in Suphan Buri and Chainat provinces, Thailand. *Curr. Res. Environ., 9*, 1–14.

Suryanarayanan, T. S., Govinda, R. M. B., & Vidal, S., (2018). Biological control through fungal endophytes: Gaps in knowledge hindering success. *Curr. Biotechnol., 7*, 185–198.

Torres, M. S., White, Jr. J. F., Zhang, X., Hinton, D. M., & Bacon, C. W., (2012). Endophyte-mediated adjustments in host morphology and physiology and effects on host fitness traits in grasses. *Fungal Ecol., 5*, 322–330.

Waqas, M., Khan, A. L., Kamran, M., Hamayun, M., Kang, S. M., Kim, Y. H., & Lee, I. J., (2012). Endophytic fungi produce gibberellins and indoleacetic acid and promote host-plant growth during stress. *Molecules, 17*, 10754–10773.

Waqas, M., Khan, A. L., Shahzad, R., Ullah, I., Khan, A. R., & Lee, I. J., (2015). Mutualistic fungal endophytes produce phytohormones and organic acids that promote japonica rice plant growth under prolonged heat stress. *J. Zhejiang Univ. Sci. B., 16*, 1011–1018.

www.fao.org (accessed on 2 June 2022).

www.ourworldindata.org (accessed on 2 June 2022).

Xie, Y., Han, S., Li, X., Amombo, E., & Fu, J., (2017). Amelioration of salt stress on bermudagrass by the fungus *Aspergillus aculeatus*. *Mol. Plant-Microbe Interact., 30*, 245–254.

Xu, L., Li, X., Han, L., Li, D., & Song, G., (2017). *Epichloe* endophyte infection improved drought and heat tolerance of tall fescue through altered antioxidant enzyme activity. *Eur. J. Hortic. Sci., 82*, 90–97.

Xu, L., Wang, A., Wang, J., Wei, Q., & Zhang, W., (2017). *Piriformospora indica* confers drought tolerance on *Zea mays* L. through enhancement of antioxidant activity and expression of drought-related genes. *Crop J., 5*, 251–258.

Yamaji, K., Watanabe, Y., Masuya, H., Shigeto, A., Yui, H., & Haruma, T., (2016). Root fungal endophytes enhance heavy-metal stress tolerance of *Clethra barbinervis* growing naturally at mining sites via growth enhancement, promotion of nutrient uptake and decrease of heavy-metal concentration. *Plos One*, e0169089. doi: 10.1371/journal.pone.0169089.

Zahoor, M., Hussain, A., Rahman, H., & Irshad, M., (2017). Alleviation of heavy metal toxicity and phytostimulation of *Brassica campestris* L. by endophytic *Mucor sp.* MHR-7. *AGRIS, 142*, 139–149.

CHAPTER 2

Comparative Study of Physiological and Biochemical Response of *Tribulus terrestris* L. and *Pedalium murex* L. to Salt Stress

VARSHA SANJAY KHUDE

Associate Professor, Botany Department, Devchand College, Arjunnagar, Maharashtra, India, E-mail: drvarshakhude81@gmail.com

ABSTRACT

In the present investigation, an attempt has been made to study the salinity tolerance of *Tribulus terrestris* L. and *Pedalium murex* L. very promising medicinal plants used in cough, cold and urinary disorders. To study the effect of NaCl salinity, the plants were treated with increasing concentration of NaCl (100, 200, 300, and 400 mM), and the results obtained are compared with those of normal or control plants based on data collected. These plants were critically analyzed for proline, TAN, and activity of some enzymes related to oxygen toxicity, e.g., superoxide dismutase, catalase, and peroxidase. The significant findings of the present work are proline accumulation is more in *Pedalium* as compared to that in *Tribulus.* TAN is also more in *Pedalium* as compared to that in *Tribulus,* so *Pedalium* appears to be more tolerant than *Tribulus*. There was an enhancement in the activity of enzymes like superoxide dismutase, catalase, and POX both on fresh weight and protein basis up to 200 mM salinity stress in *Tribulus* and up to 300 mM in *Pedalium.* Such behavior of enzymes is related to plant adaptability towards saline conditions. However, such an increased activity of these enzymes might help in minimizing the hazardous effects of ROS in these plants. From these observations,

Plant Metabolites under Environmental Stress: Mechanisms, Responses, and Adaptation Strategies.
Nivas M. Desai, Manasi Patil, Umesh R. Pawar (Eds.)

it was concluded that both the plants are salt-tolerant species, but *Pedalium* was more salt-tolerant than that of *Tribulus terrestris* L.

2.1 INTRODUCTION

Abiotic stresses such as drought, salinity, low or high temperature, and other environmental extremes are the major cause of poor plant growth and reduced crop yield in the world. Soil salinity is one of the important abiotic stresses, which has become a problem of great concern to the environment and is expanding very rapidly throughout the world. More and more agricultural land becomes extinct because of the accumulation of salt, and the problem of salinity arises, because of salt stress there are several effects takes place on the plants, such as physical drought due to osmotic pressure, specific ion toxicity because of Na^+ and Cl^-, seed germination and seedling growth affected, changes in hormonal balance ABA and ethylene, uptake of K^+ and Ca^+ affected, photosynthetic enzymes are affected, and stimulates respiratory enzymes, i.e., salt respiration and Nitrogen metabolism is also affected. To face the saline problem, it is necessary to understand the effect of salinity on plants, different adaptations developed by plants growing under salt stress conditions and the mechanism of salt tolerance in them. Therefore, it is highly essential to test some important medicinal plants for their salinity tolerance so that such plants can be tried for their growth in such problem soils which have become rather useless for most of the crops. It is also essential to understand the mechanisms underlying this phenomenon. Keeping this view in mind, the present investigation focuses on the comparative study of the physiology of salt tolerance in two promising medicinal plants *Tribulus terrestris* L. and *Pedalium murex* L.

Tribulus terrestris L. belonging to the family zygophylaceae is useful in the treatment of all types of cold disorders like cold stress disorders, arthritis, first stage dropsy, diseases of serous fluid, dysuria, eczema, cooling diuretic, tonic, aphrodisiac, used in calculus affection, urinary discharges, and impotence, it increase the sperm content.

Pedalium murex L. belonging to the family Pedaliaceae is growing in the field as a common weed but has several medicinal properties embodied in them. Particularly those associated with diuretic, aphrodisiac, antispasmodic, anti-inflammatory, digestive, carminative, tonic, rejuvenating. It is also useful in renal and vesicle calculi, puerperal diseases, spermatorrhea, spasmodic affections, amenorrhea, dysmenorrheal, varied conditions of pita, inflammation, dyspepsia, flatulence, ulcers, fever, and general debility.

There is a very little work on medicinal plants. Most of the medicinal plants have been studied for their content of medicinal component. The medicinal principle is associated from almost each and every part of this plant. Presently, this plant is highly demanded in the world of medicine. In fact, they have been extensively studied for the phytochemical aspects, particularly for alkaloids.

1. **Salinity and Plant Metabolism:** A number of workers (Bernstein, 1964; Strogonov, 1964; Mass and Hoffman, 1977) have screened different crops for their salt tolerance and have observed differential response of crops to salinity; each crop has its own threshold capacity for salinity. Beyond, which the productivity linearly decreases with increasing salinity. According to them, barley, Bermuda grass, cotton, sugar beet, wheat grass, etc., are salt-tolerant, while almond, apple, onion bean, carrot, etc., are the most sensitive and most of the remaining crops are intermediates.

   Plants resist salinity by four different methods (Fitter and Hay, 1987):
   i. Phenological escape;
   ii. Exclusion;
   iii. Amelioration; and
   iv. Tolerance.
2. **Proline:** It is an important amino acid whose biosynthesis is generally increased in salt-tolerant plants (Ro Dang, 1982; Imamul and Larher, 1985) acts as an osmoticum.
3. **TAN (Titrable Acid Number):** i.e., acidity status of the living cell is important for smooth functioning of the cell Increase in organic acid content due to salinity has been observed by Strogonov (1964); and (Popp and Kinzel, 1971). It has been considered that the accumulation of organic acids in plant is relevant to the adjustment of the cation-anion balance in plant sap and to facilitate the transport of metabolic cations in plants.
4. **Antioxidative Enzymes:** When plants are exposed to abiotic stresses such as high light intensity, drought, high salinity, herbicide treatment or mineral deficiencies. These stresses act as a catalyst in producing free radicals where reactive oxygen species (ROS) such as superoxide (O_2^-), hydroxyl (OH), and hydrogen peroxide (H_2O_2) are produced.

Superoxide radicals are generated in the mitochondria of all aerobic organisms primarily at the NADH-dehydrogenase and Ubiquinone-cytochrome b region of the cyanide sensitive respiratory pathway. It can also be produced as a metabolic byproduct of a number of oxidative enzymes such as aldehyde oxidase, xanthine oxides and lipoxygenase (LOX). Similarly, H_2O_2 and O_2^- are known to be produced in nuclei, glyoxysomes, and peroxisomes under conditions of saturating reducing equivalents. This ROS initiate membrane lipid peroxidation (LPO), destroy chlorophylls, inhibit photosynthesis, in activate key metabolic enzymes such as NADH-oxidase, glutamine synthetase and mark proteins for proteolysis. The changes in the primary and secondary structure of DNA and RNA such as base destruction, single, and double-strand breaks and cross-linking were shown to occur in the presence of H_2O_2. Superoxide radicals appear to be an effective oxidant of α, β dehydroxyethyl thiamine intermediate of the transketolase oxidized reactions.

To mitigate and repair damage initiated by ROS, plants have developed a complex antioxidant system. Most of the higher plants have elaborate non-enzymatic and enzymatic defense systems to counter the inevitable oxidative stress. The non-enzymatic defense systems include vitamin C, vitamin E, glutathione, carotenoids, flavonoids, phenolics acids, alkaloids, and chlorophyll derivatives which play an important role as scavengers of singlet oxygen ($^1O_2$) and superoxide radicals. Antioxidant enzymes play a significant role in plants to protect them against the damaging effect of these ROS. These cytotoxic Superoxide radicals are rapidly dismuted by superoxide dismutase (SOD) to H_2O_2, a product which is relatively stable and can be detoxified further by catalase or peroxidase (POX).

Therefore, keeping this view in mind, in the present chapter, focus is given on the effect of NaCl salinity on Proline content and TAN of these two medicinal plants.

2.2 EXPERIMENTAL METHODS AND MATERIALS

To study the effect of NaCl on physiological and biochemical activities the seedlings of *Tribulus terrestris* L. and *Pedalium murex* L. were raised separately from seeds in pot soil culture and after complete establishment of seedlings for 10 days with normal watering the salinity treatment (100, 200, 300, and 400 mM NaCl) were commenced. The control plants received regular non-saline water. When the plants were 20 days old, were used for analysis.

2.2.1 PROLINE

Free proline concentrations of fresh and dried parts of the plant were estimated following the method by Bate's et al. (1973). The material was homogenized in 3% sulphosalicylic acid and filtered through Whatman No. 1 filter paper. About 2 ml of extract was mixed with 2 ml acid ninhydrin reagent (2.5 g ninhydrin in 60 ml glacial acetic acid and 40 ml orthrophosphoric acid were mixed and heated for 5 to 10 minutes and then cooled) and 2 ml glacial acetic acid. The contents were boiled on water bath for 1 hour and immediately cooled in ice bath. 4 ml of toluene was added and the mixture was vigorously shaken and kept at room temperature for few minutes. The absorbance of the toluene chromatophore developed was read at 520 nm on a spectrophotometer against toluene blank. Standard curve of proline (0.1 mg per ml) was prepared and proline contents were calculated.

2.2.2 TITRATABLE ACID NUMBER (TAN)

The method described by Thomas and Beevers (1949) was used to determine TAN. The fresh leaf material was washed and rinsed with distilled water and blotted to dry. It was accurately weighed (2 g) and cut into small bits. Then it was immersed in boiling distilled water and boiled for half an hour with occasional addition of hot water to prevent drying. It was cooled and filtered through double-layered cheesecloth and volume was made with distilled water (20 ml). The filtrate was then titrated against standardized NaOH (approximate N/40) using phenolphthalein as an indicator. NaOH was standardized against N/40 oxalic acid using the same indicator titratable acid number (TAN) represents the number of ml of decinormal NaOH required to neutralize the acids present in 100 g of fresh tissue.

2.2.3 ENZYMES RELATED TO DEFENSE MECHANISM AGAINST OXYGEN TOXICITY

i. **Super Oxide Dismutase (E.C.1.15.1.1.):** Activity was determined by the method described by Giannopolits and Ries (1977). 0.5 g of fresh material was homogenized in mortar with pestle by using K-phosphate buffer (pH 7.8) 5 ml, containing 1% PVP. Under cold condition, the extract was squished through layered muslin cloth. PVP is added to protect the enzyme from the action of Polyphenols. The filtrate so obtained was centrifuged at 15,000 ×g for 10 minutes at 0–2°C in refrigerator centrifuge. The supernatant was carefully decanded and saved at 0–4°C temperature. This preparation used as an enzyme source.

For enzymatic reaction an assay mixture for SOD contained 2 ml K-phosphate buffer (pH 7.8), 0.2 ml methionine (13 mM) 0.1 ml nitro blue tetrazolium (77 uM), EDTA 0.5 ml (0.1 mM), enzyme 0.1 ml and Riboflavin 0.1 ml (2 uM) is added lastly. The total volume of reaction mixture was 3 ml. The reaction started by the light (sunlight). Change in optical density was recorded after 30 minutes at 560 nm on UV-double beam spectrophotometer (Shimadzu UV-190). The activity of enzyme SOD is expressed as Delta O.D. $h^{-1}g^{-1}$ fresh tissue and Delta O.D. $h^{-1}mg^{-1}$ Protein.

ii. **Catalase (E.C.1.11.1.6):** Catalase activity was determined by the method described by Maehly (1954). 0.5 g fresh leaf material was homogenized in 10 ml of 0.1 M phosphate buffer (pH 7.0) then it was filtered through a four-layered muslin cloth. The filtrate so obtained was centrifuged at 10,000 rpm for 10 minutes at 0–4°C. The supernatant was used as an enzyme source.

An enzyme assay mixture contained 2.8 ml H_2O_2 phosphate buffer (0.64 ml of H_2O_2 diluted to 100 ml with 0.1 M phosphate buffer pH 7.0) and 0.2 ml of enzyme. The components were mixed rapidly and change in optical density was recorded per minute at 240 nm on UV-double beam spectrophotometer (Shimadzu UV-190). The activity of enzyme peroxidase is expressed as Delta O.D. $min^{-1}g^{-1}$ fresh tissue and Delta O.D. $min^{-1}mg^{-1}$ protein.

iii **Peroxidase (E.C.1.11.1.7):** POX activity was determined by the method described by Maehly (1954). Enzyme source was prepared in similar way as mentioned above for catalase.

An enzyme assay mixture contained 2 ml phosphate buffer (0.1 M, pH 7), 1 ml guaicol (20 mM) and 1 ml of enzyme source. The reaction was initiated by addition of 0.05 ml H_2O_2 (1 mM). Change in optical density due to oxidation of guaiacol was recorded per minute at 470 nm on spectrophotometer with frequent stirring of reaction mixture with glass rod. The activity of enzyme peroxidase is expressed as Delta O.D. $min^{-1}g^{-1}$ fresh tissue and Delta O.D. $min^{-1}mg^{-1}$ protein.

2.2.4 ENZYME PROTEINS

The proteins from enzyme were determined by the method described by Lowry et al. (1951). In the assay mixture, 0.05 ml preparation was diluted to 1 ml with distilled water and mixed with 5 ml of freshly prepared reagent 'c'(50 ml 2% Na_2CO_3 in 0.1 N aqueous NaOH mixed with 1 ml 0.5% $CuSo_4$ $5H_2O$ in 1% Na-K tartrate). After 15 minutes, 0.5 ml Folin phenol reagent(100 g sodium tungstate mixed with 25 g sodium molybdate in 700 ml distilled water, 50 ml 80% Phosphoric acid and 100 ml concentrated HCl were refluxed gently for 10 hours using water condenser. To that, 150 g lithium sulfate, 50 ml distilled water, and a few drops of bromine water were added. The mixture was boiled for 15 minutes without water condenser to remove excess bromine. The mixture was cooled and then adjusted 1 N by titrating it against 1 N NaOH) was added. The color was allowed to develop for 30 minutes and then absorbance was recorded at 660 nm. Albumin solution (0.1 mg ml^{-1}) was used for the preparation of the standard curve of protein. Blank was prepared by mixing distilled water, reagent C and Folin phenol reagent.

2.3 RESULT AND DISCUSSION

2.3.1 PROLINE

Proline (pyrrolidine-2-carboxylic acid) is an important secondary amino acid. Generally, it is accumulated in the tissue of plants growing under salt or water stress. The accumulated proline helps in osmotic adjustment of such plants.

Effect of NaCl salinity on proline content of leaves of *Tribulus terrestris* L. and *Pedalium murex* L. are shown in Table 2.1 and Figure 2.1. From the observations, it can be seen that the values of proline from fresh leaves of both *Tribulus* and *Pedalium* are increased with increased level of salinity. The correlation between proline content and salinity is positive and not significant. The MSI value for *Tribulus* is more than that of *Pedalium.*

TABLE 2.1 Effect of NaCl Salinity on Proline Concentration (mg 100^{-1} g Fresh Tissue) in the Leaves of *Tribulus terrestris* L. and *Pedalium murex* L.

NaCl Treatment (mM)	Proline Concentration	
	***Tribulus terrestris* L.**	***Pedalium murex* L.**
0 (control)	25	41.2
100	44	72.7
200	94	112.0
300	142	162.0
400	170	185.0
MSI	450	322.63
t	–1.3826	–1.1296
r	0.9907	0.9948

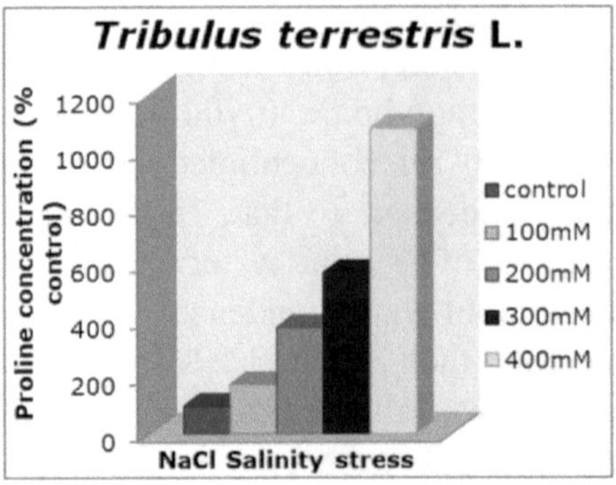

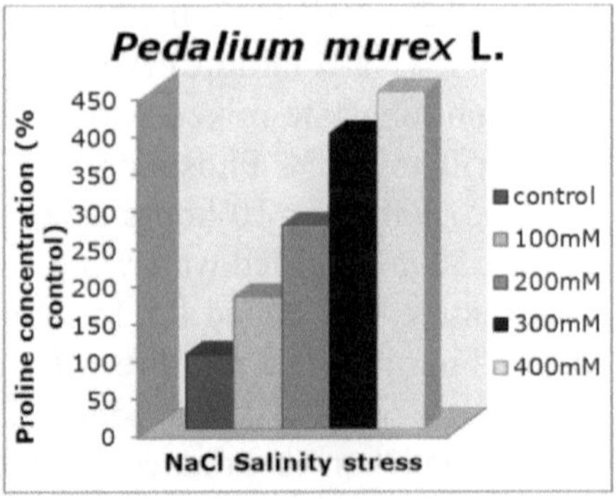

FIGURE 2.1 Effect of NaCl salinity on proline concentration (mg 100^{-1} g fresh tissue) in the leaves of *Tribulus terrestris* L. and *Pedalium murex* L.

The above findings are supported by the observations made by Krishnamoorthy and Siddque (1985) in *Vigna unguiculata;* Salunkhe and Karadge (1989) in *Dodonaea viscosa*; Singh et al. (1994) in *Vigna radiate*; Berteli et al. (1995) in tomato; Durgaprasad et al. (1996) in soybean and Muthukumarswami and Pannerselvum (1997) in green gram. They all reported the accumulation of proline in these plants under saline condition. Rao and Rao (1979) noticed that an increase in proline content under saline conditions may be due to conversion of keto acids into amino acids. Sahoo et al. (2001) observed that salinity caused to increase the rate of proline accumulation in rice. Panda (2001) found an increase in proline content of green gram due to salinity stress. Toro (2000) also noticed an increase in proline content due to salt treatment in *Setaria italica* CV's. Maighany et al. (2002) observed an increase in proline content of *Triticum* due to salt treatment. Gaikwad (2003) has made similar type of observations in *Catharanthus.* The findings of Bandeoglu et al. (2004) *lentil* and Ahmad and John (2005) in *Pisum sativum* are also on a similar line. An increase in Proline content of *Tribulus* and *Pedalium* might be for osmotic adjustment under salinity stress.

2.3.2 TITRABLE ACID NUMBER (TAN)

Effect of NaCl salinity on TAN of leaves of *Tribulus terrestris* L. and *Pedalium murex* L. are shown in Table 2.2 and depicted Figure 2.2. It is evident from the observations that, TAN in the leaves of both *Tribulus* and *Pedalium* are increased considerably with increasing NaCl salinity. It indicates that this plant has the ability to develop the mechanism of salt tolerance when grown under saline conditions. The correlation between TAN and salinity is positive but not significant. The MSI value for *Tribulus* is more than that of *Pedalium.*

TABLE 2.2 Effect of NaCl Salinity on Leaf Juice Acidity (TAN*) in *Tribulus terrestris* L. and *Pedalium murex* L.

NaCl Treatment (mM)	TAN	
	***Tribulus terrestris* L.**	***Pedalium murex* L.**
0 (control)	22.72	45.45
100	44.19	60.60
200	79.54	79.54
300	99.74	109.84
400	108.78	116.16
MSI	365.59	201.39
t	–1.777	–1.6339
r	0.9819	0.9856

*ml decinormal NaOH required to neutralize the acids present in 100 g plant material.

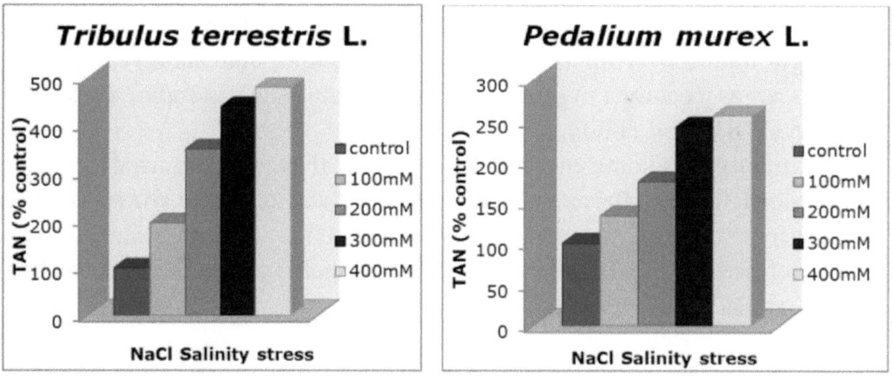

FIGURE 2.2 Effect of NaCl salinity on leaf juice acidity (TAN*) in *Tribulus terrestris* L. and *Pedalium murex* L.

Organic acids are regarded as important metabolites in living cells because many of these compounds are intermediates of a central metabolic pathway, TCA cycle; some organic acids like malic acid are involved in photosynthetic carbon assimilation pathway. They provide carbon skeleton for the synthesis of a number of important metabolites. Increase in organic acid content due to salinity has been observed by Strogonov (1964) in maize.

It has been considered that the accumulation of organic acids in plant is relevant to the adjustment of the cation-anion balance in plant sap and to facilitate the transport of metabolic cations in plants (Popp and Kinzel, 1971). Tolerance of Bermuda grass may be due to increased concentration of organic acids in the cell sap. According to Krishnamurthy et al. (1987) an increase in TAN may be an indication of salt tolerance in rice varieties.

Opposite to this a reduction in organic acid content under salt stress conditions has been reported by Perez-Alfocea et al. (1993) in *Lycopersicum esculantum* and in *L. pinnellii* leaves. Wadkar (1989) observed that in *Crotalaria retusa* and in *C. verrucosa,* TAN was decreased under saline conditions. Gaikwad (2003) also showed a decreased TAN in *Catharanthus* due to salinity.

An increment in the TAN in both the plants species investigated might have added to their mechanism of salt tolerance. Therefore, it is suggested that both *Tribulus terrestris* L. and *Pedalium murex L.* is likely to be salt-tolerant species.

2.3.3 ENZYMES RELATED TO OXIDATIVE STRESS

Plants in the environment are exposed to a variety of abiotic stresses, such as high light intensity, temperature extremes, drought, high salinity, etc. These stresses act as a catalyst in producing free radicals such as superoxide (O2–), hydroxyl (OH), and H2O2.

The plants have some enzymes to overcome the damage caused by oxygen toxicity or ROS, and these enzymes are called antioxidative enzymes. In the present investigation, the enzymes like superoxide dismutase, catalase, and POX were studied for their antioxidative property in the presence of NaCl salinity stress.

2.3.3.1 SUPER OXIDE DISMUTASE

Superoxide dismutase (EC.1.15.1.1.) is one of the important enzyme acts as antioxidants within a cell. The SOD constitutes the first line of defense mechanism against ROS (O_2^-). In a plant oxygen is produced at any location of a cell, where an electron transport chain (ETC) is present and hence O_2 activation may occur in different compartment of the cell including mitochondria, chloroplast, glyoxysomes, peroxisomes, apoplast, and the cytosol. Therefore, it is crucial that SOD's are present for the removal of O_2^- in the compartments where O_2^- radicals are formed. Superoxide dismutase converts the O_2^- or superoxide radical into H_2O_2 which is less lethal as compare to superoxide radicals.

Effect of NaCl salinity on the activity and soluble protein content of enzyme Super oxide dismutase of *Tribulus terrestris* L. and *Pedalium murex* L. is shown in Table 2.3 and depicted Figure 2.3. From the observations, it is clear that, the activity of enzyme Superoxide dismutase both on fresh tissue and protein basis in leaves of *Tribulus* Increases up to 200 mM NaCl concentration and later on as the level of salinity increases the activity decreases, while in *Pedalium* the activity of enzyme both on fresh tissue and protein basis increases up to 300 mM NaCl concentration and then decreases at higher, i.e., 400 mM NaCl salt concentration. In both the plants, the correlation between the activity of enzyme Superoxide dismutase and salinity is negative and not significant. The MSI value for *Pedalium* is more than that of *Tribulus*. The observation is supported by Agarwal and Pandey (2004) in *Cassia angustifolia*. Jungklang et al. (2004) in *Sesbania*; and Demiral et al. (2005) in *Hordeum*. From the present results, it appears that NaCl stress act as a catalyst in producing free radical reactions resulting in oxidative stress where ROS are produced. These ROS are dismutated by SOD whose activity was increased with the increasing levels of salinity. This increased activity of SOD might be helping in minimizing the hazardous effects of ROS in these plants under such adverse conditions.

TABLE 2.3 Effect of NaCl Salinity on Soluble Proteins and Activity of Enzyme Superoxide Dismutase in the Leaves of *Tribulus terrestris* L. and *Pedalium murex* L.

NaCl Treatment (mM)	**Superoxide Dismutase**			
	***Tribulus terrestris* L.**		***Pedalium murex* L**	
	Enzyme Activity	**Soluble Protein**	**Enzyme Activity**	**Soluble Protein**
0 (control)	5	8.3	4	5.8
100	5	8.3	5	8.3
200	6	12.0	6	11.0
300	4	5.8	7	13.2
400	2	3.6	3	5.1
MSI	85.0	89.45	131.25	162.06
t	–2.766	–2.7204	–2.7575	–2.7050
r	–0.7298	–0.5981	0	0.1613

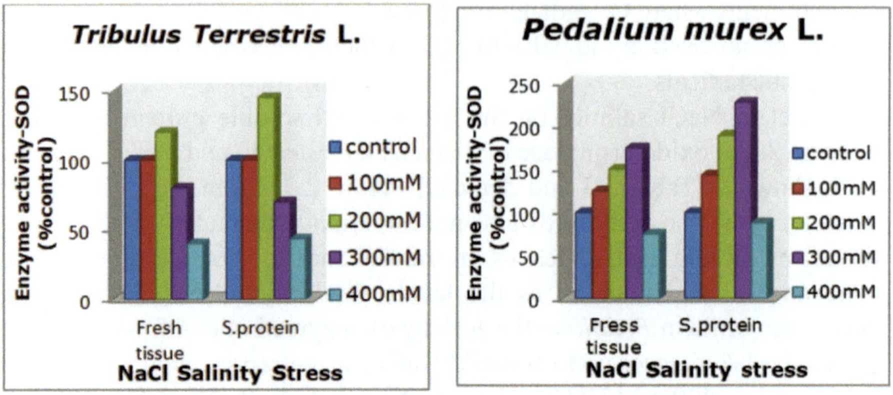

FIGURE 2.3 Effect of NaCl salinity on soluble proteins and activity of enzyme superoxide dismutase in the leaves of *Tribulus terrestris* L. and *Pedalium murex* L.

Each value is a mean of three determinations.

2.3.3.2 CATALASE

Catalase (E.C.1.11.1.6) represents one of the most important enzymes involved in protective process in the cell as it brings about degradation of harmful metabolite, H_2O_2. H_2O_2 is a form of ROS generated as a consequence of oxidative stress; H_2O_2 is generated via superoxide in rather an uncontrollable manner during electron transfer processes which are part and parcel of respiration and photosynthesis. According to Bartosoz (1997) H_2O_2 generation during electron transport is increased as a result of exposure to environmental stress such as excess drought and cold. Neill et al. (2002) recorded that H_2O_2 generation is also promoted in plants following exposure to a wide variety of abiotic and biotic stimuli such as extremes of temperature, UV radiations, ozone exposure, and excess excitation energy, phytohormone like ABA, dehydration, wounding, and pathogenesis. Bolwell et al. (2002) suggested that potential source of H_2O_2 generate NADPH oxidase, cell wall POX, amine oxidase, oxalate oxidase, and flouin-containing oxidase. H_2O_2 is a relatively long-lived molecule that can diffuse some distances from its production site. According to Dat et al. (2000), H_2O_2 serve as a signaling molecule in plants under stress condition. It may inactivate enzymes by oxidizing their thiol groups. Vranova et al. (2000) reported that H_2O_2 could bring about programmed cell death (PCD). Willekens et al. (1997) indicated that Catalase can serve as a sink for H_2O_2.

The effect of NaCl salinity on the activity and soluble protein content of enzyme Catalase of *Tribulus terrestris* L. and *Pedalium murex* L. is shown in Table 2.4 and depicted in Figure 2.4. From the observation, it is clear that in both the plants, the activity of enzyme Catalase both on fresh tissue and protein basis increased up to 200 mM NaCl concentration in *Tribulus* and up to 300 mM NaCl concentration in *Pedalium* but later on at higher salinity concentration its activity is decreased.

TABLE 2.4 Effect of NaCl Salinity on Soluble Proteins and Activity of Enzyme Catalase in the Leaves of *Tribulus terrestris* L. and *Pedalium murex* L.

NaCl Treatment (mM)	Catalase			
	Tribulus terrestris L.		*Pedalium murex* L	
	Enzyme Activity	Soluble Protein	Enzyme Activity	Soluble Protein
0 (control)	6	12.2	6	12.2
100	8	16.1	7	14.1
200	9	18.3	9	18.3
300	6	12.2	11	20.9
400	4	7.9	5	10.4
MSI	112.50	111.68	131.25	162.06
t	–2.766	–2.7204	–2.7206	–2.6127
r	–0.4866	–0.4925	0.1313	0.1165

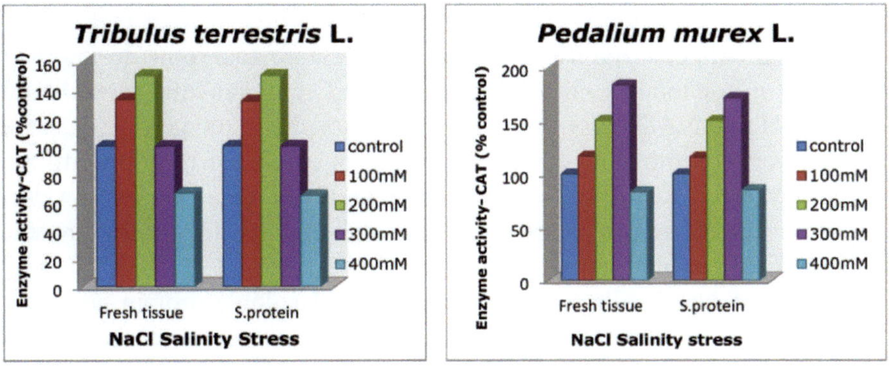

FIGURE 2.4 Effect of NaCl salinity on soluble proteins and activity of enzyme catalase in the leaves of *Tribulus terrestris* L. and *Pedalium murex* L.

In both the plants, the correlation between the activity of enzyme Catalase and salinity is negative and not significant. The MSI value for *Pedalium* is more than that of *Tribulus*.

The above findings are supported by Saha and Gupta (1997) he reported that the lower doses of NaCl (5 and 10 mmhos/cm) promoted Catalase activity in sunflower but further increase in NaCl concentration Catalase activity reduces. Similarly, Venkateshan et al. (1999) in Ipomoea pes-caprae reported that concentration of NaCl up to 20 mM stimulates the activity of enzyme Catalase, but an extreme level of salinity, the activity of these enzymes inhibited. The same trend is observed by Manikanandan and Venkatesan (2004) in *Aegiceros corniculatum.* Jungklang et al. (2004) in *Sesbania.* also observed increase in catalase activity.

Opposite to this view, the inhibitory effect of salinity on the activity of enzyme Catalase by Cavalcanti et al. (2004) in cowpea and Tejera et al. (2004) in *Phaseolus vulgaris*.

Each value is a mean of three determinations.

2.3.3.3 PEROXIDASE (POX)

Peroxidase (E.C.1.11.1.7) is one of the most extensively studied enzymes in plant. It is a heme-containing enzyme ubiquitous in the plant kingdom and belonging to a superfamily which comprises class I enzymes from mitochondria, chloroplast, and bacteria. Class II from fungi and class III from classical plant POX Welider (1992). Plant POX are glycoproteins characterized by the presence of oligosaccharide chains linked to the protein moiety which are having an effect on the stability of the enzyme, Hu and Huystee (1989). Ivanova et al. (1967) investigated that POX is associated with electron transfer from $NADH_2$ to cytochrome during respiration in mitochondria, in the absence of H_2O_2; it can catalyze oxidation of NADH and NADPH with the help of atmospheric oxygen. H_2O_2 serves as electron acceptor and a redox reaction is catalyzed by POX utilizing aromatic amino acid ferro cytochrome 'c' and $NADPH_2$. Pelvi et al. (1972) reported that the reduction of nitrate is catalyzed by POX in the presence of specific electron donors.

Effect of NaCl salinity on the activity and soluble protein content of enzyme POX of *Tribulus terrestris* L. and *Pedalium murex* L. is shown in Table 2.5 and depicted in Figure 2.5. From the observation, it is clear that in both the plants, the activity of enzyme POX both on fresh tissue and protein basis increases up to 200 mM NaCl concentration in *Tribulus* and up to 300 mM NaCl concentration in *Pedalium* but later on at higher salinity level its activity decreases. In both the plants, the correlation between the activity of enzyme POX and salinity is negative and not significant. The MSI value of *Tribulus* is more than that of *Pedalium*.

TABLE 2.5 Effect of NaCl Salinity on Soluble Proteins and Activity of Enzyme Peroxidase in the Leaves of *Tribulus terrestris* L. and *Pedalium murex* L.

NaCl Treatment (mM)	Catalase			
	***Tribulus terrestris* L.**		***Pedalium murex* L**	
	Enzyme Activity	**Soluble Protein**	**Enzyme Activity**	**Soluble Protein**
0 (control)	2	1.4	3	1.7
100	3	1.7	4	2.9
200	5	3.5	5	3.5
300	3	1.7	6	5.8
400	3	0.8	2	1.4
MSI	150.0	137.50	141.66	200.00
t	–2.7887	–2.8026	–2.7717	–2.7849
r	–0.2132	–0.1881	0	0.2071

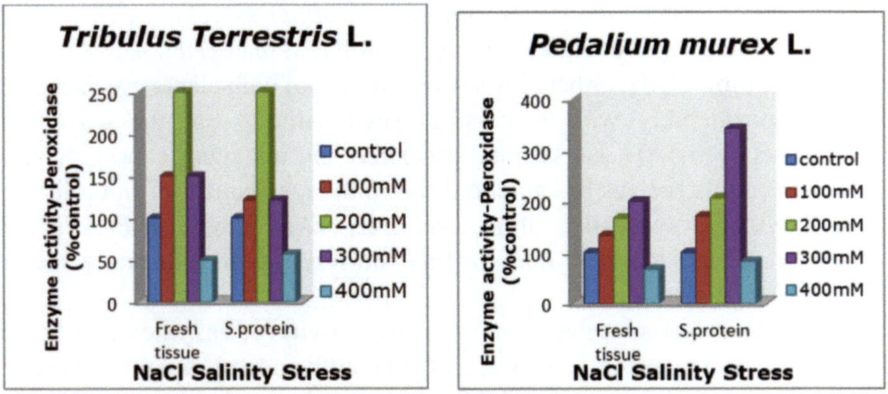

FIGURE 2.5 Effect of NaCl salinity on soluble proteins and activity of enzyme peroxidase in the leaves of *Tribulus terrestris* L. and *Pedalium murex* L.

The above findings are supported by El-Fouly and Jang (1972) in wheat seedlings, the lower level of salinity had no adverse effect on POX activity but the higher level decreased enzyme activity. Enhancement in POX activity was recorded by Singh et al. (2001) in *lentil* genotype; Singh et al. (2001) in chickpea. Jungklang et al. (2004) in *Sesbania, and* Demiral et al. (2005) in *Hordeum*, also observed an increase in POX activity. On the other hand, decrease in enzyme activity was recorded by Gopalkrishnan et al. (1996) in cotton and Papineli and Pannerselvum (1996) in *Vigna mungo*.

Each value is a mean of three determinations.

In the present investigation, there is enhancement in the activity and soluble protein content of enzymes, which are related to oxygen toxicity in the plant, i.e., superoxide dismutase, catalase, and POX. Such behavior of enzymes related to plant adaptability towards saline condition. From this it can be concluded that both *Tribulus terrestris* L. and *Pedalium murex* are salt-tolerant species.

2.4 CONCLUSION

1. In both the plants the correlation between proline content and salinity is positive. Increased proline content due to salinity is one of the important features so that both the plants are able to tolerate the saline conditions. This indicates that probably Proline plays a significant role in osmotic adjustment of these plants under saline conditions. Proline accumulation is more in *Pedalium* as compared to that in *Tribulus,* so *Pedalium* is more salt-tolerant than *Tribulus*.
2. Titrable acid number (TAN) in the leaves of both plants was increased considerably with increasing NaCl salinity. So, the correlation between TAN and salinity is positive but not significant, indicating that this species has able to develop the mechanism of salt tolerance when grown under saline conditions. TAN is more in *Pedalium* as compared to that in *Tribulus* So *Pedalium* appears to be more tolerant than *Tribulus*.
3. There was an enhancement in the activity of enzymes like Superoxide dismutase, Catalase, and POX both on fresh weight and protein basis up to 200 mM salinity stress in *Tribulus* and up to 300 mM in *Pedalium.* Such behavior of enzymes is related to plant adaptability towards saline conditions. However, such an increased activity of these enzymes might help in minimizing the hazardous effects of ROS in these plants. From these observations, it was concluded that both the plants are salt-tolerant species, but *Pedalium* was more salt-tolerant than that of *Tribulus terrestris* L.

KEYWORDS

- **hydrogen peroxide**
- ***Pedalium***
- **reactive oxygen species**
- **superoxide**
- **titratable acid number**
- ***Tribulus terrestris* L.**

REFERENCES

Agarwal, S., & Pandey, V., (2004). Antioxidant enzyme responses to NaCl stress in *Cassia angustifolia. Biol. Plant., 48*(4), 555–560.

Ahmad, P., & John, R., (2005). Effect of salt stress on growth and biochemical parameters of *Pisum sativum* L. Archives Agron. *Soil Sci., 51*(6), 665–672.

Bandeoglu, E., Eyidogen, F., Yucel, M., & Oktem, H. A., (2004). Antioxidant responses of shoots and roots of lentil to salinity stress. *Plant Growth Regulation, 42*(1), 69–77.

Bartosoz, G., (1997). Oxidative stress in plants. *Acta Physio. Plantarum, 19*, 47–64.

Bates, L. S., Waldren, R. P., & Teare, I. D., (1973). Rapid determination of free proline for water-stress studies. *Plant and Soil, 39*, 205–207.

Bernstein, L., (1964). In: Maybe, A. P., & Mayer, A. M., (eds.), *Physiological Basis of Salt Tolerance in Plants*. Transl. IPSI. Jerusalem.

Berteli, E., Corrales, E., Guerrero, C., Ariza, M. J., Pliego, F., & Vaipuesta, V., (1995). Salt stress increases ferredoxin-dependent glutamate synthase activity and protein level in the leaves of tomato. *Physiol. Plant., 93*, 259–264.

Bolwell, G. P., Bindschedler, L. V., Blee, K. A., Butt, V. S., Davies, D. R., Gardner, S. L., Gernish, C., & Minibayeva, F., (2002). The apoplastic oxidative burst in response to biotic stress in plants. A three component system *J. Exp. Bot., 53*, 1367–1376.

Cavalcanti, F. R., Oliveira, J. T. A., Martins-Miranda, A. S., Viegas, R. A., & Silveira, J. A. G., (2004). Superoxide dismutase, catalase and peroxidase activity do not confer protection against damage in salt-stressed cowpea leaves. *New Phytol., 163*(3), 563–571.

Dat, J., Vandenbeele, S., Vranova, E., Vanmiontaqu, M., Inze, D., & Van-Breusegem, F., (2000). Dual-action of the active oxygen species during plant stress responses. *Cellular and Molecular Life Science, 57*, 779–795.

Demral, M. A., Aydin, M., & Yorulmaz, A., (2005). Effect of salinity on growth chemical composition and antioxidative enzyme activity of two malting barley (*Hordeum vulgare* L.) cultivars. *Turkish J. Biology, 29*(2), 117–123.

Durgaprasad, K. M. R., Muthukumarasamy, M., & Panneerselvam, R., (1996). Changes in protein metabolism induced by NaCl salinity in soybean seedlings. *Indian J. Plant Physiol., 1*(2), 98–101.

Elfouly, & Jung, (1972). Activity of acid phosphatase in wheat seedlings associated with soil stress condition. *Plant Soil, 36*, 497–500.

Fitter, A. H., & Hay, R. K. M., (1987). In: *Environmental Physiology of Plants* (2nd edn.) Academic Press London; Chapter 6: Ionic toxicity, pp. 225–259; Chapter 9: An ecological perspective, pp. 330–346.

Gaikwad, P. V., & Karadge, B. A., (2003). Influence of sodium chloride salinity on growth and organic constituents of *Catharanthus roseus* G. Don. *Indian J. Plant Physiol., 8*(4), 392–397.

Giannopolits, C. N., & Ries, S. K., (1977). Superoxide dismutase I. Occurrence in higher plants. *Plant Physiology, 59*, 309–314.

Gopalkrishnan, N., Khader, S. E. S. P. A., & Nathan, A. R. S., (1996). Enzymatic aspects of alleviation of salinity stress by phytohormones in cotton. In: *National Seminar on 'Role of Crop Physiology in Improving Agricultural Productivity: New Vistas for Plant Physiol. Research* (p. 58). Coimbatore (India).

Hu, C., & Van, H. R. B., (1989). Role of carbohydrate moieties in peanut peroxidases. *Biochem. J., 263*, 129–135.

Immamul, H. S., M., & Larher, F., (1985). Dynamics of sodium ion, potassium ion and proline accumulation in salt-stressed *Vigna sinensis* L. and *Phaceolus Aureus* L. *J. Plant Physiol., 119*, 133–137.

Ivanova, T. M., Davydova, M. A., & Fubin, B. A., (1967). On catalytic function of peroxidase *Biokimiya., 32*, 607–611.

Jungklang, J., Sunohara, Y., & Matsumoto, H., (2004). Antioxidative enzymes response to NaCl stress in salt-tolerant *Sesbania rostrata. Weed-Biology-and-Management, 4*(2), 81–84.

Jungklang, J., Sunohara, Y., & Matsumoto, H., (2004). Antioxidative enzymes response to NaCl stress in salt-tolerant *Sesbania rostrata. Weed-Biology-and-Management, 4*(2), 81–84.

Krishnamoorthy, H. N., & Siddique, S., (1985). *Effect of Salinity and Betanine on Vigna unguiculata (Cowpea Var. HFC-42-1) NaH.* Seminar on plant physiology. BHU Varanasi 1985. Abstracts of papers, 66–67.

Krishnamurthy, R., Anbazhangan, M., & Bhagwat, K. A., (1987). Chlorophyll breakdown and sodium chloride toxicity of rice. *Ind. J. Agric. Sci., 57*, 567–570.

Lowry, O. H., Rosebrough, N. J., Farr, A. L., & Randall, R. J., (1951). Protein measurement with the folin phenol reagent. *J. Biol. Chem., 193*, 265–275.

Maehly, A. C., (1954). In: Gick, D., (ed.), *Methods in Biochemical Analysis* (Vol. 1, p. 385, 386). Interscience Publishers Inc. New York.

Maighany, F., & Ebrahimzadeh, H., (2002). The effect of salinity stress on proline metabolism in two wheat (*Triticum aestivum*) cultivars. *Rostaniha., 3*(1–4), 35–37.

Manikanandan, T., & Venkatesan, A., (2004). Influence of NaCl on growth, organic constituents and in certain antioxidant enzymes of *Aegiceros corniculatum* Blanco. *Geobios, 31*(1), 21–24.

Mass, E. V., & Hoffman, G. J., (1977). Crop tolerance current assessment. *J. Irrig. Drainage De. Am. Soc. Civ. Eng., 103*, 115–134.

Muthukumarasamy, M., & Panneerselvam, R., (1997b). Growth and biochemical changes in green gram under NaCl stress. *Geobios, 24*(2, 3), 107–111.

Neill, S. T., Desikan, P., Clarke, A., Hurst, R. D., & Hancock, J. F., (2002). Hydrogen peroxide and nitrate oxide as signaling molecules in plants *J. Exp. Bot., 53*, 1237–1247.

Panda, S. K., (2001). Oxidative response of green gram seeds under salinity stress. *Ind. J. Plant Physiol., 6*(4), 438–440.

Papinehi, S., & Pannecrelvum, R., (1996). Effect of calcium chloride on the metabolism of NaCl stressed *Vigna mungo* seedling. In. *National Seminar on "Role of Crop Physiology in*

Improving Agricultural Productivity: New Vistas for Plant Physiol. Research," Coimbatore (India).

Pelvi, Y. A. V., Narova, N. N., & Drohysillivia, N. I., (1972). Nitrate reducing activity of plant peroxidase. *Fiziologia Rasteini, 19,* 340–347.

Perez-Alfocea, F., Estan, M. T., Caro, M., & Guerrier, G., (1993). Osmotic adjustment in *Lycopersicon esculentum* and *Lycopersicon pennellii* under sodium chloride and polyethylene glycol 6000 isoosmotic stresses. *Physiol. Plant, 87*(4), 493–498.

Popp, M., & Kinzel, H., (1971). Chemical composition of Australian mangroves I. Inorganic ions and organic acids. *Z. Pflanzenphysiol., 113,* 395–409.

Rao, G. G., & Rao, R. G., (1979). Sodium chloride salinity induced keto-acid-amino acid changes in groundnut (*Arachis hypogaea* L.) leaves. *Indian J. Plant Physiol., 22*(2), 121–126.

Ro-Dong, P., (1982). Changes in the contents of some metabolites and ions and in some metabolites and ions and in some enzyme levels in rice plants grown under water and salt-stressed condition. *Hanguk Nonghwa Hakhoe Chi., 25,* 135–141.

Saha, K., & Gupta, K., (1997). Effect of NaCl salinity on ethylene production and metabolism in sunflower seedling. *Indian J. Plant Physiol., 2,* 127–130.

Sahoo, S. K., Sahoo, N., & Sahu, A. C., (2001). Sodium chloride stress-induced proline accumulation in attached senescing leaves of rice cultivars. *Ind. J. Plant Physiol., 6*(4), 423–426.

Salunkhe, S. S., & Karadge, B. A., (1989). Influence of salinity and water stress on growth and organic constituents of *Dodonaea viscosa* L. *Geobios, 16,* 252–258.

Sing, A. K., Singh, R. A., & Sharma, S. G., (2001). Influence of salinity on activity of hydrolytic and oxidative enzymes in chickpea seedlings. *Indian J. Plant Physiol., 6,* 84–86.

Singh, S. P., Singh, B. B., & Singh, M., (1994). Effect of kinetin on chlorophyll, nitrogen and proline in mungbean (*Vigna radiata*) under saline conditions. *Indian J. Plant Physiol., 37*(1), 37–39.

Strogonov, B. P., (1964). Salt tolerance of plants. *Agric. Inf. Bul U.S Dept. Agric*., 283.

Tejera, N. A., Campos, R., Sanjuan, J., & Lluch, C., (2004). Nitrogenase and antioxidant enzyme activities in *Phaseolus vulgaris* nodules formed by *Rhizobium tropici* isogenic strains with varying tolerance to salt stress. *J. Plant Physiol., 161*(3), 329–338.

Thomas, R., & Beevers, M., (1949). Physiological studies in acid metabolism in green plants. II. Evidence of CO_2 fixation in *Bryophyllum calycium* and the study of diurnal fluctuation in the genus. *New Phytol., 48,* 421–447.

Toro, S. V., (2000). *Studies in Physiology of Salt Tolerance in Setaria italica L. Beauv.* A PhD Thesis submitted to Shivaji University, Kolhapur, India.

Venkatesan, A., Chellappan, K. P., & Vankatesalu, V., (1997). Salinity stress on mineral nutrition and growth of *Ipomoea pes-caprae* sweet. *Geobios, 24,* 112–118.

Vranova, E., Inze, D., & Van Breusegem, F. (2002). Signal transduction during oxidative stress. *J. Exp. Bot. 53*(372), 1227–1236.

Wadkar, P. R., (1989). *Studies in Salt Tolerance of Crotalaria L. Species.* A Ph.D. Thesis Submitted to Shivaji University, Kolhapur (India).

Welinder, K. G., (1992). Superfamily of plant-fungal and bacterial peroxidases. *Curr. Opinion Structure Biol., 2,* 388–392.

Willekens, H., Chamnongpol, S., Davey, M., Schraudner, M., Langebartels, C., Montagu, M. V., Inz'e, D., & Camp, W. V., (1997). Catalase is a sink for H_2O_2 and is indispensable for stress defense in C_3 plants. *The EMBO Journal, 16,* 4806–4816.

CHAPTER 3

Adaptation Strategies of *Rhododendron* Under Environmental Stress

SHREYASI BISWAS and MAHUYA MUKHOPADHYAY

Department of Microbiology, Lady Brabourne College, Kolkata–700017, West Bengal, India, E-mails: mahuya.mukhopadhyay@ladybrabourne.com; moumahuya1@yahoo.com; hornbillgorumara@gmail.com (M. Mukhopadhyay)

ABSTRACT

Natural resources, especially flora and fauna are disturbed due to natural and anthropogenic stresses. With the diminishing green vegetation everywhere, *Rhododendrons* are also experiencing adverse effects of environmental stress. Several species have been extinct in the wild, many are considered as threatened with extinction, and they require a critical conservation. It is one of the most important genera in horticulture plants which have several medicinal and economic values. These woody flowering plants are distributed throughout the mountains in the world. These plants normally grow in temperate climates in acidic soil. Cold and drought stresses due to climatic and edaphic factors, limit the growth and development of several plants, including *Rhododendrons*. The successful growth of the plants is the resultant of combating the power of the plants with the stress factors. The endophytic microorganisms play a great role in the stress response by the plant and help their survivability in extreme conditions. This review deals with a few aspects of stress management mechanism and conservation of these important ornamental plants.

Plant Metabolites under Environmental Stress: Mechanisms, Responses, and Adaptation Strategies.
Nivas M. Desai, Manasi Patil, Umesh R. Pawar (Eds.)

3.1 INTRODUCTION

Plants of Ericaceae are widespread in all continents except Antarctica (Luteyn, 2002). *Rhododendrons* belong to the family Ericaceae, distributed in Himalayan ranges with a wide variety of beautiful flowers in India. These are evergreen shrubs or trees. The word "*Rhododendron*" is derived from Greek word "rhodo" means "rose" and "dendron" means "tree." In West Bengal they are found in Darjeeling and Sikkim Himalaya and its most popular name is 'Gurans.' Plant species of this genus have great medicinal and economic importance (Prakash et al., 2016; Negi et al., 2013; Gautam et al., 2020; Kashiwada et al., 2001). *Rhododendrons* are rich in phytochemicals (Kiruba et al., 2011). Fresh flowers of *Rhododendron arboreum* are used to make jams, jellies, and local 'brew' which is very efficacious in the prevention of altitude sickness. As we know various stress conditions either biotic or abiotic have shown effects on plants and plants also survive under those situations via adaptation (Riverola et al., 2019).

3.2 HABITATS OF RHODODENDRONS

This is the largest genus that belongs to the Ericaceae family and widely distributed in Sino-Himalayas (areas of Eastern Himalayas and western China). Apart from this, *Rhododendrons* are found in Japan, Thailand, Malaysia, Indonesia, Myanmar, Philippines, Australia, Pakistan, Afghanistan, northern America, and southern Europe. *Rhododendrons* occupy the mountains at high altitudes in northeast India, and about 121 taxa have been reported from India (Mao, 2010). In another report 109 taxa were found in India, diversity was found in Arunachal Pradesh, Sikkim, and Darjeeling, few were reported in Jammu and Kashmir (Spethmann, 2010). *Rhododendrons* usually prefer well-drained acidic soil with humans and some of its species can grow in sheltered wet sand along with mountain rivers. In West Bengal, this genus is widely extended with its ineffable attractive flowers in Sandakphu range. *Rhododendron arboreum* is most dominating species of the Himalayas (Tiwari and Chouhan, 2006). A new variety of *Rhododendron grande* Wight named *Rhododendron grande* var *singalense* was reported in the year 2014 from Singalila National Park, India (Rai et al., 2014). In 2016, another new species was found in Singalila National Park was named as *Rhododendron apradae*, which has a close resemblance to *Rhododendron arboreum* subsp. *cinnamomeum* and *Rhododendron grande* Wight (Rai et al., 2016). *Rhododendron* is considered as 'keystone element' in the Himalayan range as it has an impact on other species, forest management and succession. Well, this was observed in case of Sikkim falls which is a center of biodiversity and endemism (Figures 3.1 and 3.2) (Shenga, 1994; Kumar, 1993).

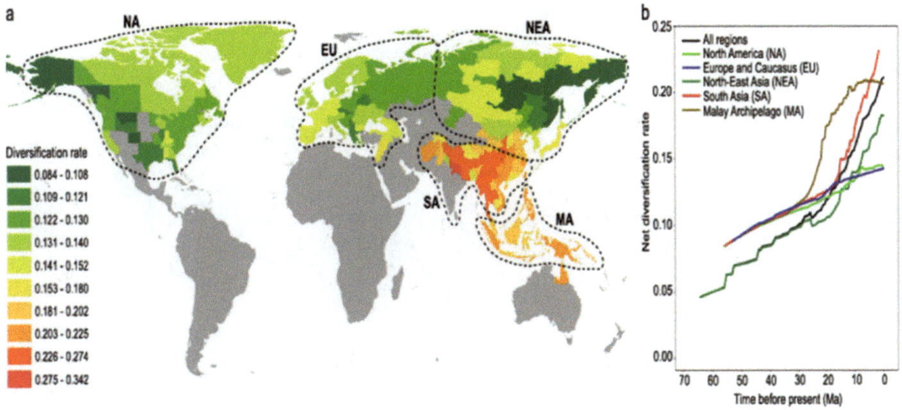

FIGURE 3.1 Global distribution and diversification of *Rhododendron.*

Source: Reprinted from Shrestha et al., 2018. Open access.

3.3 WHAT IS STRESS FOR PLANTS AND HOW THEY SURVIVE UNDER THESE CONDITIONS?

Stress for plants means any external condition that adversely effect on plants by reducing their growth, development, and reproduction rate. Plants in higher altitudes or in desert areas face different types of environmental stresses like oxidative stress, water stress, drought stress, stress due to UV radiation, and metal ion stress (Foyer et al., 1994; Connolly et al., 2002). Plants survive under these conditions by adopting various mechanisms like altered gene expression, producing secondary metabolites, changes in metabolic pathways, etc. (Foyer et al., 2006; Divi et al., 2016). Brassinosteroid is reported as a naturally occurring plant steroid in response to various stress conditions (Krishna, 2003).

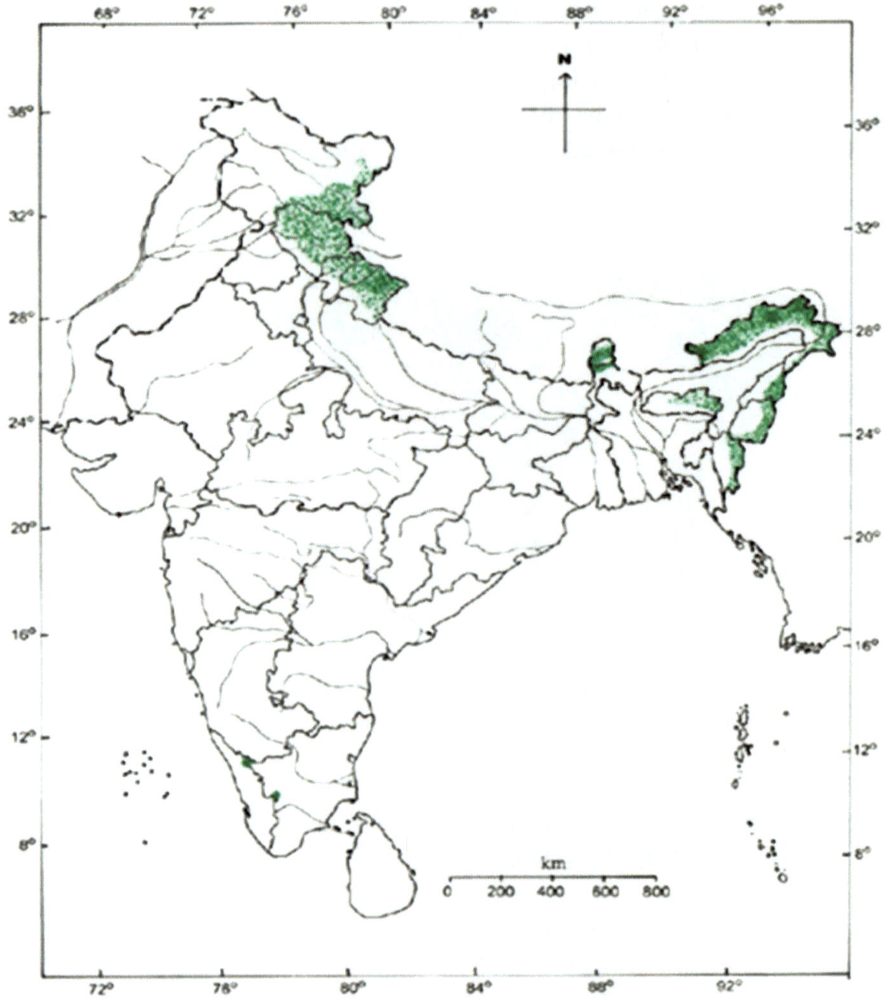

FIGURE 3.2 Habitats of *Rhododendrons* in India.

Source: National Atlas and Thematic Mapping Organization, Department of Science and Technology, Government of India, Kolkata – 700064; Map outline drawn from India–Physiography, Published in *Rhododendron* habitats in India, Debjyoti Bhattacharyya and M. Sanjappa Botanical Survey of India Salt Lake City, Kolkata, West Bengal, India. *Journal American Rhododendron Society*, 2008, Vol. 1, p. 62).

3.4 TYPES OF STRESS

There are various types of stresses for plants:

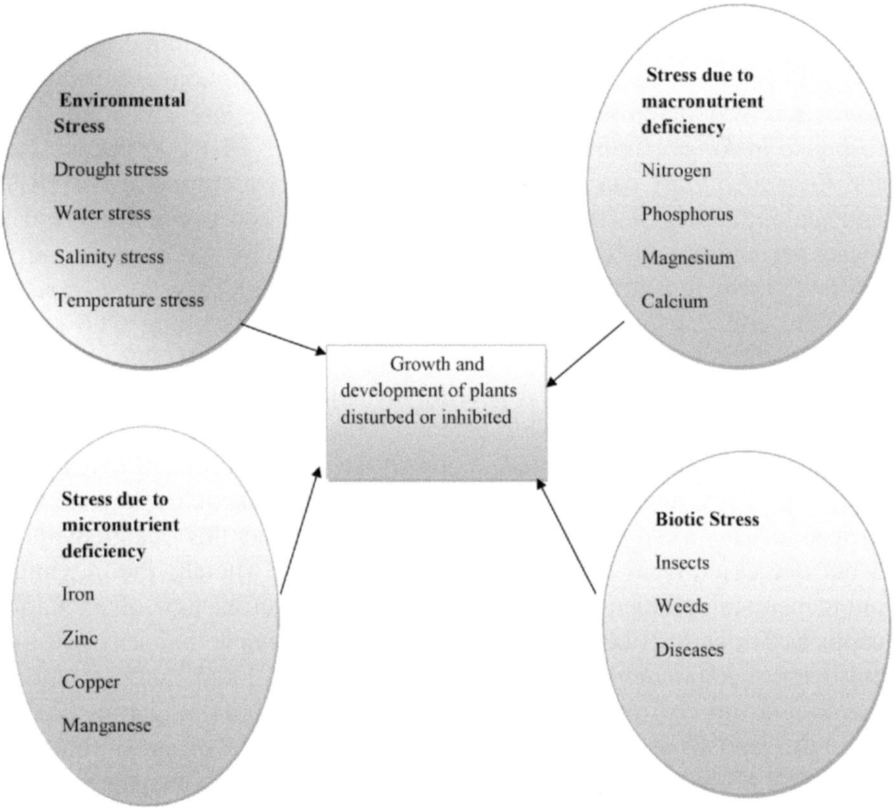

3.4.1 DROUGHT AND HEAT TOLERANCE OF RHODODENDRONS

Drought stress is limitation of water which causes reduced growth rate, reduced photosynthesis, and accumulation of reactive oxygen species (ROS), etc. Basically, there are genetic regulations behind the survival strategies of *Rhododendrons* to withstand such stress conditions. One such finding suggested that differentially expressed genes (DEGs) encoding photosystem proteins, chlorophyll protein, antioxidants were down regulated under drought conditions but immediately restore their functions upon re-watering (Cai et al., 2019). Recently in another report there are 310 and 1,452 DEGs were expressed under moderate and severe drought conditions, respectively, and show protein protection activity, metabolic process in *Rhododendron pulchrum* (Wang et al., 2020). Thermal stress tolerance of *Rhododendrons* were investigated but not completely explored yet. The malondialdehyde, hydrogen peroxide (H_2O_2), superoxide radicals have been used to study the heat stress tolerance of this plant(Wang et al., 2011). Illumina sequencing and qRT-PCR analysis resulted in various transcription factors (TFs) were involved in crosstalk between hormonal signaling and heat stress response in *Rhododendron hainanense* (Zhao et al., 2018). Excessive heat for *Rhododendrons* can cause electrolyte leakage, accumulation of H_2O_2 and malondialdehyde, reduction of chlorophyll a and b contents. Due to these peroxidases (POXs) and superoxide dismutase (SOD) respond to malondialdehyde and reduced its level in leaves of two *Rhododendron* cultivars and it was also proved that stomatal factors along with xylem vessels, chloroplasts are important under heat stress of *Rhododendron* cultivars (Shen et al., 2017). Measuring the chlorophyll fluorescence kinetic parameters is a useful method to detect the relation between photosynthetic activities and heat stress tolerance. Four *Rhododendron* cultivars were treated under 38°C/35C and their chlorophyll fluorescence kinetic parameters were measured. The initial fluorescence was increased under this treatment at day 3 but decreased from day 6 to 9 in four cultivars, whereas the maximum photochemical efficiency of photosystem II and efficiency of excitation capture of open PSII center decreased in an undetectable manner in all four cultivars of *Rhododendron*, suggesting the structural damage to thylakoid membrane and PSII reaction center at high temperature (Fang et al., 2017). Salicylic acid most important phenolic phytohormone of stress tolerance in plants, always had a positive impact on growth, photosynthesis, and transpiration of plants. Combined application salicylic acid and $CaCl_2$ had superior result on increasing chlorophyll content, enzymatic antioxidant activities and total soluble proteins under heat stress treatment *in vitro* (Shen et al., 2016). Net photosynthetic rate (P_N) of Rhododendrons affected by high temperatures. The work by Ranney et al. showed a higher P_N at 40°C in *Rhododendron hyperythrum* which proves its well adaptability under excessive high temperature but the opposite results was found in case of *R. russatum* (Ranney et al., 1995). Brassinolide is an important plant hormone which reduced the heat damage and improving the heat tolerance at 0.50 mg/L concentration at *R. lapponicum* seedlings (Figure 3.3) (Li et al., 2017).

3.4.2 COLD STRESS TOLERANCE IN RHODODENDRONS

There are two strategies adopted by plants to combat low temperature stress, one is tolerance and other one is avoidance. Tolerance means acclimatization indicating the functional and structural changes acquired

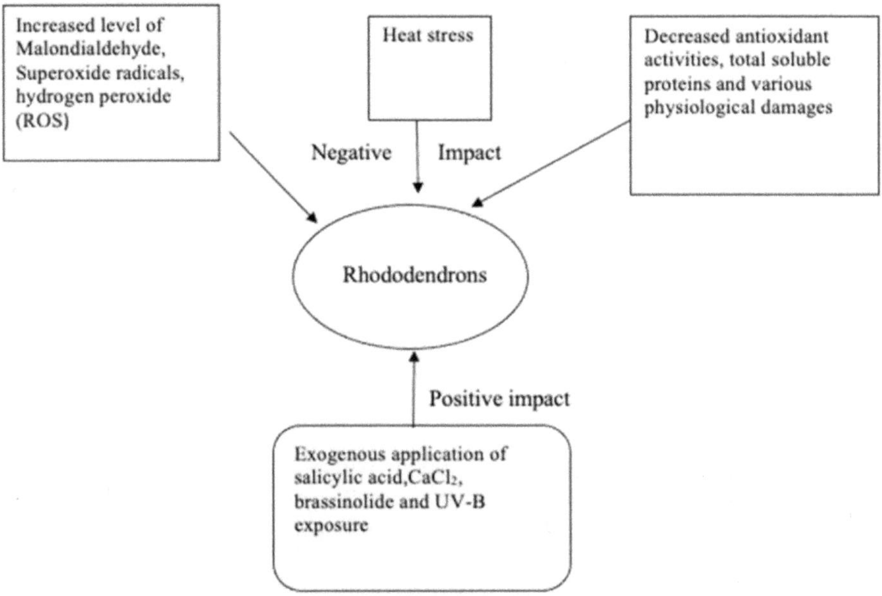

FIGURE 3.3 Schematic presentation of heat stress on *Rhododendrons*.

by plants to minimize the harmful effects and induce fitness (Kacperska, 1999). Avoidance is something, which includes the prevention of freezing of sensitive tissues, some plants rather succulents accumulate heat and release it slowly at night during cold (Nilsen and Orcutt, 1996). Woody perennials survive under freezing temperature via several adaptive mechanisms. Normally these plants have juvenile (pre-flowering) and mature stages, cold hardiness increases with the age during the juvenile stage of plants. One such finding based on T_{max} (the temperature at which maximum rate of freezing injury occurs) showed that significant annual increase in T_{max} value ranged from 3.7 to 6.4°C in the seedlings of age 3 to 5 years of old (Lim et al., 2014). In cold-acclimated *Rhododendron*s, few genes were expressed at a high level, and some were downregulated. So, it is evident from the studies that genetic regulation is the most important factor in the adaptation process. As for example, structural genes as well as genes for glycolysis and stress response were much enriched rather than energy-related genes during cold acclimatization, though some housekeeping genes were reported with the same expression in both cold-acclimated and non-acclimated *Rhododendrons* (Wei et al., 2005). Generally, during autumn, acclimation occurs in temperate plant species, in which different heat shock proteins, dehydrins, cold-regulated proteins help to stabilize membrane proteins and phospholipids, cytoplasmic membranes, and cryoprotectants like soluble sugars low–molecular weight nitrogenous compounds were produced as well (Iba, 2002; Gusta et al., 2004; Hare et al., 1998). Dehydrins are stress proteins and belong to the late embryogenesis abundant (LEA) protein family. These proteins basically expressed in different stress conditions in plants (Hanin et al., 2011). Analysis of expressed sequence tags (ESTs) from leaves of *Rhododendron catawbiense* revealed that there were five dehydrin genes detected in up regulated form, when they were isolated from cold-acclimated leaves during winter and subsequent down-regulation was reported in summer as well; that proves the stress response-ability of dehydrin proteins (Wei et al., 2019). The dehydrins can protect and rescue enzymatic activities. RcDhn5, a SK_2 type acidic dehydrin from *Rhododendron catawbiense* showed an upregulation and downregulation during cold acclimation (winter) and downregulation in non-acclimation state (summer) respectively (Peng et al., 2008). Cold acclimation results in seasonal up-regulation of many proteins including carbohydrate metabolism proteins, cell membrane proteins, RuBisCo, and many more were reported in literature (Die et al., 2017). Another well-known strategy of avoidance is supercooling in which ice nucleation is inhibited by preventing ice nucleators formation at the temperature below – 30°C. This supercooling is associated with water content in florets. In the year 1981, one such finding suggested that supercooling of flower buds was acquired by migration of water from florets to scales, which lowers the water content in florets and peduncle but results in an increase in scales of flower buds *in vitro*. The same process was followed in that experiment where flower buds are used without florets but that gave just opposite result and cause damage to flower buds and scales (Masaya Ishikawa and Sakai, 1981). Apart from this, structural adaptation is also an important strategy of plants in response to change in temperature. Thermonasty (non-directional response due to temperature) is a much-related term in this adaptation. Anatomical features of leaves reported in studies showed increased thickness in leaf blades, differences in stomatal density give protection against photoinhibition in winter (Wang et al., 2008). Field grown *Rhododendron catawbiense* is a cold-hardy plant reported in many literatures. One such study has shown that in this *Rhododendron* species phospholipid biosynthesis, desaturation, downregulation of aquaporin and upregulation of gated outward rectifying K^+ channel are associated with freezing tolerance ability (Wei et al., 2006). *Rhododendrons* can survive at cold climate with the help of UV-B cross-protection. *In vitro* analysis was done to show that UB-B (280–320 nm) irradiated leaves of *Rhododendrons* can survive temperatures below – 8°C but control plant leaves died at – 6°C and this UV-B exposure may promote phenolic compounds production and helped in heat tolerance (Scott and Scott, 2004; Dunning et al., 1994).

3.5 HOW *RHODODENDRONS* SURVIVE UNDER METAL STRESS CONDITIONS?

Heavy metal toxicity is increasing day by day due to industrialization and development of population and is the main hindrance for plant development. Endophytes in plants play several roles for growth and defense in plants. Ericoid mycorrhizal (ERM) fungi played an important role in this stress tolerance. Three ericaceous plant species *Rhododendron ponticum, Calluna vulgaris, and Vaccinium macrocarpon* were grown in mycorrhizal and non-mycorrhizal conditions *in vitro*, with copper and zinc of various concentrations, mycorrhizal plants grow well at all metal concentrations, but non-mycorrhizal plants failed to grow, which means these endophytic fungi are capable to withstand the mental stress and promote growth of these plant species (Bradley et al., 1982). *Rhododendron*s not only survive in stress conditions but also help other plants to grow and combat with stresses. Methanol leaf extracts of rhododendron restored normal cell growth, pigment contents, stomatal structures and also reduces Chromium accumulation in *Vigna radiata* plants (Gautam et al., 2021).

3.6 SALINITY STRESS

Salinity stress or salt stress is an accumulation of excessive salt in soil which eventually ceases plant growth and development. Salt accumulation can occur in everywhere. In a case study, growth of *Rhododendron* was inhibited under $NaHCO_3$ stress, but exogenous application of β-aminobutyric acid resume its growth and significantly enhanced chlorophyll a, chlorophyll b, carotenoid, net photosynthetic rate, activities of some antioxidant enzymes like glutathione reductase, POX, SOD, ascorbate peroxidase (APX) (Xu et al., 2018). Salt tolerance and antioxidants are correlated in *Rhododendrons*. Recently it was observed in 3 subcellular compartments of *Rhododendron* cultivars. This was investigated by accumulation of malondialdehyde due to high superoxide radicals and H_2O_2 under NaCl stress conditions. In one *Rhododendron* cultivar named 'Yanzhi Mi' leaves showing no significant difference in malondialdehyde content in 3 subcellular compartments-mitochondria, cytosol, and chloroplasts which indicates its good salt-tolerant ability, apart from this the antioxidant enzymes were also high in cytosol of these three cultivars than the other two subcellular compartments (Liu et al., 2020). Injury to plants due to salt stress resulted in membrane damage, malondialdehyde production though the protein content increased two folds in NaCl treatment in *Rhododendron arboreum* (Chhetri, 2014). To investigate stress tolerance, *in vitro* stress imposition is the simplest method, as for example, due to increasing stress condition proline content and carbohydrate production were enhanced in *R. arboreum* (Chhetri and Lepcha, 2015).

3.7 VARIOUS METHODS TO IMPROVE GROWTH AND DEVELOPMENT OF *RHODODENDRONS*

Methods indicated in this include several techniques for improvement of these ornamental plants. It could be soil treatment, association of soil micro-biota with *Rhododendrons*, or endophytes. A survey was conducted with soil treatment in which different types of soil like rhizosphere soil, degraded soil, sterile soil were used for seed germination of rare and endangered species of rhododendrons of Sikkim Himalaya. Best seed germination was found in rhizosphere soil mixed with degraded soil at the ratios of 1:1 and 3:1 and mean net photosynthetic rate was also higher in rhizosphere soil (Singh et al., 2008). Soil biota always have positive and negative impacts on plants, but in one study, it was reported live soil biota can confer resistance to disease caused by *Phytophthora cinnamommi* (Liu et al., 2020). Endophytes are microorganisms which enter into the plants from the soil and promote plant growth and development without showing any harmful effects (Hurek and Hurek, 2011). Mycorrhizal fungi isolated from *Rhododendron fortunei* had quite impressive effects on it, as for example, these fungi inoculated *R. fortunei* showed higher fresh and dry weight, absorbed nitrogen, and upregulated genes encoding nitrate transporters, glutamine synthetase, ammonium transporters, glutamate synthetase (Wei et al., 2016). Adventitious root formation is an important factor in plant reproduction, and mycorrhizal fungi can promote this rooting along with the ability to produce tryptophan and indole-3-acetic acid (IAA), and indole-3-pyruvic acid in *Rhododendron* (Wei et al., 2020). Plant microbes help in stress tolerance and the development of plants, but this is also true that plants can improve soil microbes and soil fertility (Zhao et al., 2016). So, the beneficial relationship between plants and soil microbes is bidirectional (Figure 3.4).

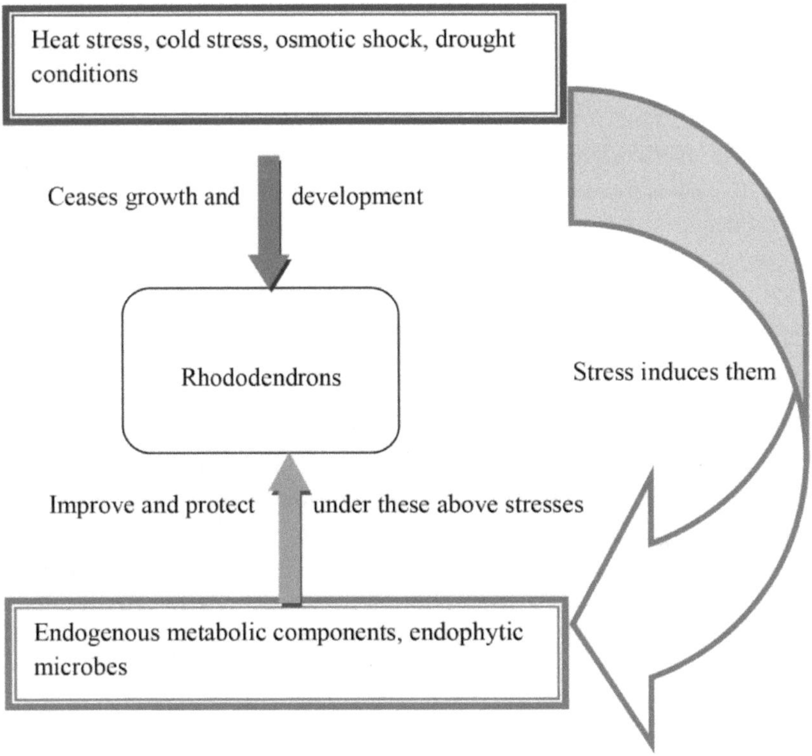

FIGURE 3.4 Relation between various stresses and stress-tolerant mechanisms of *Rhododendron.*

3.8 HOW WE CAN SAVE RHODODENDRONS?

Rhododendrons are economic ornamental woody plants that need conservation as many of the species of this genus become threatened and endangered. It is important to know more about their development, metabolism, hormonal regulations to sustain their growth. A case study has shown that major endogenous phytohormones resulted in variations in their expression after cross-pollination of two male parents' genotype of *Rhododendrons* and jasmonic acid reported with highest level in this (Tantawy et al., 2020). Biological diversity is based on genetic diversity, which is the most important aspect of conservation (Gordon et al., 2012). As these woody plants face several biotic, abiotic, and anthropogenic stresses, we have to conserve it. There are various molecular and biotechnical tools that can help to save these woody plants. Micropropagation is a very well-known process in tissue culture, through which multiple progeny plants can be grown rapidly from plantlets. In this technique ERM fungi can help to promote the growth of the plants. *Rhododendrons* inoculated with these ERMs post *vitro* have shown growth and few strains showed good results (Jansa and Vosatka, 2000). Endophytic *Streptomyces* have the ability to induce defense against diseases in *Rhododendron*s. Surprisingly antagonistic *Streptomyces* have shown to induce resistance against *Pestalotiopsis sydowiana,* causal agent of Pesailotia disease in seedlings of *Rhododendron* (Shimizu et al., 2006).

3.9 CONCLUSION

Due to different biotic, abiotic stresses and increased population and anthropogenic threats, the vegetation of different geographical regions whether temperate or tropical forests need conservation. It is only possible through the investigation to know more about phytohormones, different secondary metabolites, and the microorganisms that help to survive plants under various environmental conditions, and last but not least, to make people aware of the devastating effects of their own activities like deforestation, constructions on environment and ecology. In the case of *Rhododendrons*, there are many areas to investigate as many hormonal regulations, and actions of secondary metabolites under stress tolerance remain unexplored. It can be a new site in the future for research.

ACKNOWLEDGMENTS

The financial support for this study by the University Grants Commission under NET-JRF scheme is duly acknowledged.

KEYWORDS

- **beneficial microbes**
- **cold stress**
- **drought and heat stress**
- **embryogenesis**
- **rhododendron**
- ***Rhododendrons***

REFERENCES

Bradley, R., Burt, A. J., & Read, D. J., (1982). The biology of mycorrhiza in the Ericaceae, VIII. The role of mycorrhizal infection in heavy metal resistance. *New Phytol., 91*, 197–209.

Cai, Y. F., Wang, J. H., Zhang, L., Song, J., Peng, L. V., & Zhang, S. B., (2019). Physiological and transcriptomic analysis highlight key metabolic pathways in relation to drought tolerance in *Rhododendron delavayi*. *Physiology and Molecular Biology of Plants, 25*, 991–1008.

Chhetri, D. R., & Lepcha, D. L., (2015). Stimulated osmotic stress induces metabolic effects in *Rhododendron arboreum* smith. *Indian Journal of Plant Sciences, 4*(2), 67–70.

Chhetri, D. R., (2014). Metabolic changes *in Rhododendron arboreum* Smith in response to abiotic stress-induced by salinity and abscisic acid. *International Journal of Research in Applied, Natural and Social Sciences, 2*(6), 91–102.

Connolly, E. L., & Guerinot, L., (2002). Iron stress in plants. *Genome Biology, 3*(8), 1024.1–1024.4.

Die, J. V., Arora, R., & Rowland, L. J., (2017). Proteome dynamics of cold-acclimating *Rhododendron* species contrasting in their freezing tolerance and thermonasty behavior. *PLoS One, 12*(5), e0177389.

Dunning, C. A., Scott, L. C., & Scott, J. D., (1994). Exposure to ultraviolet-B radiation increases cold hardiness in *Rhododendron*. *Physiologia Plantarum, 92*(3), 516–520.

Fang, L., Tong, J., Dong, Y., Xu, D., Mao, J., & Zhou, Y., (2017). De novo RNA sequencing transcriptome of *Rhododendron obtusum* identified the early heat response genes involved in the transcriptional regulation of photosynthesis. *PLoS One, 12*(10), e0186376.

Foyer, C. H., Delgado, H. L., Dat, J. F., & Scott, I. M., (2006). Hydrogen peroxide-and glutathione-associated mechanisms of assimilatory stress tolerance and signaling. *Physiologia Plantarum, 100*(2), 241–254.

Foyer, C. H., Leiandais, M., & Kunert, K. J., (1994). Photooxidative stress in plants. *Physiologia Plantarum, 92*, 696–717.

Gautam, S., & Chaudhary, K., (2020). Evaluation of antidiabetic and antihyperlipidemic activity of *Rhododendron arboreum* bark extract against streptozocin induced diabetes in rats. *Journal of Medicinal Herbs and Ethnomedicine, 6*, 11–16.

Gautam, V., Kohli, S., Kapoor, D., Bakshi, P., Sharma, P., Arora, S., Bhardwaj, R., & Ahmad, P., (2021). Stress protective effect of *Rhododendron arboreum* leaves (MEL) on chromium-treated *Vigna radiata* plants. *Journal of Plant Growth Regulation, 40*, 423–435.

Gordon, S. P., Sloop, C. M., Davis, H. G., & Cushman, J. H., (2012). Population genetic diversity and structure of two rare vernal pool grasses in central California. *Conservation Genetics, 13*, 117–130.

Gusta, L. V., Wisniewski, M., Nesbitt, N. T., & Gusta, M. L., (2004). The effect of water, sugars, and proteins on the pattern of ice nucleation and propagation in acclimated and nonacclimated canola leaves. *Plant Physiology, 135*, 1642–1653.

Hanin, M., Brini, F., Ebel, C., Toda, Y., Takeda, S., & Masmoudi, K., (2011). Plant dehydrins and stress tolerance versatile proteins for complex mechanisms. *Plant Signaling & Behavior, 6*(10), 1503–1509.

Hare, P. D., Cress, W. A., & Van, S. J., (1998). Dissecting the roles of osmolyte accumulation during stress. *Plant, Cell and Environment, 21*, 535–553.

Hurek, B. R., & Hurek, T., (2011). Living inside plants: Bacterial endophytes. *Current Opinion in Plant Biology, 14*, 435–443.

Iba, K., (2002). Acclimative response to temperature stress in higher plants: approaches of gene engineering for temperature tolerance. *Annual Review of Plant Biology, 53*, 225–245.

Ishikawa, M., & Sakai, A., (1981). Freezing avoidance mechanisms by supercooling in some *Rhododendron* flower buds with reference to water relations. *Plant & Cell Physiol.*, *22*(6), 953–967.

Jansa, J., & Vosátka, M., (2000). In vitro and post vitro inoculation of micropropagated *Rhododendrons* with ericoid mycorrhizal fungi. *Applied Soil Ecology, 15*, 125–136.

Kacperska, A., (1999). Plant response to low temperature: Signaling pathways involved in plant acclimation. In: Margesin, R., & Schinner, F., (eds.), *Cold-Adapted Organisms–Ecology, Physiology, Enzymology and Molecular Biology* (pp. 79–103). *Springer-Verlag*, Berlin, Germany.

Kashiwada, Y., Yamazaki, K., Ikeshrio, Y., Yamagishi, T., Fujioka, T., Mihashi, K., Mizuki, K., et al., (2001). Isolation of rhododaurichromanic acid B and the anti-HIV principles rhododaurichromanic acid A and rhododaurichromanic acid from *Rhododendron dauricum. Tetrahedron, Science Direct, 57*(8), 1559–1563.

Kiruba, S., Mahesh, M., Nisha, S. R., Miller, P. Z., & Jeeva, S., (2011). Phytochemical analysis of the flower extracts of *Rhododendron arboreum* Sm. ssp. *nilagiricum* (Zenker) tagg. *Asian Pacific Journal of Tropical Biomedicine*, S284–S286.

Krishna, P., (2003). Brassinosteroid-mediated stress responses. *Journal of Plant Growth Regulation, 22*, 289–297.

Kumar, V., (1993). *Biological Diversity in Eastern Himalaya: A Status Report.* University of Delhi and WWF-India, New Delhi.

Li, X. L., Hua, Z. R., & Li, J., (2017). Effects of brassinolide on tolerance of *Rhododendron lapponicum* in Qinling to heat stress. *Journal of Henan Agricultural Sciences, 46*(8), 126–130.

Lim, C. C., Krebs, S. L., & Arora, R., (2014). Cold hardiness increases with age in juvenile *Rhododendron* populations. *Frontiers in Plant Science, 5*(542), 1–7.

Liu, P., Geng, X., Ozaki, U., Zhao, H., & Zheng, F., (2020). Subcellular localization and responses of antioxidant systems of three *Rhododendron* cultivars under salt stress. *Journal of the Faculty of Agriculture, Kyushu University, 65*(1), 45–55.

Liu, Y., Medeiros, J. S., & Burns, J. H., (2020). The soil biotic community protects *Rhododendrons* spp. across multiple clades from oomycete *Phytophthora cinnamomi* at a cost to plant growth. *Oecologia, 195*, 1–12.

Luteyn, J. L., (2002). Diversity, adaptation and endemism in neotropical *Ericaceae*: Biogeographical patterns in the vaccinieae. *Biol. Rev., 68*, 55–87.

Mao, A. A., (2010). The genus *Rhododendron* in north-east India. *Journal of Plant Science, 7*, 26–34.

Negi, V. S., Maikhuri, R. K., Rawat, L. S., & Chandra, A., (2013). Bioprospecting of *Rhododendron arboreum* for livelihood enhancement in Central Himalaya, India. *Environ. We Int. J. Sci. Tech., 8*, 61–70.

Nilsen, E. T., & Orcutt, D. M., (1996). *The Physiology of Plants Under Stress–Abiotic Factors*. John Wiley & Sons, Inc., New York.

Pedranzani, H., Racagni, G., Alemano, S., Miersch, O., Ramirez, I., Cortes, H. P., Taleisnik, E., et al., (2003). Salt tolerant tomato plants show increased levels of jasmonic acid. *Plant Growth Regulation, 41*, 149–158.

Peng, Y., Reyes, J. L., Wei, H., Yang, Y., Karlson, D., Covarrubias, A. A., Krebs, S. L., et al., (2008). RcDhn5, a cold acclimation-responsive dehydrin from *Rhododendron catawbiense* rescues enzyme activity from dehydration effects in vitriol and enhances freezing tolerance in *RcDhn5*-overexpressing *Arabidopsis* plants. *Physiologia Plantarum, 134*, 583–597.

Prakash, V., Rana, S., & Sagar, S., (2016). Studies on antibacterial activity of leaf extracts of *Rhododendron arboreum* and *Rhododendron campanulatum*. *Int. J. Curr. Microbiol. App. Sci., 5*(4), 315–322.

Rai, U., & Lama, D., (2016). *Rhododendron apradae* U. Rai et D. Lama sp. Nov. from Darjeeling Himalaya in India. *Pleione, 10*(2), 370–374.

Rai, U., Lama, D., Thapa, N., & Baraily, S., (2014). A new variety of *Rhododendron grande* wight [Ericaceae] from Darjeeling Himalaya in West Bengal, India. *Pleione, 8*(1), 159–162.

Ranney, T. G., Blazich, F. A., & Warren, S. L., (1995). Heat tolerance of selected species and populations of *Rhododendron*. *J. Amer. Soc. Hort. Sci., 120*(3), 423–428.

Riverola, A. P., Gupta, A., Putze, I. B., Bosch, N., Ibañes, M., & Delgado, A. I. C., (2019). Brassinosteroid signaling in plant development and adaptation to stress. *Development, 146*, dev151894. doi: 10.1242/dev.151894.

Scott, L. C., & Scott, J. D., (2004). Elevated ultraviolet-B radiation induces cross-protection to cold in leaves of *Rhododendron* under field conditions. *Photochem. Photobiol., 79*(2), 199–204.

Shen, H. F., Zhao, B., Xu, J. J., Liang, W., Huang, W. M., & Li, H. H., (2017). Effects of heat stress on changes in physiology and anatomy in two cultivars of *Rhododendron*. *South African Journal of Botany, 112*, 338–345. Elsevier.

Shen, H., Zhao, B., Xu, J., Zheng, X., & Huang, W., (2016). Effects of salicylic acid and calcium chloride on heat tolerance in *Rhododendron* 'Fen Zhen Zhu'. *J. Amer. Soc. Hort. Sci., 141*(4), 363–372.

Shenga, N. C., (1994). *Status Paper on Biodiversity in Sikkim Panda*, 5–10.

Shimizu, M., Meguro, A., Nishimura, S. H. T., & Kunoh, H., (2006). Disease resistance induced by nonantagonistic endophytic *Streptomyces* spp. on tissue-cultured seedlings of *Rhododendron*. *J. Gen Plant Pathol., 72*, 351–354.

Shrestha, N, Wang, Z, Su, X, et al. Global patterns of Rhododendron diversity: The role of evolutionary time and diversification rates. Global Ecol Biogeogr. 2018; 27: 913– 924. https://doi.org/10.1111/geb.12750

Singh, K. K., Kumar, S., & Pandey, A., (2008). Soil treatments for improving seed germination of rare and endangered Sikkim Himalayan *Rhododendrons*. *World Journal of Agricultural Sciences, 4*(3), 288–296.

Spethmann, W., Michaelis, G., & Schepker, H., (2010). *Rhododendron*s in Germany and the German *Rhododendron* gene bank. In: Mainra, A., Badola, H. K., & Mohanty, B., (eds.), *Proceedings, International Conference Rhododendrons: Conservation and Sustainable Use* Forest Environment & Wildlife Management Department, Government of Sikkim Saramsa, Gangtok-Sikkim, India.

Tantawy, A. A. E., Xie, W., Li, S., Wang, J., Peng, L., Song, J., Zhang, L., Cai, Y., & Yang, X., (2020). Changes of endogenous hormones during ovary development in *Rhododendron delavayi* × *Rhododendron sinofalconeri* and *Rhododendron delavayi* × *Rhododendron cyanocarpum*. *Eur. J. Hortic. Sci., 85*(4), 248–257.

Tiwari, O. N., & Chouhan, U. N., (2006). *Rhododendron* conservation in Sikkim Himalaya. *Current Science, 90*, 532–541.

Uday, K., Divi, U. K., Rahman, T., & Krishna, P., (2016). Gene expression and functional analyses in brassinosteroid-mediated stress tolerance. *Plant Biotechnology Journal, 14*, 419–432.

Wang, H., Mu, K., Liu, C., Guo, Y., & Deng, X., (2020). Gene expression profiling of *Rhododendron pulchrum* leaves under drought stress. *Tree Genetics &Genome, 16*, 58.

Wang, K. H., Liu, X. P., Zhang, L. H., Ling, J. H., & Li, L., (2011). Physiological-biochemical response of five species in *Rhododendron* L. to high temperature stress and comprehensive evaluation of their heat tolerance. *Journal of Plant Resources and Environment, 20*(3), 29–35.

Wang, X., Arora, R., Hoener, H. T., & Krebs, S. L., (2008). Structural adaptations in overwintering leaves of thermonastic and nonthermonastic *Rhododendron* species. *J. Amer. Soc. Hort. Sci., 133*(6), 768–776.

Wei, H., Dhanaraj, A. L., Rowland, L. J., Krebs, Y. F. S. L., & Arora, R., (2005). Comparative analysis of expressed sequence tags from cold-acclimated and non-acclimated leaves of *Rhododendron catawbiense* michx. *Planta, 221*, 406–416.

Wei, H., Dhanraj, A. L., Arora, R., Rowland, L. J., Fu, Y., & Sun, L., (2006). Identification of cold acclimation-responsive *Rhododendron* genes for lipid metabolism, membrane transport and lignin biosynthesis: Importance of moderately abundant ESTs in genomic studies. *Plant, Cell and Environment, 29*, 558–570.

Wei, H., Yang, Y., Himmel, M. E., Tucker, M. P., Ding, S. Y., Yang, S., & Arora, R., (2019). Identification and characterization of five cold stress-related *Rhododendron* dehydrin genes: Spotlight on FSK-type with multiple F-segments. *Frontiers in Bioengineering and Biotechnology, 7*(30), 1–14.

Wei, X., Chen, J., Zhang, C., & Pan, D., (2016). A new *Oidiodendron maius* strain isolated from *Rhododendron fortunei* and its effects on N uptake and plant growth. *Frontiers in Microbiology, 7*(1327), 1–11.

Wei, X., Chen, J., Zhang, C., Liu, H., Zheng, X., & Mu, J., (2020). Ericoid mycorrhizal fungus enhances microcutting rooting of *Rhododendron fortunei* and subsequent growth. *Horticulture Research, 7*, 140.

Xu, Q., Li, H. X., Xian, X. L., Lin, L., Chen, R., Song, W., Fu, W., & Pan, Y. Z., (2018). Effects of BABA on photosynthetic characteristics and antioxidative system in *Rhododendron* under $NaHCO_3$ stress. *Forest Research, 31*(2), 133–140.

Zhao, W., Qi, X., Lyu, J., Yu, Z., & Chen, X., (2016). Characterization of microbial community structure in rhizosphere soils of cowskin azalea (*Rhododendron aureum* Georgi) on northern slope of Changbai Mountains, China. *Chinese Geographical Science, 26*, 78–89.

Zhao, Y., Yu, W., Hu, X., Shi, Y., Liu, Y., Zhong, Y., Wang, P., et al., (2018). Physiological and transcriptomic analysis revealed the involvement of crucial factors in heat stress response of *Rhododendron hainanense*. *Gene, 660*, 109–119. Elsevier.

CHAPTER 4

Waterlogging Stress: ROS Generation and Its Effects on Plant

SWARUPA AGNIHOTRI[1] and VIJAYA NIKAM[2]

[1]*Department of Botany, VIVA College, Virar (W), University of Mumbai, Maharashtra, India, E-mail: Swarupaagnihotri13@gmail.com*

[2]*Department of Botany and Plant Protection, S.G. M. College, Karad, Maharashtra, India*

ABSTRACT

Waterlogging is biotic and sometimes manmade stress to the plants. Waterlogging causes a reduction in the solubility of oxygen in the water. Microflora and fauna suffer from hypoxic and anoxic conditions, which result in certain metabolic changes in these plants. In this present work, such metabolic changes related to the production of reactive oxygen species are discussed.

4.1 INTRODUCTION

Waterlogging is one of the harmful abiotic stresses which change the distribution pattern of plant species and the yield of crops. Different natural and anthropogenic causes such as highly variable weather conditions, soil erosion, lack of soil drainage, irregular irrigation, and leakage in dams are enlisted (FAO, 2007; Hasanuzzaman et al., 2012), which are responsible for creating a waterlogging situation and damaging 20–30 million hectares irrigated field. During waterlogging stress, plant roots are primarily subjected to stress conditions such as hypoxia and anoxia. Hypoxia is considered as a short-term flooding; during this stress condition roots are exposed to low

Plant Metabolites under Environmental Stress: Mechanisms, Responses, and Adaptation Strategies.
Nivas M. Desai, Manasi Patil, Umesh R. Pawar (Eds.)

oxygen while anoxia as a long-term flooding where there is total absence of oxygen (Ponnamperuma, 1972). Various morphological, physiological, and biochemical changes are noticed in plants which are exposed to waterlogging stress. Morphologically waterlogged plants exhibit aerenchyma and adventitious roots formation. Plant hormones, auxin, and ethylene interact to cause the development of adventitious roots. At the same time, aerenchyma is formed due to ethylene, acidosis, and NO formation (McNamara and Mitchell, 1989). The formation of aerenchyma in the roots is one of the adaptive features shown by the stress adaptive plants (Evans, 2004; Kulkarni and Chavan, 2014). Initiation of adventitious roots instead of basal roots is one of the waterlogging stress adaptive characteristics of the plants (Malik et al., 2001; Steffens et al., 2006). Stomata of waterlogged plants show increased resistance and decreased conductance which caused deficiency in water content in plant body (Parent et al., 2008). Reduction in oxygen level causes disturbance in transmembrane transportation in roots which resulted in a decline in water transport (Else et al., 2001). Due to a decline in oxygen level in plant, a decrease in photosynthesis and transpiration rates were observed, which ultimately caused stomatal closure (Ashraf et al., 2011; Ashraf and Arfan, 2005). Smethurst et al. (2005) reported that waterlogging stress caused alteration in PSII mechanism which leads to inadequacy of macroelements such as N, P, K, Ca, and Mg in the plants. Commonly, it was observed that increased content of sodium and iron in the waterlogged plants is harmful. But it was proved that deficiency of the macronutrients (N, P, K, Ca, and Mg) in the plants create toxic effects not Na and Fe (Sharma and Swarup, 1989; Smethurst et al., 2005). Biochemically plants which are exposed to waterlogging stress exhibit changes in their metabolic pathways to get maximum energy. At a low concentration of oxygen, under waterlogging stress plants shift their normal carbohydrate metabolism to fermentative mode to obtain ATP in less time period. Enzymes related to fermentative metabolism are also reported to increase (Davies, 1980; Vartapetian, 1991). γ-Amino butyric acid and Alanine are the two important amino acids were synthesized rapidly to cope with the waterlogging stress (Kulkarni and Chavan, 2013).

For living organism, oxygen is very vital molecule for conducting normal metabolic processes. Reactive oxygen species (ROS) are generated under low oxygen conditions as well as during the total absence of oxygen (Ashraf, 2009). These ROS are present in various forms such as superoxide (O_2^-), hydroperoxyl radical ($HO_2^•$), hydrogen peroxide (H_2O_2), alkoxy radical ($RO^•$), peroxy radical ($ROO^•$), singlet oxygen ($^1O_2$) and excited carbonyl (RO*), and the hydroxyl radical (OH). After receiving energy excited chlorophyll reacts with the oxygen to produce $^1O_2$, which is highly toxic ROS (Ashraf and Akram, 2009; Karuppanapandian and Manoharan, 2008; Vellosillo et al., 2010). These ROS are highly reactive and irreversibly damage the cellular organelles, metabolic processes, membranes, and genetic material (Karuppanapandian et al., 2011; Ashraf, 2009). ROS are generated and scavenged efficiently during the normal process of growth of an organism. But various biotic and abiotic stresses over produce them, which sometimes cause lethal effects (Mittler, 2002; Mittler et al., 2004). In the present article, various causes of ROS generation and their effects on growth and metabolism of plants during waterlogging stress are discussed.

4.2 CHEMICAL NATURE OF REACTIVE OXYGEN SPECIES (ROS)

ROS are classified as radicals and non-radicals. Under free radicals, superoxide anion ($O_2^{\bullet-}$), hydroxyl radical ($^{\bullet}OH$), alkoxy radical ($RO^{\bullet}$), and peroxyl radical ($ROO^{\bullet}$) are included while non-radicals are H_2O_2, singlet oxygen ($^1O_2$), and ozone (O_3) (Maurya, 2020). According to Kapoor et al. (2015), hypochlorous acid (HOCl), hydroperoxides (ROOH), and excited carbonyls (RO*) as ROS are present in plants. Reactive oxygen intermediates (ROI) are generated due to partial reduction of molecular oxygen which is also included in ROS. At the same time, hypobromous acid (HOBr) and hypoiodous acid (HOI) and carbonate ($CO_3^{\bullet-}$) as well as semiquinone ($SQ^{\bullet-}$) termed as ROS (Nathan et al., 2010; Farnese et al., 2016; Waszczak et al., 2018). Various forms of these ROS are formed in different cellular organelles during metabolic processes. In the present article, various locations of ROS generation and their effects on plant cell structure and metabolism are noted.

4.3 SITES OF PRODUCTION OF THE ROS

4.3.1 CHLOROPLAST

Gaseous exchange through stomata is hampered due to stress conditions. As a result of which CO_2 assimilation in dark reaction is declined. This decrease will further reduce the oxidized $NADP^+$ formation which is an acceptor of electrons from donor-reduced ferredoxin during the light reaction. According to Mehler reaction, these electrons are accepted by oxygen molecule to generate $O_2^{\bullet-}$. Internal reactions in photosystem caused formation of stable triplet chlorophyll ($^3Chl^*$) from singlet chlorophyll ($^1Chl^*$). Release of O_2 was noticed (Zolla and Rinalducci, 2002) after spin restriction of $^1O_2$. This $^1O_2$ is formed after reaction between triplet chlorophyll and triplet oxygen. The reaction center of photosystem (II) is damaged due to genesis of $^1O_2$. This radical also affects membranous proteins and lipids which are present near to site of $^1O_2$ production.

4.3.2 MITOCHONDRIA

During electron transport chain (ETC), reduced ubiquinone transfers electron to acceptor cytochrome c1 and thereby generates highly reactive, reducing semiquinone species, which later on donates electron to cytochrome b. This reduced Cyt b donates electron to O_2 which turns into $O_2^{\bullet-}$. During further metabolic reactions this relatively ROS in aqueous environment is further dismutation to H_2O_2 (Moller, 2001). Reaction between H_2O_2 and reduced Fe^{2+} and Cu^+ generates hydroxyl radical (OH^-). This free radical is toxic and reacts with membrane molecules (Moller, Loc.Sit.).

4.3.3 PEROXISOME

Enzyme xanthine oxidase (XOD) catalases the reaction of formation of uric acid from xanthine or hypoxanthine. This reaction occurs in the matrix of peroxisomes during which $O_2^{\bullet-}$ is created. Peroxisomal membrane also exhibit production of O_2^- through the transport of electron from flavoprotein NADH and Cyt b to the O_2 (LópezHuertas et al., 1999; del Rio et al., 2002). Another form of reactive species, i.e., H_2O_2 which is also produced by glycolate oxidase through oxidation of glycolate in the peroxisome, β-oxidation of fatty acids, the flavin oxidase cycle and the oxidation-reduction of $O_2^{\bullet-}$ (Corpas et al., 2001).

4.3.4 APOPLAST

Abscisic acid (ABA) induced stomatal closure is facilitated by apoplastic NADPH oxidases, which are a major source of ROS generation (Kwak et al., 2003). Cell wall-located oxidases, pH-dependent cell wall POXs, oxalate oxidases and polyamine oxidases are other forms apoplastic enzymes which are responsible for producing apoplastic ROS (Mittler, 2002).

4.3.5 PLASMA MEMBRANE

NADPH oxidase is a key enzyme for the generation of ROS (Pei et al., 2000; Kwak et al., 2003). Transfer of electron from cytoplasmic NADPH to O_2 is catalyzed by this enzyme. The product of this reaction is $O_2^{\bullet-}$ which later gets dismutase to form H_2O_2.

4.3.6 CELL WALL

H_2O_2 is formed in the cell wall due to NADH which is produced by the cell wall malate dehydrogenase. This reaction is mediated by NADH oxidase, present in plasmalemma. Peroxides are formed through oxidation of quinones. These quinones are formed during the reduction reaction which is catalyzed by cell wall located enzymes diamine oxidases (Spiteller, 2003; Higuchi, 2006).

4.3.7 CHEMICAL NATURE OF ROS

ROS are formed from the atmospheric O_2 which is present in the form of triplet oxygen, $^3O_2$ with two single parallel spin electrons. Later on, these electrons lose their potential. But various radiations, chemical, and metabolic processes help to free this oxygen moiety from spin restriction, which eventually creates ROS (Mailloux, 2016).

All these ROS are formed during regular growth and metabolism, but their production and scavenging is under control. But during stress, plant show enhancement in ROS generation and antioxidant system is not efficient to get rid of them. Mechanism of formation of ROS and their effects on cellular metabolism during waterlogging stress are reported.

4.4 WATERLOGGING STRESS AND ROS PRODUCTION

Limited amount of oxygen in the plant body caused generation of ROS in cellular organelles. During waterlogging stress, plants are exposed to anoxic situation where total absence of oxygen hampered mitochondrial and chloroplastic ETCs which ultimately resulted in formation of mitochondrial and chloroplastic ROS (Chang et al., 2012). As soluble oxygen is less in water, under waterlogging condition plants are not efficient in photosynthesis and their photorespiration rate increases which give rise to the production of ROS in peroxisomes during daytime (Mommer and Visser, 2005). Photorespiration also triggers the generation of H_2O_2 (Del Río and López-Huertas, 2016). As roots are primarily exposed to the waterlogging stress conditions, it affects metabolic processes of ribulose bisphosphate carboxylase (RuBPC), glycolate oxidase, and phosphoglycolate adversely. These altered metabolic pathways disturb photosynthetic ETS and photosystem II and such changes boosted the formation superoxide ($O_2^{\bullet-}$), singlet oxygen ($^1O_2$), H_2O_2, and hydroxyl radicals ($OH^{\bullet}$) (Ashraf, 2012, 2009). ROS scavenging is very important during stress conditions. But imbalance between ROS production and their removal through poor antioxidant system hampered plant growth and metabolism under waterlogging stress (Lasanthi-Kudahettige et al., 2007). Demidchik et al. (2007, 2010) observed that waterlogging stress at higher dose creates negative impact on oxygen level and antioxidant system of the plant. Level of ROS was also increased during this period. After post-flooding stress also plant exhibits an increase in lipid peroxidation (LPO) and membrane permeability. Transportation of ions (Ca^{2+} and K^+) through ion channels was also regulated by superoxide radicals (H_2O_2 and $OH^{\bullet}$). ROS create disruption in intracellular organization and metabolism, which result into programed cell death (PCD). Respiratory burst oxidase homolog D (RBOHD) gene is induced during waterlogging stress. Under total as well as less absence of oxygen, RBOHs are stimulated, which caused formation of ROS (Figure 4.1) (Mustroph et al., 2009; Pucciariello et al., 2012; Yao et al., 2017).

Kalashnikov et al. and Biemelt et al. (1994, 2000) noticed an increase in the level of H_2O_2 content in waterlogged leaves and roots of *Hordeum vulgare* and roots of wheat, respectively. Enhancement in ROS production during waterlogging stress was reported in *Triticum aestivum* and *Zea mays* L. (Škute et al., 2019; Salah et al., 2019). Seven days waterlogging caused an increase in H_2O_2 in *Deschampsia antarctica* (Park and Lee, 2019). Increasing waterlogging treatment (2, 4, 6, and 8 d) caused elevation in H_2O_2 content in waterlogging sensitive *Sesamum indicum* L. cv. BARI Til-4. Noticeable increase in MDA and O_2^- contents was observed in waterlogging sensitive variety of *Hordeum vulgare* (TF57) compared to the tolerant variety (TF58) on 21st day of waterlogging treatment (Luan et al., 2018). Flooding stress for 15 days resulted in increased MDA (54%) and H_2O_2 (208%) levels in *S. lycopersicum* (cv. Roma) compared to the control (Rasheed et al., 2018). A detailed study of the effect of waterlogging on tolerant as well as susceptible verities of pigeon pea (*Cajanus cajan* L. Halls) genotypes ICP 301 and Pusa 207, respectively, showed that as compared to the control at a higher dose of waterlogging (4 and 6 days) there was elevation in superoxide radical, H_2O_2 and thiobarbituric acid reactive substances (TBARS) contents. Authors suggested that these enhancements in oxidative metabolites are due to stimulation of DPI-sensitive NADPH-oxidase enzyme working under stress conditions (Kumutha et al., 2009). Due to waterlogging, rice plant suffers from reoxygenation and dehydration stresses which caused to generate malondialdehyde (MDA), superoxide anion, and H_2O_2 (Fukao et al., 2011). According to the authors, to cope with these stresses plant must extract ROS and ABA responsiveness from plant.

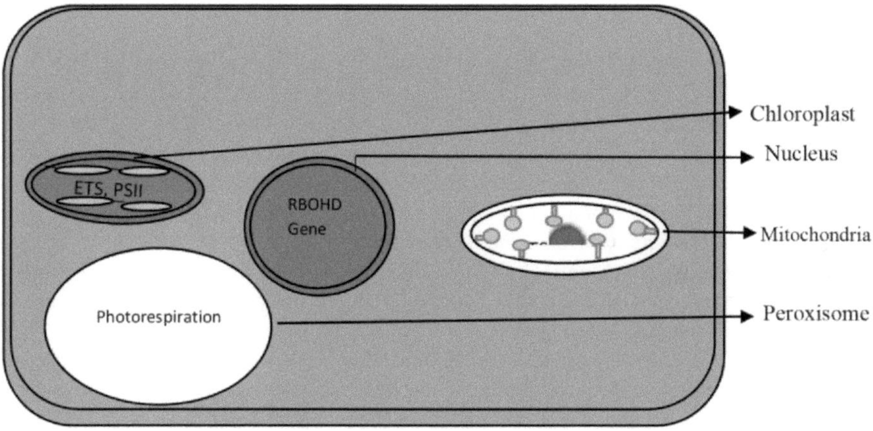

FIGURE 4.1 Sites and processes of ROS formation in anoxic/hypoxic plant cell.

For scavenging ROS plant possess various enzymatic (Catalase, POX, superoxide dismutase, ascorbate peroxidase (APX), glutathione reductase, etc.), and nonenzymatic (ascorbic acid (AA), glutathione, and phenolic compounds, etc.), antioxidants. It is observed that waterlogging susceptible plants show less activity of all these antioxidants compared to tolerant plants (Mubeen et al., 2017). In the case of waterlogging stress, ROS also plays a role as a signaling molecule which is considered as an adaptive character (Pucciariello et al., 2012; Yamauchi et al., 2014).

KEYWORDS

- **abscisic acid**
- **hypobromous acid**
- **reactive oxygen species**
- **respiratory burst oxidase homolog D**
- **thiobarbituric acid reactive substances**
- **xanthine oxidase**

REFERENCES

Ashraf, M. A., (2012). Waterlogging stress in plants: A review. *Afr. J. Agril. Res., 7*, 1976–1981.

Ashraf, M. A., Ahmad, M. S. A., Ashraf, M., Al-Qurainy, F., & Ashraf, M. Y., (2011). Alleviation of waterlogging stress in upland cotton (*Gossypium hirsutum* L.) by exogenous application of potassium in soil and as a foliar spray. *Crop Pasture Sci., 62*, 25–38.

Ashraf, M., & Akram, N. M., (2009). Improving salinity tolerance of plants through conventional breeding and genetic engineering: An analytical comparison. *Biotech. Adv., 27*, 744–752.

Ashraf, M., & Arfan, M., (2005). Gas exchange characteristics and water relations in two cultivars of *Hibiscus esculentus* under waterlogging. *Biol. Plant, 49*, 459–462.

Ashraf, M., (2009). Biotechnological approach of improving plant salt tolerance using antioxidants as markers. *Biotech. Adv., 27*, 84–93.

Biemelt, S., Keetman, U., Mock, H. P., & Grimm, B., (2000). Expression and activity of isoenzymes of superoxide dismutase in wheat roots in response to hypoxia and anoxia. *Plant Cell Environ., 23*, 135–144.

Chang, R., Jang, C. J., Branco-Price, C., Nghiem, P., & Bailey-Serres, J., (2012). Transient MPK6 activation in response to oxygen deprivation and reoxygenation is mediated by mitochondria and aids seedling survival in *Arabidopsis*. *Plant Mol Biol., 78*(1, 2), 109–122.

Corpas, F. J., Barroso, J. B., & Del Río, L. A., (2001). Peroxisomes as a source of reactive oxygen species and nitric oxide signal molecules in plant cells. *Trends Plant Sci., 6*, 145–150.

Davies, D. D., (1980). Anaerobic metabolism and the production of organic acids. In: Davies, D. D., (ed.), *The Biochemistry of Plants* (Vol. 2, pp. 581–611). Academic Press, NY, U.S.A.

Del Río, L. A., & López-Huertas, E., (2016). ROS Generation in peroxisomes and its role in cell signaling. *Plant Cell Physiol., 57*(7), 1364–1376.

Del Rio, L. A., Corpas, F. J., Sandalio, L. M., Palma, J. M., Gomez, M., & Barroso, J. B., (2002). Reactive oxygen species, antioxidant systems and nitric oxide in peroxisomes. *J. Exp. Bot., 53*, 1255–1272.

Demidchik, V., Cuin, T. A., & Svistunenko, D., (2010). *Arabidopsis* root K^+-efflux conductance activated by hydroxyl radicals: Single-channel properties, genetic basis and involvement in stress-induced cell death. *J. Cell Sci., 123*, 1468–1479.

Demidchik, V., Shabala, S. N., & Davies, J. M., (2007). Spatial variation in H_2O_2 response of *Arabidopsis thaliana* root epidermal Ca^{2+} flux and plasma membrane Ca^{2+} channels. *Plant J., 49*, 377–386.

Else, M. A., Coupland, D., Dutton, L., & Jackson, M. B., (2001). Decreased root hydraulic conductivity reduces leaf water potential, initiates stomatal closure and slows leaf expansion in flooded plants of castor oil (*Ricinus communis*) despite diminished delivery of ABA from the roots to shoots in the xylem sap. *Physiol. Plant, 111*, 46–54.

Evans, D. E., (2004). Aerenchyma formation. *New Phytol., 161*, 35–49.

Farnese, F. S., Menezes-Silva, P. E., Gusman, G. S., & Oliveira, J. A., (2016). When bad guys become good ones: The key role of reactive oxygen species and nitric oxide in the plant responses to abiotic stress. *Front. Plant Sci., 7*, 471.

Fukao, T., Yeung, E., & Bailey-Serres, J., (2011). The submergence tolerance regulator SUB1A mediates crosstalk between submergence and drought tolerance in rice. *Plant Cell, 23*, 412–427.

Hasanuzzaman, M., Hossain, M. A., Teixeira Da, S. J. A., & Fujita, M., (2012). Plant responses and tolerance to abiotic oxidative stress: Antioxidant defense is a key factor. In: Bandi, V., Shanker, A. K., Shanker, C., & Mandapaka, M., (eds.), *Crop Stress and its Management: Perspectives and Strategies* (pp. 261–316). Springer: Berlin, Germany.

Higuchi, T., (2006). Look back over the studies of lignin biochemistry. *J. Wood Sci., 52*, 2–8.

Kalashnikov, Y. E., Balakhnina, T. I., & Zakrzhevsky, D. A., (1994). Effect of soil hypoxia on activation of oxygen and the system of protection from oxidative destruction in roots and leaves of *Hordeum vulgare*. *Russ. J. Plant Physiol., 41*, 583–588.

Kapoor, D., Sharma, R., Handa, N., Kaur, H., Rattan, A., Yadav, P., Gautam, V., et al., (2015). Redox homeostasis in plants under abiotic stress: Role of electron carriers, energy metabolism mediators and proteinaceous thiols. *Front. Environ. Sci., 13*(3), 3.

Karuppanapandian, T., & Manoharan, K., (2008). Uptake and translocation of tri-and hexavalent chromium and their effects on black gram (*Vigna mungo* L. Hepper cv. Co4) roots. *J. Plant Biol., 51*, 192–201.

Karuppanapandian, T., Wang, H. W., Prabakaran, N., Jeyalakshmi, K., Kwon, M., Manoharan, K., & Kim, W., (2011). 2,4-dichlorophenoxyacetic acid-induced leaf senescence in mung bean (*Vigna radiata* L. Wilczek) and senescence inhibition by co-treatment with silver nanoparticles. *Plant Physiol. Biochem., 49*, 168–177.

Kudahettige, N. R., Magneschi, L., Loreti, E., Gonzali, S., Licausi, F., Novi, G., Beretta, O., et al., (2007). Transcript profiling of the anoxic rice coleoptile. *Plant Physiol., 144*, 218–231.

Kulkarni, S. S., & Chavan, P. D., (2013). Study of some aspects of anaerobic metabolism in roots of finger millet and rice plants subjected to waterlogging stress. *Int J. Bot., 9*(2), 80–85.

Kulkarni, S. S., & Chavan, P. D., (2014). Study of effect of waterlogging on root anatomy of ragi and rice. *American J. Plant Physiol., 9*(2), 46–51.

Kumutha, D., Ezhilmathi, K., Sairam, R. K., Srivastava, G. C., Deshmukh, P. S., & Meena, R. C., (2009). Waterlogging induced oxidative stress and antioxidant activity in pigeon pea genotypes. *Biol. Plant., 53*, 75–84.

Kwak, J. M., Mori, I. C., Pei, Z. M., Leonhardt, N., Torres, M. A., Dangl, J. L., Bloom, R. E., et al., (2003). NADPH oxidase AtrbohD and AtrbohF genes function in ROS-dependent ABA signaling in *Arabidopsis*. *EMBO. J., 22*, 2623–2633.

Lopez-Huertas, E., Corpas, F. J., Sandalio, L. M., Del Río, L. A., (1999). Characterization of membrane polypeptides from pea leaf peroxisomes involved in superoxide radical generation. *Biochem. J., 337*, 531–536.

Luan, H., Shen, H., Pan, Y., Guo, B., Lv, C., & Xu, R., (2018). Elucidating the hypoxic stress response in barley (*Hordeum vulgare* L.) during waterlogging: A proteomics approach. *Sci. Rep., 8*, 1–13.

Mailloux, J. R., (2016). Application of mitochondria-targeted pharmaceuticals for the treatment of heart disease. *Curr. Pharm. Des., 22*, 4763–4779.

Malik, A. I., Colmer, T. D., Lamber, H., & Schortemeyer, M., (2001). Changes in physiological and morphological traits of roots and shoots of wheat in response to different depths of waterlogging. *Aust. J. Plant. Physiol., 28*, 1121–1131.

Maurya, A. K., (2020). Oxidative stress in crop plants. In: Hasanuzzaman, M., (ed.), *Agronomic Crops* (pp. 349–380). Springer: Singapore.

Mittler, R., (2002). Oxidative stress, antioxidants and stress tolerance. *Trends Plant Sci., 7*, 405–410.

Mittler, R., Vanderauwera, S., Gollery, M., & Van, B. F., (2004). Reactive oxygen gene network of plants. *Trends Plant Sci., 9*, 490–498.

Moller, I. M., (2001). Plant mitochondria and oxidative stress: Electron transport, NADPH turnover, and metabolism of reactive oxygen species. *Annu. Rev. Plant Physiol. Plant Mol. Biol., 52*, 561–591.

Mommer, L., & Visser, E. J., (2005). Underwater photosynthesis in flooded terrestrial plants: A matter of leaf plasticity. *Ann Bot., 96*(4), 581–589.

Mubeen, A. H., Khan, S. P., Singh, A. K., Singh, A. R., Gauta, M., Khan, N., & Khan, M., (2017). Effect of water logging and salinity stress on physiological and biochemical changes in tolerant and susceptible varieties of *Triticum aestivum* L. *Int. J. Curr. Microbiol. App. Sci., 6*(5), 975–981.

Mustroph, A., Zanetti, M. E., Jang, C. J., Holtan, H. E., Repetti, P. P., Galbraith, D. W., Girke, T., & Bailey-Serres, J., (2009). Profiling translatomes of discrete cell populations resolves altered cellular priorities during hypoxia in *Arabidopsis*. *Proc Natl Acad Sci U S A., 106*(44), 18843–18848.

Nathan, C., & Ding, A., (2010). SnapShot: Reactive oxygen intermediates (ROI). *Cell, 140*, 951.

Paltaa, J. A., Ganjealic, A., Turnerb, N. C., & Siddique, K. H. M., (2010). Effects of transient subsurface waterlogging on root growth, plant biomass and yield of chickpea. *Agric. Water Manag., 97*, 1469–1476.

Parent, C., Berger, A., Folzer, H., Dat, J., Crevecoeur, M., Badot, P. M., & Capelli, N., (2008). A novel nonsymbiotic hemoglobin from oak: Cellular and tissue specificity of gene expression. *New Phytol., 177*, 142–154.

Park, J. S., & Lee, E. J., (2019). Waterlogging induced oxidative stress and the mortality of the Antarctic plant, *Deschampsia antarctica. J. Ecol. Environ., 43*, 29.

Pei, Z. M., Murata, Y., Benning, G., Thomine, S., Klusener, B., Allen, G. J., Grill, E., & Schroeder, J. I., (2000). Calcium channels activated by hydrogen peroxide mediate abscisic acid signaling in guard cells. *Nature, 406,* 731–734.

Ponnamperuma, F. N., (1972). The chemistry of submerged soils. *Advances in Agronomy, 24*, 29–96.

Pucciariello, C., Parlanti, S., Banti, V., Novi, G., & Perata, P., (2012). Reactive oxygen species-driven transcription in *Arabidopsis* under oxygen deprivation. *Plant Physiol., 159*(1), 184–196.

Rasheed, R., Iqbal, M., Ashraf, M. A., Hussain, I., Shafiq, F., Yousaf, A., & Zaheer, A., (2018). Glycine betaine counteracts the inhibitory effects of waterlogging on growth, photosynthetic pigments, oxidative defense system, nutrient composition, and fruit quality in tomato. *J. Hortic. Sci. Biotechnol., 93*, 385–391.

Salah, A., Zhan, M., Cao, C., Han, Y., Ling, L., Liu, Z., & Jiang, Y., (2019). γ-Aminobutyric acid promotes chloroplast ultrastructure, antioxidant capacity, and growth of waterlogged maize seedlings. *Sci. Rep., 9*(1), 484.

Setter, T. L., Waters, I., Sharma, S. K., Singh, K. N., Kulshreshtha, N., & Yaduvanshi, N. P. S., (2009). Review of wheat improvement for waterlogging tolerance in Australia and India: The importance of anaerobiosis and element toxicities associated with different soils. *Ann. Bot., 103*, 221–235.

Sharma, D. P., & Swarup, A., (1989). Effect of nutrient composition of wheat in alkaline soils. *J. Agric. Sci., 112*, 191–197.

Škute, N., Savicka, M., Kulbachna, A., Petjukevičs, A., & Harlamova, N., (2019). Influence of flooding on leaf cell membranes of three Latvian wheat cultivars (*Triticum aestivum* (L)). In: *Proceedings of the 12th International Scientific and Practical Conference I, 287*, 290.

Smethurst, C. F., Garnet, T., & Shabala, S., (2005). Nutrition and chlorophyll fluorescence responses of lucerne (*Medicago sativa*) to waterlogging subsequent recovery. *Plant Soil, 270*(1, 2), 31–45.

Vartapetian, B. B., (1991). Flood-sensitive plants under primary and secondary anoxia: Ultrastructural and metabolic responses. In: Jackson, M. B., Davies, D. D., & Lambers, H., (eds.), *Plant Life under Oxygen Deprivation* (pp. 201–216). SPB Academic Publ. Hague, Netherlands.

Vellosillo, T., Vicente, J., Kulasekaran, S., Hamberg, M., & Castresana, C., (2010). Emerging complexity in reactive oxygen species production and signaling during the response of plants to pathogens. *Plant Physiol., 154*, 444–448.

Waszczak, C., Carmody, M., & Kangasjärvi, J., (2018). Reactive oxygen species in plant signaling. *Ann. Rev. Plant Biol., 69*, 209–236.

Yamauchi, T., Watanabe, K., Fukazawa, A., Mori, H., Abe, F., Kawaguchi, K., Oyanagi, A., & Nakazono, M., (2014). Ethylene and reactive oxygen species are involved in root aerenchyma formation and adaptation of wheat seedlings to oxygen-deficient conditions. *J Exp Bot., 65*(1), 261–273.

Yao, Y., He, R. J., Xie, Q. L., Zhao, X. H., Deng, X. M., He, J. B., Song, L., et al., (2017). Ethylene response factor 74 (ERF74) plays an essential role in controlling a respiratory burst oxidase homolog D (RbohD)-dependent mechanism in response to different stresses in *Arabidopsis*. *New Phytol., 213*(4), 1667–1681.

CHAPTER 5

Remediation of Organic Textile Dyes and Effluent by Freshwater Filamentous Cyanobacteria: A Review

MADHULIKA GUPTA

Assistant Professor, Department of Microbiology, Lady Brabourne College, P1/2 Suhrawardy Avenue, Kolkata–700017, West Bengal, India,
E-mail: madhulikajhum@gmail.com

ABSTRACT

Organic chemical coloring compounds called dyes form an important constituent of the textile industry. Among the diverse range of dyes, azo dyes are the oldest and major group of dyes. These textile industries utilize large volumes of water for their various operation processes and release wastewater as effluents. The textile effluent consists of several toxic inorganics and organic ingredients along with heavy metals present in ionic and complex forms. Liberation of untreated dye containing textile effluents affects the environment and human health. Elimination of colored compounds released from textile industries can be carried out by a number of conventional modes of physicochemical methods, which has a high cost, and produce a large amount of sludge and harmful products. A better substitute to conventional methods is the degradation of dyes by microbes which involves low cost, less sludge build-up, and breaks down toxic dyes into simple inorganic compounds which cause less damage to living organisms. This review deals with the classification of dyes, environmental hazards encountered, conventional modes versus biological methods of treatment of textile dyes and effluent and the role of filamentous Cyanobacteria (also known as

Plant Metabolites under Environmental Stress: Mechanisms, Responses, and Adaptation Strategies.
Nivas M. Desai, Manasi Patil, Umesh R. Pawar (Eds.)

blue-green algae) in the decolorization of textile dyes by the mechanism of biosorption and enzymatic approach involving reductive and oxidative enzymes under various conditions.

5.1 INTRODUCTION

The principal and the chemically essential organic compounds which imparts color to the drapery are known as the textile dyes. Commonly used dyes in textile industries are azo dyes, anthraquinones, and indigo dyes (Manu and Sanjeev, 2002). The different stages of processing in textile fabric involve large volumes of water as well as generates wastewater (ISPCH, 1995). Among the different types of dyes, approximately the dyes used in bulk and on a mass scale in the textile dyeing industries are the azo dyes. The advantages of using such dyes are for its attributes such as chemical stability, easy synthesis, and availability of varieties of colors (Chang et al., 2004). Discharge of dye containing textile effluent affects water transparency with unpleasant appearance, elevated levels of BOD cause speedy exhaustion of dissolved oxygen, decrease in photosynthetic rate, toxic effect on flora, fauna, and humans due to high levels of COD and their end-products may act as deadly carcinogens and mutagens for the biotic life (Mughda and Usha, 2012; Xu et al., 2005; Metcalf and Eddy Inc., 1991). Treatment of dye-containing textile effluents by various physical and chemical methods can be employed but have certain drawbacks such as high cost, low efficiency, produces wastes difficult to dispose and may not be appropriate for all categories of dyes (Shah, 2014; Robinson et al., 2001; Singh and Singh, 2017; Rafu et al., 2009; Mehta et al., 2011; Nora, 2011). The job of microbes in the breakdown of discharge from the textile industries have attracted attention as an eco-friendly and cost-effective alternative method compared to that of other methods (Chen et al., 2003). Blue-green algae also known as Cyanobacteria have gained an important position in the remediation of textile dyes and effluents apart from carbon dioxide fixation and providing oxygen for aerobic bacterial degraders (Huang, 2010; Mallick, 2002).

5.2 CLASSIFICATION OF DYES

Complex unsaturated organic compounds that absorb light (chromophore) in the visible spectrum (380–750 nm), gives color to the visible region are called dyes (Rehman et al., 2020; Mounia et al., n.d.). On the basis of the technique of application, Direct (ex. Direct Red1), Vat (ex. Vat Blue 1), Sulphur (ex. Sulphur Black), and Reactive (ex. Cibacron F) dyes are appropriate and utilized for Cotton and melded fabrics. Disperse dyes (ex. Disperse Red 1) are exclusively used on synthetic textiles while Acidic (ex. Acid Orange 7), and Basic (ex. Basic Blue 9) are applied on wool and silk and Azoic dyes (ex. Direct Brown 78) are useful for printing inks and pigments respectively (Chequer et al., 2013).

Recent classification of dyes is based on Chromophore groups present in chemical structure such as azo dyes, anthraquinone dyes, etc., method of application of the dye and the Color index designated as C.I. containing a generic name and a chemical constitution number (Gürses et al., 2016; Benkhaya et al., 2020).

5.3 ENVIRONMENTAL MENACES

In modern-day adulteration of aquatic habitats has become one of the prime setbacks. The cost-effective use of this environmental reserve has received particular consideration as scarcity of water could pose a situation of great anxiety in the near future (Chequer et al., 2013).

The highest aquatic pollution is caused due to the dyeing process in textile industries amongst the other industries. The final step in the textile industries is the procedure of Dyeing, which comprises of washing in baths to remove surplus of the primary dye residue which do not adhere to the fibers. A large quantity of dyes is lost in this process of dying, which eventually pollutes the environment released as effluent (Chequer et al., 2013; Gupta, 2020). Often in wastewater treatment plants, activated sludge is availed, which causes an upsurge in the release of dyes resulting in the further toxicity of aquatic bodies.

Usage of reducing agents causes huge loss of dyes in the aquatic system due to imperfect adherence of dyes to the textile fibers as well as its addition in the bleaching process to remove the effluent color in order to abide by the legislation. As a result, this causes the creation of mutagenic substances when the degraded colorless dye effluent is sent to sewage treatment plants in various municipalities where chlorine (Cl) is added and the said effluent is released into the aquatic system. This would severely affect the health of the living organisms (Chequer et al., 2013).

The strong color of the textile effluent even below 1 ppm pose a great effect on the aquatic environment due to its turbidity, ability to cause huge strength of pollution and dye concentration ranging from 10–200 mg/l

(Chequer et al., 2013; Rajendran and Gunasekaran, 2006; Murugesan and Kalaichelvan, 2003). High levels of BOD in effluents due to accumulation of organic matter leading to shortfall in the dissolved oxygen concentration, intense color, resistant to degradation affects photosynthesis and oxygen-evolving mechanisms by hindering the path of sun rays which inhibits growth and activity of microorganisms and leads to toxicity of aquatic life. Health disorders such as nausea, hemorrhage, ulceration of skin and mucous membranes are caused due to the presence of dyes and their degraded products. Highest rates of toxicity were observed in case of basic, diazo, and direct dyes (Chequer et al., 2013; Chang et al., 2001).

The permanency in the nature of the toxic azo dyes and their side products has made the water treatment plants unsuccessful in their elimination which may be designated as carcinogenic or mutagenic imperiling the human well-being. The fabric color stays in the aquatic bodies for a prolonged period as they are manufactured in a consumer-friendly manner in order to prevent degradation by maintaining the stability of the colors. For example, azo dye Acid Violet 7 elicits a number of deleterious effects such as chromosomal abnormalities, peroxidation of lipids and cessation of enzyme activity (acetylcholinesterase) (Mansour et al., 2010; Singh and Singh, 2017). Compounds such as Methyl Red and its degradation product N,N-dimethylphenylenediamine (DMPD) are toxic and mutagenic (Wong and Yu, 1999) and remains unchanged in culture (Ayed et al., 2011). Therefore, there is an urgent requirement to confine their access in the ecosystem.

5.4 TREATMENT METHODS: CONVENTIONAL VS. BIOLOGICAL

The synthetic foundation and complex aromatic molecular structures of textile dyes make them durable, recalcitrant to degrade. The simplicity in the process of yearly production indicates azo dyes (also containing more than one azo bond) to be the most profitable one among other dyes manufactured globally. In India, 80,000 tons of dyes and pigments are produced and 10,000 different textile dyes are available (Lalnunhlimi and Krishnaswamy, 2016; Carliell et al., 1995).

The most important marker of aquatic pollution is the colored industrial effluent during the synthesis of dyes and other processes where pigment loss is quite conspicuous causing impairment to life underwater and is esthetically and visually disagreeable (Easton, 1995; Maguire, 1992; Chequer et al., 2013).

Conventional modes consisting of physical and chemical methods involves ozonation, adsorption, flocculation, bleaching, electrochemistry, irradiation, membrane filtration, etc., are usually carried out for handling and management of textile wastewater. These methods are highly specific, not economical, but to some extent effective but the complex molecular structure of dyes makes quite problematic to be treated by these conventional methods. The conventional modes of treatment often land up in excess cost, accumulation of sludge in large volumes along with the synthesis of detrimental side products. Traditional treatments of physical and chemical methods include coagulation with the help of lime, iron or aluminum salts, adsorption with activated carbon, ion exchange, filtration, etc., cause decolorization but involves the formation of sludge and requirement for revival of the applied adsorbent ingredients. In order to remove the intense color of the textile effluent oxidizing agent sodium hypochlorite is added but eventually it aggravates the problem due to the release of a toxic compound called aromatic amines (Rafu et al., 2009; Mehta et al., 2011; Nora, 2011; Schliephake et al., 2000; Shah, 2014).

Importance is therefore now engrossed on the microbial biodegradation of textile dyes as a superior substitute (An et al., 2002). The wastewater of the textile industry contains heavy metals in ionic and complex form. So, bioaccumulation and biosorption has a better potential than other conventional methods (Rahman et al., 2016). A variety of microorganisms play a crucial role in the degradation and absorption of a large number of dyes involving prokaryotes and eukaryotes (Robinson et al., 2001; Shah, 2014). The handling of dye containing effluent employing biological approach displays certain discrete benefits over treatment involved in standard traditional methods of treatment. It is more economical and harmless sludge is accumulated less, eco-friendly, converts dyes to simple inorganic compounds by mineralization which are less harmful to living forms. The basic step in the process of color removal involves the breakdown of azo bonds, the principal mechanism behind the degradation and decolorization process. Reductive cleavage occurs in the case of azo dyes while under aerobic conditions, the processing of the aromatic amines formed are carried out (Kapdan et al., 2003). So, under aerobic as well as anaerobic environment the microbial population should function competently in this respect (Tripathi and Srivastava, 2011). Another alternative approach for decolorization and degradation of textile dyes and effluents is offered by enzymatic methodology over conventional modes of treatment as it involves less sludge making and cost-effective as well. A number of enzymes such as azo-reductases (El-Sheekh et al., 2009), polyphenol oxidases, manganese peroxidase (POX), lignin peroxidase, laccases, tyrosinase, N-demethylase (Oturkar et al., 2011; Martorell et al., 2012), dye decolorizing POXs and cellobiose dehydrogenase are actively involved in the degradation procedure (Singh and Singh, 2017; Liers et al., 2010; Tilli et al., 2011).

Thus, microorganisms play an important role in the remediation of heavy metals as well as decolorization of textile effluents which also contains heavy metals by biological processes of biosorption, biodegradation, and enzymatic approach which is stable and profitable.

5.5 ROLE OF CYANOBACTERIA IN REMEDIATION OF TEXTILE DYES AND EFFLUENT

Biological treatment of textile dyes has been considered to be the best method as it involves the ability to degrade most of the dyes, eco-friendly, and economical method (Singh and Singh, 2017; Vitor and Corso, 2008; Pajot et al., 2011). Cyanobacteria can effectively decolorize textile dyes and effluents by biosorption and by enzymatic methods.

5.5.1 BIOSORPTION

The accumulation or uptake of chemical compounds by Cyanobacteria is attributed to its cell wall components like different functional groups like as carboxyl, hydroxyl phosphate amino, etc., which confers strong attractive forces between the cell wall and the dye concerned (Singh and Singh, 2017; Srinivasan and Viraraghavan, 2010; Das and Charumathi, 2012). The considerable surface area availability in both marine and freshwater habitats and pronounced binding properties indicate higher possibilities of biosorption in them (Singh, 2018; Donmez and Aksu, 2002). The physical process underlying the mechanism of uptake being electrostatic affinity and the capacity to form a complex between the dye and cell wall of Cyanobacteria (Satiroglu et al., 2002). Aspects like pH, temperature, dye concentration, contact time, ionic strength, molecular structure of the dye and kind of Cyanobacteria used determines the efficiency of biosorption (Singh and Singh, 2017; Reganathan et al., 2006; Vijayaraghavan and Yun, 2007; Erden et al., 2011; Ambrosio et al., 2012). Species of Cyanobacteria such as *Anabaena variabilis, Nostoc muscorum and Lyngbya majuscule* showed effective uptake of pollutants from textile industry when the pH was raised from 7.3 to 9.6 (Kotteswari et al., 2007). From the reference of Priscila et al. (2017), it has been shown that Cyanobacteria with filamentous forms acquires the time to build up the interaction between the dye and the concerned organism. Several species of *Oscillatoria* were found to convert the aromatic amines generated into simple organic compounds or CO_2 and H_2O and proved to be efficient and successful degraders of azo dye. Some of them also used dyes as their carbon and nitrogen source (Jinqi and Haitian, 1992). *Phormidium autumnale* UTEX 1580 efficiently degraded the indigo dye, the resultant products being isatin and anthranilic acid which neither increased toxicity nor showed mutagenic effect in mutagenicity test with *Allium cepa* (Priscila, 2017) Currently, immobilized Cyanobacteria such as *Oscillatoria brevis* and *Westiellopsis prolifica* are used to reduce the color from the textile effluent containing dyes Remazol and vinyl sulfone. They were not only efficient in decolorization of the dyes but also reduced BOD, COD, and removed ammonia, nitrate, and total organic and inorganic phosphates from the effluent. The matrix recently used in immobilization of Cyanobacteria is polyvinyl and polyurethane foams compared to that of commonly used agar, alginate gels as it has high mechanical stability for long time use in bioreactors, not disrupted by phosphate ions (Vijayakumar and Manoharan, 2012). Immobilized Cyanobacterial strain *Phormidium* has well decolorizing activity under thermophilic circumstances. The dead cyanobacterial cells compared to the free-living ones are beneficial and strategic as they do not have nutrient requirement, can be put in storage, and used for long time and can be revived by organic solvents (Singh and Singh, 2017; Fu and Viraraghavan, 2001). The vital and significant benefits of biosorption by Cyanobacteria involve high selectivity, efficiency, economic, operating at low concentration, and good removal accomplishment (Aksu and Donmez, 2003).

5.5.2 ENZYMATIC METHOD

Azo dyes are less prone to degradation by microorganisms due to their azo linkage (–N N), in some cases due to the presence of sulphonic (SO_3^-) and other electron-withdrawing groups that result in electron deficiency. Reductases play a significant role in the degradation of azo dyes under proper conditions. In case of the anaerobic process of degradation of azo dye by microbes commences with the breakage of the azo linkage with the help of anaerobic azo-reductase into respective amines and transfer of electron by a redox mediator that behaves as a transporter of electron between the dye located extracellularly and the reductase present intracellularly. Azo-reductases (El-Sheekh et al., 2009) catalyze this reaction only in the presence of reducing equivalents, e.g., FADH, NADH, and NADPH. Dye reducing activity of most azo dyes is not dependent on the uptake of the dye within the cell as they are not likely to pass through the cell membrane due to their high molecular weight and presence of sulphonate groups as substituents. Various enzymes that reduce azo linkages have been identified: FMN-dependent reductases, FMN-independent reductases, NADH-dependent reductases, NADPH-dependent reductases and NADH-DCIP reductases. The oxidative degradation of azo dyes is catalyzed by POXs and phenoloxidases, such as N-demethylase, manganese POX, lignin POX laccase, tyrosinase, etc. (Singh and Singh, 2017; Oturkar et al., 2011; Martorell et al., 2012), dye-decolorizing POXs (Liers et al., 2010) and cellobiose dehydrogenase (Tilli et al., 2011).

Cyanobacteria possess several enzymes which play an important role in the decolorization of azo dyes. *Oscillatoria curviceps* degraded Acid Black dye with the enzymes such as azoreductase, polyphenol oxidase and laccase (Singh and Singh, 2017; Priya et al., 2011). Under azo dye stress an induction of azo reductase was reported in Cyanobacteria *Anabaena subcylindrical* when treated with Tartrazine (mono azo) and Ponceau (diazo) dyes (Urushigawa and Yonezawa, 1977). Moreover, compared to bacteria and fungi, which needs the involvement of carbon and other increments for the removal of dyes, Cyanobacteria have no such requirements (Omar, 2008). Cyanobacterial laccase could be used as an eco-friendly and competent approach in the decolorization of textile dyes (Rao et al., 2010).

5.6 CONCLUSION

The textile industries emit considerable dye in wastewater effluent, which contaminates the aquatic bodies causing deterioration of flora, fauna, and human health. A number of shortcomings are encountered in the treatment of textile wastewater while following physical and chemical approaches. To combat such obstacles, the role of microbes in the current scenario as an advantageous alternative has been recommended. This review concentrates on the capability of freshwater filamentous Cyanobacteria to remediate the textile dyes and effluent effectively and economically by biosorption and enzymatic approach under varied conditions to less harmful products. Further, due to their no involvement of other add-ons and carbon for the elimination of dyes compared to that of bacteria and fungi, Cyanobacteria can be considered as a promising option in the field of bioremediation against the stress caused by textile dyes and effluent.

ACKNOWLEDGMENTS

I express my heartfelt thanks to the Department of Microbiology, Lady Brabourne College, Kolkata for providing me the opportunity to perform the research work.

KEYWORDS

- **biosorption**
- **cyanobacteria**
- **dyes**
- **enzymatic approach**
- **textile effluent**
- ***Westiellopsis prolifica***

REFERENCES

Aksu, Z., & Donmez, G., (2003). A comparative study on the biosorption characteristics of some yeast for Remazol blue reactive dye. *Chemosphere, 50*, 1075–1083. doi: 10.1016/S0045-6535(02)00623-9.

Ambrosio, S. T., Vilar, J. C., Da Silva, C. A. A., Okada, K., Nascimento, A. E., & Longo, R. L., (2012). A biosorption isotherm model for the removal of reactive azo dyes by inactivated mycelia of *Cunninghamella elegans* UCP542. *Molecules, 17*, 452–462. doi: 10.3390/molecules17010452.

An, S. Y., Min, S. K., Cha, I. H., Choi, Y. L., Cho, Y. S., Kim, C. H., & Lee, Y. C., (2002). Decolorization of triphenylmethane and azo dyes by *Citrobacter* sp. *Biotechnol. Lett., 24*, 1037–1040.

Ayed, L., Mahdhi, A., Cheref, A., & Bakhrouf, A., (2011). Decolorization and degradation of azo dye methyl red by an isolated *Sphingomonas paucimoboilis*: Biotoxicity and metabolites characterization. *Desalination, 274*, 272–277. doi: 10.1016/j.desal.2011.02.024.

Benkhaya, S., et al., (2020). A review on classifications, recent synthesis and applications of textile dyes inorganic *Chemistry Communications, 115*, 107891.

Carliell, C. M., Barclay, S. J., Naidoo, N., Buckley, C. A., Mulholland, D. A., & Senior, E., (1995). Microbial decolorization of a reactive azo dye under anaerobic conditions. *Water SA, 21*(1), 619.

Chang, J. S., Chen, B. Y., & Lin, Y. S., (2004). Stimulation of bacterial decolorization of an azo dye by extracellular metabolites from *Escherichia coli* Strain NO_3. *Bioresour. Technol., 91*, 243–248. doi: 10.1016/S0960-8524(03)00196-2.

Chang, Jo-Shu, Chou, Chien, & Chen, Shan-Yu, (2001). *Process Biochem., 36*, 757–763.

Chen, K. C., Wu, J. Y., Liou, D. J., & Hwang, S. C. J., (2003). *J. Biotechnol, 101*, 57–68.

Chequer, F. M. D., Oliveira, G. A. R., Ferraz, E. R. A., Cardoso, J. C., & Zanoni, M. V. B., (2013). Oliveira DP textile dyes: Dyeing process and environmental impact. In: Günay, M., (ed.), *Eco-Friendly Textile Dyeing and Finishing* (pp. 151–168). IntechOpen. doi: 10.5772/53659.

Das, N., & Charumathi, D., (2012). Remediation of synthetic dyes from wastewater using yeast a review. *Ind. J. Biotechnol., 11*, 369–380.

Donmez, G., & Asku, Z., (2002). Removal of chromium (VI) from saline wastewater by *Dunaliella* species. *Process Biochem., 38*, 751–762. doi: 10.1016/S0032- 9592(02)00204-2.

Easton, J., (1995). The dye maker's view. In: Cooper, P., (ed.), *Color in Dyehouse Effluent* (p. 11). Bradford, UK: Society of Dyers and Colorists.

El-Sheekh, M. M., Gharieb, M. M., & Abou-El-Souod, G. W., (2009). Biodegradation of dyes by some green algae and cyanobacteria. *Int. Biodeter. Biodegr., 63*, 699–704. doi: 10.1016/j.ibiod.2009.04.010.

Erden, E., Kaymaz, Y., & Pazarlioglu, N. K., (2011). Biosorption kinetics of a direct azo dye Sirius blue K-CFN by *Trametes versicolor. Electron. J. Biotechnol., 14*, 1–10. doi: 10.2225/vol14-issue2-full text-8.

Fu, Y., & Viraraghavan, T., (2001). Fungal decolorization of dye wastewaters: A review. *Bioresour. Technol., 79*, 251–262. doi: 10.1016/S0960-8524(01)00028-1.

Gupta, M., (2020). *Remediation of Textile Dyes (Direct Red and Acid Orange) by Freshwater Cyanobacteria, Research Interventions and Advances in Plant Science* (pp. 123–130). Bhumi Publishing.

Gürses, A., et al., (2016). *Classification of Dye and Pigments, in Dyes Pigments* (pp. 31–45). Springer.

Huang, G., Chen, F., Wei, D., Zhang, X., & Chen, G., (2010). *Appl Energ., 87*, 38–46.

ISPCH, (1995). *Industrial Safety and Pollution Control Handbook* (2nd edn., pp. 45–446). A Joint Publication of National Safety Council and Associate (Data) Publishers Pvt. Ltd., Hyderabad, 2nd reprint.

Jinqi, L., & Haitian, L., (1992). *Environ. Pollut.,* 75, 279–283.

Kapdan, I. K., Tekol, M., & Sengul, F., (2003). Decolorization of simulated textile wastewater in an anaerobic-aerobic sequential treatment system. *ProcBiochem., 38*, 1031–1037.

Kotteswari, M., Murugesan, S. J., Kamaleswari, & Veeralakshmi, M., (2007). *Indian Hydrobiology, 10*(1), 109–116.

Lalnunhlimi, S., & Krishnaswamy, V., (2016). Decolorization of azo dyes (direct blue151 & direct red 31) by moderately alkaliphilic bacterial consortium *Brazilian Journal of Microbiology, 7*(1), 39–46.

Liers, C., Bobeth, C., Pecyna, M., Ullrich, R., & Hofrichter, M., (2010). DyP-like peroxidases of the jelly fungus *Auricularia auriculajudae* oxidize nonphenolic lignin model compounds and high-redox potential dyes. *Appl. Microbiol. Biotechnol., 85*, 1869–1879. doi: 10.1007/s00253-009-2173-7.

Maguire, R. J., (1992). Occurrence and persistence of dyes in a Canadian river. *Water Sci. Technol., 25*, 265–270.

Mallick, N., (2002). A review. *Biometals, 15*, 377–390.

Mansour, H. B., Ayed-Ajmi, Y., Mosrati, R., Corroler, D., Ghedira, K., & Barillier, D., (2010). Acid violet 7 and its biodegradation products induce chromosome aberrations, lipid peroxidation and cholinesterase inhibition in mouse bone marrow. *Environ. Sci. Pollut. Res. Int., 17*, 1371–1378. doi: 10.1007/s11356-010-0323-1.

Manu, B., & Sanjeev, C., (2002). *Biores. Technol., 82*, 225–231.

Martorell, M. M., Pajot, H. F., & De Figueroa, L. I. C., (2012). Dye decolorizing yeasts isolated from las Yungas rainforest dye assimilation and removal used as selection criteria. *Int. Biodeter. Biodegr., 66*, 25–32. doi: 10.1016/j.ibiod.2011.10.005.

Mehta, P., Surana, M., Mehta, R., & Kabra, B. V., (2011). *Der Chemica Sinica, 2*(5), 37–46.

Metcalf and Eddy Inc, (1991). *Wastewater Engineering, Treatment, Disposal and Reuse* (3rd edn.). Mc Graw Hill Publishing Company, Mc Graw Hill International editions, Civil Engineering series, Singapore.

Mounia, M. C., et al., (n.d.). *Analyze, Synthese et Etude Solvatochromique de Composes Tinctoriaux.*

Mugdha, A., & Usha, M., (2012). Enzymatic treatment of waste containing dyestuffs using different delivery systems. *Sci. Rev. Chem. Commun., 2*, 31–40.

Murugesan, K., & Kalaichelvan, P. T., (2003). *IJEB, 41*, 1076–1087.

Nora, M., (2011). Hilal, *Der Chemica Sinica, 2*(4), 262–273.

Omar, H. H., (2008). Algal decolorization and degradation of monoazo and diazo dyes. *Pak. J. Biol. Sci., 11*, 1310–1316. doi: 10.3923/pjbs.2008.1310.1316.

Oturkar, C. C., Nemade, H. N., Mulik, P. M., Patole, M. S., Hawaldar, R. R., & Gawai, K. R., (2011). Mechanistic investigation of decolorization and degradation of reactive red 120 by *Bacillus lentus* BI377. *Bioresour. Technol., 102*, 758–764. doi: 10.1016/j.biortech.2010.08.094.

Pajot, H. F., Delgado, O., De Figueroa, L. I. C., & Farina, J. I., (2011). Unraveling the decolorizing ability of yeast isolates from dye polluted and virgin environment: An ecological and taxonomical overview. *Antonie Van Leeuwenhoek, 99*, 443–456. doi: 10.1007/s10482-010-9495-4.

Priscila, M. D., Silva-Stenico, M. E., Moraes, L. A. B., Fiore, M. F., & Monteiro, R. T. R., (2017). Degradation of textile dyes by cyanobacteria. *Brazil Journal of Microbiology, 48*(1), 25–31.

Priya, B., Uma, L., Ahamed, A. K., Subramanian, G., & Prabaharan, D., (2011). Ability to use the diazo dye C.I. acid black 1 as nitrogen source by the marine cyanobacterium *Oscillatoria curviceps* BDU92191. *Bioresour. Technol., 102*, 7218–7223. doi: 10.1016/j.biortech.2011.02.117.

Rafu, M. A., & Ashraf, S. S., (2009). *J. Hazard Mater, 166*, 6–16.

Rahman, (2016). Research Article Open Access. *Journal of Textile Science & Engineering.* ISSN: 2165-8064 6:1 http://dx.doi.org/10.4172/2165-8064.1000242.

Rajendran, P., & Gunasekaran, P., (2006). MJP Publishers, 138–159.

Rao, X. J., Ling, E., & Yu, X. Q., (2010). He role of lysozyme in the prophenoloxidase 692 activation system of *Manduca sexta*: An in vitro approach. *Dev Comp Immunol., 34*, 264–271.

Rehman, A., et al., (2020). The application of cationic-nonionic mixed micellar media for enhanced solubilization of direct brown 2 dye. *J. Mol. Liq., 301*, 112408.

Renganathan, S., Thilagaraj, W. R., Miranda, L. R., Gautam, P., & Velan, M., (2006). Accumulation of acid orange 7, acid red 18 and reactive black 5 by growing *Schizophyllum commune*. *Bioresour. Technol., 97*, 2189–2193. doi: 10.1016/j.biortech.2005.09.018.

Robinson, T., McMullan, G., Marchant, R., & Nigam, P., (2001). Remediation of dyes in textile effluent: A critical review on current treatment technologies with a proposed alternative. *Bioresour. Technol., 77*, 247–255.

Satiroglu, N., Yalcinkaya, Y., Denizli, A., Arica, M. Y., Bektas, S., & Genc, O., (2002). Application of NaOH treated *Polyporus versicolor* for removal of divalent ions of group IIB elements from synthetic wastewater. *Process Biochem., 38*, 65–72. doi: 10.1016/S0032-9592(02)00057-2.

Shah, M. P., (2014). Microbial degradation of acid orange dye by an application of *Pseudomonas* spp. ETL-1979 isolated from textile effluent: An environmental bio-medical approach. *OA Biotechnology, 3*(1), 3.

Singh, N., (2018). Significance of bioadsorption process on textile industry wastewater. *The impact and Prospects of Green Chemistry for Textile Technology* (pp. 367–416). Woodhead Publishing. doi: 10.1016/B978-008-102491-1.00013-7.

Singh, P. K., & Singh, R. L., (2017). Bio-removal of azo dyes: A review. *Int J. Appl. Sci. Biotechnol., 5*(2), 108–126.

Srinivasan, A., & Viraraghavan, T., (2010). Decolorization of dye wastewaters by biosorbents: A review. *J. Environ. Manag., 91*, 1915–1929. doi: 10.1016/j.jenvman.2010.05.003.

Tilli, S., Ciullini, I., Scozzafava, A., & Briganti, F., (2011). Differential decolorization of textile dyes in mixtures and the joint effect of laccase and cellobiose dehydrogenase activities present in extracellular extracts from *Funalia trogii. Enzyme Microb. Tech., 49*, 465–471. doi: 10.1016/j.enzmictec.2011.08.002.

Tripathi, A., & Srivastava, S. K., (2011). Eco-friendly treatment of azo dyes: Bio decolorization using bacterial strains. *International Journal of Bioscience, Biochemistry and Bioinformatics, 1*(1).

Urushigawa, Y., & Yonejawa, Y., (1977). Chemo-biological interactions in biological purification system II-biodegradation of azo compound by activated sludge. *Bull. Environ. Contam. Toxicol., 17*, 214–218.

Vijayakumar, S., & Manoharan, C., (2012). Treatment of dye effluent using free and immobilized cyanobacteria. *J Bioremed Biodeg., 3*, 10. doi: 10.4172/2155-6199.1000165.

Vijayaraghavan, K., & Yun, Y. S., (2007). Utilization of fermentation waste (*Corynebacterium glutamicum*) for biosorption of reactive black 5 from aqueous solution. *J. Hazard. Mater., 141*, 45–52. doi: 10.1016/j.jhazmat.2006.06.081.

Vitor, V., & Corso, C. R., (2008). Decolorization of textile dye by isolated *Candida albicans* from industrial effluents. *J. Ind. Microbiol. Biotechnol., 35*, 1353–1357. doi: 10.1007/s10295-008-0435-5.

Wong, Y., & Yu, J., (1999). Laccase catalyzed decolorization of synthetic dyes. *Water Res., 33*, 3512–3520. doi: 10.1016/j.colsurfa.2005.04.028.

Xu, M., Guo, J., Cen, Y., Zhong, X., Cao, W., & Sun, G., (2005). *Shewanella decolorationis sp.* Nov., a dye decolorizing bacterium isolated from activated sludge of a wastewater treatment plant. *Int. J. Syst. Evol. Microbiol., 55*, 363–368. doi: 10.1099/ijs.0.63157-0.

CHAPTER 6

Dust Pollution and Plants: A Case Study of Mumbai, Maharashtra, India

NITESH JOSHI

Rizvi College of Arts Science and Commerce, Department of Botany, Mumbai, Maharashtra, India, E-mail: niteshcjoshi@gmail.com

ABSTRACT

This chapter is a review of literature mainly dealing with dust and plants with reference to Mumbai city. The use of plants to monitor dust in cities and their role in mitigating dust pollution is discussed. Plants ideal for selection in developing green belts in an urban region is discussed with a comprehensive approach of using a dust chamber. The characteristics useful in selecting plants in a city for reducing dust are enumerated as well as plants suitable as phytomonitors.

6.1 MUMBAI URBAN REGION AND POLLUTION

Maharashtra is one of the most industrialized states, and its capital, Mumbai, is termed as the financial capital of India. Population-wise, Mumbai is one of the largest districts in the country. The population of Mumbai is 12.48 million as per the 2011 census and its Population Density is 39,446 per sq. km. Area-wise; it is the second smallest district in Maharashtra State. Area of Mumbai above sea level as per the development plan is 453.05 km^2. Its maximum width is 17 km (east to west) and length is 42 km (north to south). The gardens and parks in the city take up to 73.67 hectares of area. Similarly, recreation grounds and playgrounds share 91.31 hectares and 69.81 hectares

Plant Metabolites under Environmental Stress: Mechanisms, Responses, and Adaptation Strategies.
Nivas M. Desai, Manasi Patil, Umesh R. Pawar (Eds.)

of areas, respectively. This city is located on the western sea coast of India at 18°53'N to 19°16'N latitude and 72°E to 72°59'E longitude (Archana et al., 2010).

Mumbai Suburban District has three Talukas/Tahsils, namely, Andheri, Borivali, and Kurla. The jurisdiction of the Mumbai Suburban District is from Bandra to Dahisar, from Kurla (Chuna Bhatti) to Mulund and from Kurla and up to Trombay Creek. This district has 4 Parliamentary Constituencies and 26 Assembly Constituencies. Many important places have adorned the beauty of this districts like Mount Mary Church, Kanheri Caves, Jogeshwari Caves, Fantasy Land, Essel World, Water Kingdom, Sanjay Gandhi National Park, Aarey Colony, Film city, Santacruz Airport, Sahar International Airport, Tulsi Lake, Vihar Lake, Powai Lake, Mithi River, Dahisar River, Poiser, and Oshiwara River, INS Hamla, Indian Institute of Technology, Powai, Bhabha Atomic Research Centre, Rashtriya Chemicals and Fertilizers, Hindustan, and Bharat Petroleum Refineries. A 3.5 km stretch passes through the Bandra-Kurla Complex, which is planned and developed by the Mumbai Metropolitan Region Development Authority. The total number of vehicles in Mumbai as of January 2011 is 19,09,804 which includes 54.5% two-wheelers, 31.2% cars, 3.5% taxis, 2.8% heavy vehicles, 7.6% three-wheeler and 0.4% of other vehicles (MPCB, 2010) which contribute significantly to air pollution. The road length in Mumbai is about 2,000 km, comprising of about 1,950 km of MCGM maintained roads and about 50 km of State Highways (23.55 km of Eastern Express Highway from Sion to Thane and 25.33 km of Western Express Highway from Bandra to Dahisar).

Pollution is the addition of undesirable change in the environment, which is prevalent in this city. This is caused by the release of products as a result of human activities. Hence these by-products can be physical, chemical, or biological and generally find their way into the environment directly or indirectly. These could be harmful to human beings and other organisms.

When pollutants are released into the atmosphere, they are called air pollutants. These can be solid, liquid, or gaseous, including noise. These pollutants can prove to be injurious to human beings, other organisms, and property CPCB (2011). Air pollutants may be either emitted into the atmosphere or formed within the atmosphere itself. Air Pollutants are of two types, i.e., primary, and secondary pollutants, former are pumped into the atmosphere and directly pollute the air, while those that are formed in the air when primary pollutants react are known as secondary pollutants. The city faces and is exposed to all types of air pollutants, some even above the permissible limits.

The pollutants that spoil the air are Primary air pollutants, Secondary air pollutants, Gaseous air pollutants, and Particulate Pollutants (WHO, 2005). The Primary air pollutants include are Sulphur dioxide, oxides of Carbon and Nitrogen, and volatile organic compounds (VOC). Ozone is a more common secondary air pollutant (Seinfeld and Pandis, 1998). Gaseous air pollutants enter the human respiratory system, either as gases or through moisture. These get quickly deposited in the upper respiratory tract and not penetrate and soon enter the lungs. The pollutants suspended in the atmosphere in the form of solid or liquid phase are Particulate air pollutants, i.e., suspended particulate matter (SPM). Particulate matter (PM) is classified as; (1) PM_{10}; (2) $PM_{2.5}$; (3) PM_1; (4) dust and sea particles 0.1 mm in diameter(Morawska and Salthammer, 2003). Large particles tend to fall due to gravity or wind-driven processes and have a short atmospheric existence. In practice there are few particles in the atmosphere exceeding 20 μm in diameter. Dust has diverse chemical compositions which are highly dependent on their sources.

The atmospheric pollutants also include hydrocarbons, oxygenates, halogenated, and other carbon compounds collectively known as VOC. The main source is typically through leakages. The exhaust pipe of a vehicle may therefore be as important as the fuel tank Cars, trucks; other vehicles are also a source of VOC emissions. Paints and adhesives contain organic solvents, which disperse in the atmosphere to allow the active ingredients to dry.

Highway and roads are the major cause of atmospheric pollution. Sources of highway runoff pollutants have been reviewed and summarized in numerous reports (Barber et al., 2006; Barrett et al., 1995). The sources of pollution can be classified into three general categories: atmospheric deposition, vehicles (including fuels and exhaust emissions). There are enough reports by government agencies to substantiate the fact that dust is a very prevalent problem in the city throughout the year.

6.1.1 SOURCES AND HEALTH HAZARDS OF AIR POLLUTION

Air pollutants are derived various from sources which may be created by man or by activities of nature. Forest fire, volcanic emissions, sandstorm, pollen scattering, etc., contribute towards pollution which are not in human control. Man-made sources on the other hand consist of substances emitted during burning of fossil fuels in factories and thermal power plants, which emit myriad of pollutants which are harmful. Most developing countries like India are faced with deteriorating air quality due to rapid industrialization and urbanization. Lack of awareness, poor quality of roads and improper legislation also contribute to poor air quality (Joshi and Chauhan, 2008). Another contributory factor is stationary and mobile emissions.

Concentrations that substantially exceed the odor detection threshold can result in annoying and discomforting physiological symptoms of headache and nausea in humans (EOHP, 2006). When a severe exposure of a pollutant takes place particularly in a closed space, in an industrial area or due to an accidental release of Hydrogen sulfide, the effects can be harmful (Bist et al., 2011).

6.1.2 EFFECTS OF AIR POLLUTANTS ON HUMAN HEALTH

Studies done all over the world have shown that there is an increase in several diseases like chronic respiratory diseases, lung cancer, tuberculosis, stomach cancer and heart diseases. These are all linked to air pollution. The roadside dust particles and heavy metals along with it were found responsible for causing diseases like asthma and chronic lung diseases. Six major air pollutants identified by USEPA (United States Environmental Protection Agency).

6.1.3 DUST POLLUTION–SOURCES AND ENVIRONMENTAL IMPACT: AIR POLLUTANTS AND HUMAN HEALTH

The plants have different level of tolerance towards the harshness of the environment. They differ considerably in their responses to pollutants (Rao, 1985). Some plants are sensitive and some are found to be tolerant as reflected in their air pollution tolerance index (APTI). Ascorbic acid (AA) plays an important role in tolerance (Beckerson and Hofstra, 1980; Laisk et al., 1988). The plant species differ in their perception and response to air pollutants due to influence of genetic and/or environmental factors. Pollutants emitted from automobile, cement factories and industrial sources adversely affect the air quality leading to health hazards of not only humans and animals but plants too (Table 6.1).

The plants growing in the ambient environment of urban-industrial areas are exposed to a mixture of gases from various sources. Therefore, in ecological studies, considering the physical presence and toxicity of a single air pollutant acting alone may not be reasonable. While individual plant responses can be demonstrated following brief exposures to different concentrations of air pollutants, for long periods, field experiments are necessary to detect any ecosystem changes.

TABLE 6.1 Effects of Hazardous Dust Types on Various Plant Species

Tree Species	Dust Type/Source	Effect	References
'Fruit trees' and *Quercus lobate*	Cement factory	Blocked stomata.	Pierce (1909)
Citrus authantium, C. limon	Cement factory and road dust	Reduced fruit set, leaf growth, starch production and increased necrosis.	Parish (1910); Pierce (1910)
Prunus avium, Malus domestica, Pyrus communis	Cement factory	Reduced fruit set.	Anderson (1914)
Tilia cordata, Acer pseudoplatanus, Prunus spinosa, Corylus avellana, Rosa canina, Aesculus hippocastanum, Salix viminalis, Sambucus nigra	Cement	Cell destruction, bark peeling, leaf necrosis.	Czaja (1961)
Piceae abies, Pinus sylvestris, Taxus baccata, Rosa canina	Cement	Cell destruction.	Czaja (1962)
Ouercus robur, Pinus sylvestris, Populus alba	Cement factory	Leaf injury, reduced growth.	Bohne (1963)
Tsuga canadenis	Limestone quarry	Chlorotic needles	Manning (1971)
Acer rubrum, Quercus prinus, Q. rubra, Liriodendron tulipifera	Limestone quarry	Reduced growth Increased growth	Brant and Rhoades (1972, 1973)
Psidium guajava	Cement factory	Increased tissue Ca, K, Na, and P	Lal and Ambasht (1982)
Populus nigra and Prunus laurocerasus	Building works	Increased leaf temperature	Steinhubel and Halas (1967)
Populus tremula	Inert silica gel	Reduced diffusive resistance	Fluckiger et al. (1979)
Populus tremula, Betula pendula, Alnus glutinosa, Fraxinus excelsior	Urban road	Blocked stomata and reduced diffusive resistance.	Fluckiger et al. (1977)
Rhododendron catawbiense	Urban road	Increased absorption of insolation	Eller (1974, 1977); Eller and Brunner (1975)
Populus tremula, Acer campestre, Betula pendula, Alnus glutinosus, Prunus avium, Quercus spp.	Urban road	Increased leaf temperature	Fluckiger et al. (1978)

TABLE 6.1 *(Continued)*

Tree Species	Dust Type/Source	Effect	References
Fraxinus excelsior	Urban road	Reduced diffusive resistance	Fluckiger et al. (1982)
Abies alba	Urban road	Reduced growth, but compounded by Pb and NOx	Braun and Fluckiger (1987)
Fraxinus excelsior	Urban road	Increased leaf temperature	Guggenheim et al. (1980)
Viburnum tinis	Motor vehicle exhaust	Reduced photosynthesis and diffusive resistance.	Thompson et al. (1984)
Mangifera indica, Citrus limon	Coal dust	Reduced growth, fruit set and leaf lesions and partial defoliation	Rao (1971)
Daphne laureola	Urban area	Blocked stomata	Pyatt (1973)
Quercus petraea	Smokeless fuel factory	Blocked stomata and reduced night time diffusive resistance.	Ricks and Williams (1974, 1975)
Pinus sylvestris	Fertilizer factory	Young trees increase growth, old trees decrease growth, leaf injury.	Havas and Huttunen (1972)

Source: Vishaka (2021).

According to Rao (1980), the effects of air pollution (gaseous pollutant, acid deposition and particulates) on various levels of ecosystem organization may be summarized as: (i) absorption and accumulation of pollutants; (ii) Injury to autotrophs and primary and secondary consumers due to pollutant accumulation; (iii) Change in quantitative characteristics of plant and animal species with a shift in competition; (iv) Loss of stability in ecosystem; (v) disruptions in associations of biotic components and biogeochemical cycle; and (vi) landscape erosion.

6.2 PLANT INDICATORS OF AIR POLLUTION

Bioindicators reveals the presence of a substance in its vicinity by showing typical symptoms which can be distinguished from the effects of other natural or anthropogenic stresses (Steubbing et al., 1989). Plants are stationary, and thus they play a very important role in indicating the changes taking place in an environment, such plants are called indicators. Use of visual symptoms of injury in plants sensitive to a pollutant and Use of chemical tissue analysis of accumulator plants are important aspects of bioindicator plants Keller (1982). Visible symptoms in plants like Necrosis, in Gladioli in presence of fluoride accumulation; bronzing and necrosis of *Urtica urens* L. in Peroxy-acetyl nitrate; Necrosis, in *Medicago sativa* L. showing intercostal chlorosis in presence of pollutants; Speckle necrosis due to O_2 on *Nicotiana tabacum* L. Intercostal necrosis in the presence of oxides of Nitrogen in Spinach are all useful hints (Posthumus, 1982).

Falla et al. (2000); and Garrec and Van Haluwyn (2002) have modified the concepts of phytomonitoring and biomonitoring and have given terms to describe them as follows:

1. **Bioindicators (Reactive or Response Indicators):** These are plants showing visible symptoms on their vegetative and reproductive parts like leaves and flowers, and fruits in the form of necrosis, chlorosis, etc., as well as reduction in growth.
2. **Biosensors or Biomarkers:** These are plants in which biochemical and physiological estimations and showing marked reduction in metabolites are used.
3. **Bioaccumulators:** These are plants that are in general tolerant to air pollutants but they accumulate airborne elements and gases into their tissues.
4. **Ecological Indicators:** These are plant population or communities which change as per changes in ecology.

Due to their very large surface area leaves of plants represent an efficient pollutant trapping device. Shelter belts of pollutant tolerant species of herbs, shrubs, and trees in urban-industrial areas may help to reduce air pollution by improving air quality (Manning and Fedder, 1980). Air pollutants may have very low ambient concentrations and are difficult to measure accurately with physical and chemical methods. In such conditions, plants can accumulate those pollutants to a level that is easier to analyze. The parameters which can be used for phytomonitoring of air quality are mentioned below:

1. **Plant Growth:** Altered growth rates, and reduction in morphological growth or growth changes or rates or growth is effectively used to monitor pollution, Manning and Fedder (1980).
2. **Macro Characters:** Visible injury symptoms and reduction in growth have been reported by many workers.

   Tobacco Bel–W3 (*Nicotiana tabacum* L.) is used to monitor ozone and the plant has been done for many years in different countries like France (Vergé et al., 2002) and The Netherlands (Posthumus, 1982).
3. **Micro Characters:** Frequency of stomata and appearance of stomatal and epidermal cells, stomatal pore size and cuticular damage, trichome density, etc., have been used as indicators of pollution in the past (Rao, 1985; Desai and Kapoor, 2013).
4. **Biochemical Characters:** Where visible symptoms are difficult to observe, biochemical parameters become effective in knowing pollution levels, peroxidase activity, total chlorophylls, catalase activity total protein contents AA content and indices like APTI are useful (Keller, 1974; Nandi et al., 1980; Rao, 1981; Singh and Rao, 1985).

   Auto-exhaust pollution showed marked alterations in photosynthetic pigments and protein content in foliar tissues were noticed by Verma and Singh (2006). Increased dust deposition, showed a decrease in chlorophyll content (Prajapati and Tripathi, 2008).
5. **Lichens:** *Lecanora conizaeoides* was used in an industrial area in Denmark to monitor heavy metals and gases (Pilegaard, 1978) because of its high tolerance to air pollution. Species surviving in an area with high SO_2 concentrations. *Xanthoria parietina* was used in north-eastern Italy to monitor airborne heavy metals (Nimis et al., 2000).
6. **Bryophytes:** These are useful bioaccumulators of inorganic and particulate organic pollutants (Thomas, 1986) and are also important in biomonitoring of heavy metal deposits (Steinnes, 1995); growth and efficiency for element uptake by bryophytes has been studied by Sucharova and Suchara (1998).

7. **Mosses:** *Pleurozium schreberi* and *Hylocomium splendens,* have been recognized as accumulators of heavy metals in Scandinavia (Rühling and Tyler, 1973) and Poland (Grodzinska, 1978), *Hypnum cupressiforme* in the UK (Goodman and Roberts, 1971; Lee and Tallis, 1973) and in Germany by Thomas (1983).
8. **Higher Plants:** Plants capture air pollutants with their above-ground plant parts. The leaves are the most important parts to absorb or adsorb pollutants, so also is the bark (Schulz et al., 2000).

Polyalthia longifolia (Ashok), *Tamarindus indica, Thuja occidentallis, Ziziphus jujube, Acacia arabica, Citrus* sps., *Diospyros* sps., *Ficus benghalensis, Ficus religiosa, Lillium* sps. etc., are some of the plant species suggested by MOEF (2005) for their possible potential in pollution control.

6.3 PHYTOMONITORING

Plants are effectively used as cheap monitoring systems (Steubing et al., 1981, 1982, 1989). Young needles of *Pinus*, chlorosis indicates sulfur dioxide pollution, necrosis indicates Fluoride pollution, bleaching indicates oxides of Nitrogen pollution while chlorosis indicates chloride pollution in the atmosphere, browning in moss leaves indicate fluoride accumulation. The plant species to be used in pollution monitoring should have certain features such as: species should be easy to identify in the field and easy to handle for damage analysis; species should have a wide range of distribution to be used in different areas; they should be Sensitive to many types of pollutants so that it can be used to monitor different types of pollutants in the area; produce highly specific damage symptoms in response to particular types and concentrations of pollutants.

Most useful and commonly used plants include sensitive species of some lower plant forms, conifers oaks and many crop plants. Lower forms are more commonly used in monitoring air pollution because their range of pollutant specificity is usually much higher than that of higher vascular plants (Chaphekar, 1972; Posthumus, 1982; Steubing, 1982; Rao, 1985; Joshi, 1990). Map of dust pollution in Mumbai was last prepared by Chaphekar et al. (1980) using dust collected from the Leaves of *Mangifera indica* L. and *Thespesia populnia* (L.). *Commelina benghalensis* L. was used as a phytomonitor and an air quality index (AQI) map was prepared using morphological parameters. A similar AQI map was prepared for the city Chaphekar et al. (1980).

6.4 CURRENT SCENARIO OF DUST POLLUTION IN MUMBAI

System of Air Quality and Weather Forecasting and Research carried out monitoring work and it reported that Friday, 29th Jan. 2016 was the most polluted day since June 2015 and its pollution mounts to record high, as bad as Delhi (TOI, 30th Jan. 2016). The tragedies like Deonar dumping ground fire has increased the AQI from 325 in the morning to 341 in the evening. The previous worst AQI of Mumbai was recorded as 313 in November 2012 (the day after Diwali). On 28th January 2015, the city was enveloped in smog, leading to an increase in pollution and decrease in temperature. IMD (Indian Meteorological Departments) recorded 14°C at Santacruz and 18.5°C at Colaba (TOI, 28th Jan. 2016). The AQI showed that PM in Mumbai stood at 183 on Sunday, 5th April and on Monday, 6th Apr. 2015 the reading was 201 in Mumbai (TOI, 7th Apr. 2015). The dusty weather has been experienced in Mumbai and many other parts of the state of Maharashtra (DNA, 5th Apr. 2015). The haze and dusty weather on Sunday, 5th Apr. 2015 was a result of the recent dust storm in the Gulf region on 1st and 2nd Apr. 2015. The visibility was between 2,000 and 4,000 meters (TOI, 6th Apr. 2015) in some areas.

NEERI in its reports along with CPCB (NEERI, 2010), concluded that vehicular pollution accounts for less than 6% of particulate matter (PM_{10}) and bakeries played a larger role in particulate pollution than the vehicles in Mumbai. The highest percentage of PM comes from power plants. PM from paved road dust and unpaved road dust are together, contribute more dust than power plants (TOI, 13th Dec. 2015). The Dust storm on 24th Mar. 2013 in Kuwait has been blowing through, finally settled as haze in Mumbai and other parts of Maharashtra (TOI, 12th Apr. 2013). On Wednesday, 21st Mar. 2012, the atmosphere was observed to be dusty morning due to a giant dust plume that stretched across the Arabian Sea from the coast of Oman to India (Rediff.com News, 21st Mar. 2012). On Wednesday, 21st Mar. 2012 there was a thick layer of dust enveloping Mumbai city which reduced visibility levels extremely lower than pollution levels by nearly 500%. The dust levels as reflected by, SPM, levels rose to 1,200 units (TOI, 22nd Mar. 2012).

PM2.5 levels—tiny pollutant particles of 2.5-micron size—were below permissible limits across all the 11 real-time ambient air quality monitoring stations (AAQMS) in Mumbai, stated the Maharashtra Pollution Control Board (MPCB) annual report of April 2019–March 2020, released last month. PM2.5 is a prominent pollutant in the city. PM2.5 are the result of fuel combustion, mainly from car engines, coal/natural gas-fired power plants, wood burning and fireplaces (Indian Express, 2020).

AAQMS, monitor pollution in Mumbai, out of which, 15 recorded average PM 10 concentration within the standard limit (60 μg/m^3) as compared to four in 2018–2019. The highest RSPM levels were recorded at Kurla (140 μg/m^3), Sion (117 μg/m^3), Sahar (116 μg/m^3), and Mulund (103 μg/m^3). Some stations recorded concentration in the range of 64–98 μg/m^3. The annual safe limit for PM10 is 60 μg/m^3. "Respirable Suspended Matter poses a greater risk to human health as these particles can enter deep into the lungs through respiration and some may even enter the bloodstream. These levels are a matter of great concern for Maharashtra, as it is one of the most persistent reasons for increasing air pollution," read the report.

MPCB also analyzed the AQI range for 2019–2020 for 11 monitoring stations in the city. Good AQI was recorded for only 21% of total observations, with most good air days at Borivli (43%) and minimum at Kurla (3.8%) (Indian Express March 1, 2021).

The primary sources of metals from vehicles are friction in engine and suspension systems, brake pad and tire wear, and rust and corrosion (Kearfott et al., 2005; Barrett et al., 1995a; Lancaster, 2005). Metals have also been linked to plating on guardrails (Lancaster, 2005), vehicle emissions, and impurities in deicing salts (Barrett et al., 1995a). In addition, the presence of rumble strips can aid in dislodging metals and rust from vehicles through increased vibration (Lancaster, 2005). Another indirect source of metals in runoff is suspended solids. In particular, several nationwide studies have shown that concentrations of zinc, copper, and chromium correlate well with TSS concentrations (Kearfott et al., 2005; Kayhanian et al., 2003). In the Pacific Northwest, concentrations of metals are often strongly correlated with TSS concentrations (Chui, 1981; Preciado and Li, 2006), and in studies where both dissolved and total metals are reported, a large fraction of total metals is associated with particles (Portele, 1981). Finally, atmospheric deposition can be an important source for some metals in highway runoff. Plant leaves acts as bio-filters as they absorb large quantities of particles from the environment (CPCB, 2007).

6.4.1 PHYTOMONITORING OF DUST IN MUMBAI CITY

Urban vegetation in Mumbai city comprises of gardens, avenue trees and ornamental plants in Gardens, road dividers, Traffic Islands, and side roads on Highways. This urban vegetation is largely maintained by Municipal Corporation or corporate sectors in their effort to keep the city green. Mumbai is a densely populated and polluted city. The pollutants range from oxides of nitrogen, carbon, sulfur, and dust particles (Joshi and Chauhan, 2008).

The city shows the high range of dust levels between 190 μg/m^3 and 500 μg/m^3 (Municipal Corporation of Greater Mumbai (MCGM), 2013). The total number of vehicles in Mumbai as on January 2011 is 1,909,804 including 54.5% two wheelers, 31.2% cars, 3.5% taxis, 2.8% heavy vehicles, 7.6% three-wheeler and 0.4% of other vehicles (Environmental Status of Brihan Mumbai, 2010–2011), which contribute significantly to air pollution. Dust pollution itself comprises 40% of total air pollution problems in India (Lone et al., 2005). In Mumbai, the MPCB monitors air pollution at 3 sites and MCGM at 6 sites. The values cross the daily permissible levels of 200 μg/m^3 in winter and summer seasons.

Heavy vehicular traffic results circulation of suspended PM in the ambient air, which accumulates on roadside plants. The plants can effectively be used to monitor dust is a well-known fact (Yunus et al., 1985; Joshi, 1990). Urban trees have effectively been used in the city in the past for monitoring air pollution (Chaphekar et al., 1980). The use of plants to reduce pollution by dust is a common practice in developed countries. Plant are regarded as biofilters as they absorb large quantities of particles from the environment (Central Pollution Control Board, 2007). A plants dust interception capacity depends on its surface geometry, phyllotaxy, and external characteristics of its leaf such as hairs, cuticle, and lengths of petioles, height/canopy of trees and the prevailing weather conditions with direction and speed of the wind (Prajapati and Tripathi, 2008). In urban environment, dust on the roads can be potentially harmful to roadside vegetation, wildlife, and the neighboring human settlements (Bhattacharya et al., 2011).

Nerium odorum Aiton. is the best plant species for phytomonitoring and phytoremediation. Also, it's easy availability and maintenance, makes it suitable for growing all over the city on the roadsides and road dividers. Since *Nerium odorum* Aiton. was the best dust capturer at maximum number of sites, a dust map was prepared to indicate the seasonal dust fall at various sites during the study period using *Nerium odorum* Aiton. *Pedilanthes tithamaloides* in 2016 (Alkama and Joshi), *Ficus benjamina* in Alibag in the year 2017 by Vishaka Vaidya, *Tithonia diverisfolia* in 2015 (Joshi and Bist, 2018) was effectively used to phytomonitors dust and ambient air quality in Mumbai and later in 2018 in Alibag (Vishaka, 2021).

6.4.2 DUST PROJECT IN MUMBAI

A survey was conducted throughout the city to prepare an inventory of ornamental plant species available on roadsides and road dividers. Western

Suburbs, Central Mumbai (Eastern suburbs) and South Mumbai were visited. The selected sites include Primary, Secondary, and Tertiary Roads. Express Highways like Eastern express highways and Western Express highways.). The work is a part of University Grant commissions under major UGC projects, India. Leaves were washed with water using spray bottles, carefully collected on pre-weighed Whatman's filter paper No. 1 and dust was estimated as given by Joshi, 1990; and Chaphekar et al. (1980). The values were expressed as g/m^2. Study was carried out between Dec-11 to May-12 and Oct-12 to Nov-12. Foliar dust was collected from the main road and a few secondary roads in the city which covered 81 sites. For the sake of convenience, the sites were broadly divided into 7 areas based on their locations within the city.

It was evident from the statistical analysis that foliar dust capturing capacities between the four plant species viz for *Ficus benjamina, Bougainvillea spectabilis, Nerium odorumm and Pedilanthus tithimyloides* and the trend is as follows; *Ficus benjamina >Nerium odorum>pedilanthus tithimyloides > Bougainvillea spectablis*.

High correlation was found in the amount of dust deposited to the prevailing traffic conditions. *Ficus benjamina, Pedilanthes tithymaloides, Nerium odorum, Ixora coccinea, Calotropis gigantea* and *Ziziphus jujube* were found to be the best dust capturers during the study.

Express highways and major arterial roads were found to have consistently high values of both SPM and dust fall values. Whereas areas with high SPM values also showed higher values of foliar dust.

Dust fall depends on local and regional climate as well as the prevailing microclimatic conditions. Hence, different sites show different values of foliar dust deposition on different plant species. There is a general increase in Dust fall on the leaves from the month of winter to Summer of 2014.

Variations in wind speed, wind direction, humidity of air, etc., play a vital role in dust deposition. Hence plants can be used for phytomonitoring dust. Air pollution due to anthropogenic sources such as traffic, industries is the major source contributing to the personal exposure to respirable particulates. In India, there are many old vehicles (more than 15 years old) still in use, these should be phased out totally and transportation system should be improved to reduce idling time of vehicles at the traffic signals and old vehicles phased out. There are many small scale industries of an unauthorized nature. They should be identified and brought under the purview of the regulatory authorities. Plants can either be placed or locally grown to monitor dust fall and its accumulation. Plants as monitors of dust have an advantage over gas analyzers and high volume samplers. Thus, use of plants can be a cost-effective substitute and can be placed at specific sites. *Calotropis gigantean, Ixora coccinea, Ficus benjamina, Pedilanthes tithymaloides, Nerium odorum* and *Ziziphus jujube* can be used as a Phytomonitor of dust in the city. These plants are easy to grow, and need relatively less care and water. Therefore, highly suitable as dust monitors (Joshi, 2014).

6.5 DUST CHAMBER STUDIES

Human activities in the field of development in India created huge amounts of air pollutants (Abbasi, 1998), with dust being in large quantities (Das, 1981; Das et al., 1981; Lone et al., 2005). These dust particles are made of different elements, and some are dangerous and their phytotoxicity increases under humid conditions (Chaphekar, 1972). Plants intercept tons of dust and absorb noise near noisy factory area (Warren, 1973). The amount of dust is never constant in the atmosphere and is ever-changing from time to time and place to place. Meteorological conditions influence particle movements (Fritschen et al., 1970; Edmonds and Driver, 1974). That trees are the best sinks for particulate pollutants like radioactive, trace elements, pollen, spore, salt particles, dust, etc., is well documented (William, 1990). The use of trees is therefore the best solution to the problems of pollution in urban and industrial areas (Agarwal and Tiwari, 1997).

6.6 VEGETATION AS SINK FOR DUST

Sink for atmospheric pollutants (Meetham, 1950, 1954; Robinson and Robins, 1971; Weinstock, 1969; Hill, 1971; Moser et al., 1991) is the vegetation around us. Vegetation in the localities also have been used to monitor dust (Yunus et al., 1985; Joshi, 1990). Foliar surfaces of plants have been regarded as biofilters as they absorb large quantities of particles from the environment (CPCB, 2007).

6.6.1 DUST POLLUTION STUDIES IN INDIA ON PLANTS

Dust pollution is the common problem in cities. High amount of respirable dust is observed around cement factories (Singh and Rao, 1978, 1983). The plant growing in the control site was healthier than the plant growing near the cement industry and an increase in the amounts of cement dust pollutant adversely affected the plant growth, productivity, photosynthetic ability, opening, and closure of leaf stomata, etc. (Chaurasia, 2013).

Also, the different plant species can never have the same dust capturing capacities due to their own morphological differences such as trichomes, phyllotaxy, etc. The plants may be very good, good, or bad in holding dust on their surfaces (Mancharkar, 2001). It was found that, under controlled wind tunnel conditions, rough pubescent sunflower leaves could catch more dust than smooth waxy tulip poplar leaves (Wedding et al., 1975). The dust is captured by leaves of plants, leaf epidermal outgrowths like hairs and scales, hairy axils of stems and leaf bases, etc. (Das, 1981).

According to Rao, 1985, some plants are sensitive and some are found to have better APTI. Ascorbic acid (AA), along with leaf pH, plays a significant role in determining the pollutant response of plants (Chaudhary and Rao, 1977). Plasma membranes and chloroplasts are considered the main sites of pollutant action (Beckerson and Hofstra, 1980; Laisk et al., 1988). The plant species differ in their perception and response to air pollutants due to influence of genetic and/or environmental factors (Taylor and Pitelka, 1992; Tingey and Anderson, 1991). A field study on dust deposition at different roadsides revealed that highways and industrial areas were dustier than residential areas (Mancharkar, 2001). The building construction-dust can adversely affect flowering, micromorphology, biochemical process, etc., of the plants (Desai and Kapoor, 2013).

Foliar dust recorded of various plant species differed at different sites, and no plant behaved the same at all sites, i.e., the plant which is the best dust capturer at one site may be the worst in holding dust at other sites. Different plants have different dust capturing capacities was established in Chapter 2. The amount of dust also varied depending upon the ecological conditions provided by an urban region. Plants growing near signals, along highways, and along seaside had different foliar dust values for the same plant species and time periods.

Trees planted in rows can provide several benefits to soils and crops. Such plantings are referred to as windbreaks, shelter belts or green belts. Urban vegetation as a network of green spaces which provide life support functions and places for recreation.

6.6.2 GREEN BELT

Trees are considered the lungs of the cities by acting as sink for some of the harmful gases released by vehicles and industries operating in the city area.

Greenbelt development plan for a particular sector mainly depends upon: (1) sinking capacity of the ecosystem, nature, and extent of pollution, (2) climatic factors, soil, and water quality (Horaginamani et al., 2012).

6.6.2.1 OBJECTIVES OF GREEN BELT DEVELOPMENT

The objectives of Green Belt development takes care of micro level air pollution abatement including enhancement of socio-economic status of the region (Gupta et al., 2008).

- The primary objective of green belt development is reducing of air and noise pollution;
- Green Belts reduce the soil erosion and improves esthetic enhancement;
- The socioeconomic status of the region increases due to green belts, by generating employment avenues and also participates in environmental protection;
- The width of the Green Belt and the choice of plants species to be used for planting varies from industry to industry of air pollutants.

Green belt serves five purposes:

- to check the unrestricted sprawl of large built-up areas;
- to avoid neighboring towns merging into one another;
- assisting in safeguarding from encroachment;
- preserving original character of towns;
- assisting in urban regeneration, encouraging the recycling of derelict lands.

Different plant species show different APTI. Studies revealed that the maximum APTI is observed in *Azadirachta indica* and minimum is *Psidium spicigera* (Horaginamani et al., 2012). Study at Varanasi revealed that *F. benghalensis L.* and *F. religiosa* are excellent performers in urban forest, *P. longifolia, F. glomerata (Roxb.), A. indicus* and *M. indica* are very good performers in the marble industrial areas of Potwar region and in the similar fashion *C. fistula L., D. roxburghii, T. arjuna, P. guajava L., M. hortensis* and *D. sissoo* are good performers in urban forest with respect to their API (Pandey et al., 2015).

Green belt literally means an area of open land around a city, on which building is restricted. The idea of Green Belt was initially suggested in the late 19th century. I Ebenezer Howard's in 1898 proposed Garden Cities were intended to be "planned, self-contained, communities surrounded by greenbelts, containing carefully balanced areas of residences, industry, and agriculture." The first Green Belts were designated in London where they were supported by both national planning, legislation, and policy. Shelterbelt and/or windbreak is a technique of plantation to provide shelter from the wind and to protect soil from erosion. Usually, it is made up of one or more rows of plants. Windbreaks and shelter belts are important soil conservation practices improving the microclimate for production of crops and protection of rangelands from wind erosion. A wind break is a barrier for protection from winds on farms. A shelter belt is longer, consisting of shrubs and trees for protection of crops.

The characters of plants, like leaf surface, size, shape, number of trichomes, waxy deposition, etc., were mainly responsible for trapping the dust (Joshi, 1990). A significant role is played by the nature of trees, type of canopy, branching pattern, surface of bark, phyllotaxy, venation, trichomes, etc., in abatement of pollution (Mancharkar, 2001). Alfalfa has been investigated for uptake of ordinary pollutants like oxides of Sulphur and Nitrogen (Hill, 1971). Leaf uniqueness, i.e., form, and size, orientation, surface texture, occurrence/ lack of leaf hairs, petioles length, meteorological conditions, leaf attributes, phenology, and height of plants affect the dust interception/accumulation capacity of different plants (Younis et al., 2013). The foliage characteristics impact dust accumulation (Somashekar et al., 1999; Garg et al., 2000).

Morphological attributes such as cuticular waxes, raised epidermal cell margins, trichomes, ledges of stomata and epicuticular lignin and waxes give roughness to leaves (Pal et al., 2002). The morphology and internal structure of leaves is altered by a heavy load of dust pollutants (Somashekar et al., 1999; Gostin, 2009; Sukumaran, 2012).

Ministry of Environment and Forest listed some (2005) criteria for selection of plants for controlling air pollution:

- Trees, shrubs should have dense foliage;
- Leaves should large surface area, because leaves absorb pollutants;
- Evergreen trees preferred;
- Resistant trees are preferred, particularly in the early stages of their growth.

According to Ministry of Environment and Forest (2005), some of the suitable plants having potential for pollution control are:

- *Acacia arabica*;
- Citrus species;
- Diospyros species;

- *Ficus bengalensis*;
- *Ficus religiosa*;
- *Lillium* spp.;
- *Polyathia lotigifolia*;
- *Tamarindus indica*;
- *Thuja occidentallis*;
- *Prosopis juliflora*;
- *Zizypus jujuba*.

6.6.3 STUDIES ON PLANTS FOR GREEN BELTS IN URBAN AREAS OF INDIA

A recent study revealed that under Indian climatic conditions *Tectona grandis* showed low scores of APTI, but high dust accumulation; hence it is unfit for green belt regions but suitable for heavy traffic and road side areas (Sahu et al., 2015). Green belt development plan for a particular sector mainly depends upon: (i) nature and extent of pollution load, (ii) sinking capacity of the ecosystem, (iii) climatic factors and (iv) soil and water quality (Horaginamani et al., 2012). Similarly, another recent study revealed that *Ficus benghalensis* L. and *Ficus religiosa* L. would be excellent in cities whereas *Polyalthia longifolia* (Sonnerat) Thw., *Ficus glomerata* Roxb., *Anthocephalus indicus* and *Mangifera indica* were very good performers in industrial areas of Potwar region. In the similar fashion *Cassia fistula* L., *Drypetes roxburghii*, *Terminalia arjuna*, *Psidium guajava* L., *Millingtonia hortensis* and *Dalbergia sissoo* were estimated to be good performers to control pollution (Pandey et al., 2015).

Screening of dust to plants under control conditions was done using a dust constructed on the lines developed by Mancharkar (2001) Plants were rated according to their dust retention indices (DRI).

Mancharkar and Chaphekar (2005) have discovered that plant species have high DRI and thus good dust capturers. Some of the plant species showing higher DRI values are:

- *Abitulon indicum*;
- *Calotropis gigantea*;
- *Ficus benghalensis*;
- *Ficus glomerata*;
- *Gmelina arborea*;
- *Mangifera indica*;

- *Terminalia bellerica*;
- *Thespesia populnea.*

A dust chamber was constructed on more or less similar lines in this work to study the dust capturing capacities of various plants found growing along the roads of the city. Using this Dust Chamber, plants were exposed to dust in still or controlled conditions and then focused to windy or simulated condition. The wind shows continuous variation in its speed and direction. The average wind speed can vary (Tyagi et al., 2011). Wind speeds vary from location to location in Mumbai city (Kumar et al., 2012).

Dust chambers have been used l in analyzing the efficiency of plants in monitoring and measuring pollutants under controlled conditions (Darley, 1966; Hill, 1971; Mancharkar, 2001) A dust chamber was constructed (Joshi, 2014) to study plant and their dust retention capacities. Different plant species have different dust capturing capacities. Thus, it became necessary to study the dust fall under standard conditions. 127 species belonging to 52 families were screened for dust fall in the dust chamber. Values like DFI (Dust Fall Index) indicating dust fall in still conditions and DRI (Dust Retention Index) indicating dust fall in windy conditions were determined to understand the same. These values were calculated in terms of percentage values, and a scale was made to indicate plants in different categories like Excellent, Good, Moderate, and Weak dust capturers and dust retainers. DFI or/and DRI values were ranging from 10% to 30%, 31% to 50%, 51 to 70% and 71% to 90% and were called weak, moderate, good, and excellent dust capturers/retainers. There were 52 tree species which showed excellent DFI but only 2 of them showed excellent DRI. Similarly, 17 shrub plant species showed excellent DFI but only 2 of them could show excellent DRI. There were 20 herb species showing excellent DFI but only 7 of them showed excellent DRI. Also, out of 6 climbers showing excellent DFI none showed excellent DRI. *Bryophyllum pinnatum* (Lam.) Oken, *Ficusbenjamina* L. var. *nuda* (Miq.) M. F. Barrett, *Nerium odorum* Aiton., *Pedilanthus tithymaloides* Poit., *Pedilanthus tithymaloides* Poit. *variegatus* and *Tithonia rotundifolia* (Mill.) S.F. Blake was found to be the excellent dust retainers. The members of Urticaceae, Asteraceae, Euphorbiaceae, Apocynaceae, and Crassulaceae families were found to be the excellent dust retainers by showing the DRI values from 71% to 90%.

Out of total 127 plant species, there were 64 plant species common in the field. Using these species, a comparison was made to establish the connection between Field study and Laboratory study. The species such as *Acalypha wilkesiana Mull. Arg., Bauhinia purpurea* L., *Canna indica L., Cassia fistula Linn., Ceiba pentandra* (L.) Gaertn., *Citrus limon* (L.) Burm.f., *Codiaeum variegatum* (L.) A. Juss., *Ficus glomerata Roxb., Operculina turpethum* (L.) Silva Manso, *Ricinus communis L., Samanea saman* (Jacq.) Merr. and *Terminalia catappa L.* were found to have more or less similar trend on the field and in the laboratory. Which means, the plants having more foliar dust deposition showed higher DFI and/or DRI and the one having less foliar dust deposition showed lower DFI and/or DRI. Whereas the remaining plants could not follow any such trend or pattern. Plants with good DRI can be used for phytoremediation of dust and development of green belts. For the selection of tree species in various localities, one has to take care as to selection of plants for green belt areas. Industrial areas plants with high DFI and its growth patterns have to be taken into consideration, also the various data available on its biochemical resistance to air pollutants viz; APTI. For residential areas, the proximity to buildings and other concrete structures makes it more difficult to select the tree or shrub species, so apart from the above mentioned guide list, one has to take into account the overall ecological availabilities in the area. The chapter on dust chamber suggested the types of plants suitable for the development of green belts.

Similarly, the excellent dust retaining tree, shrub, herb, and climber viz; *Ficus benjamina* L. var. *nuda* (Miq.) M. F. Barrett, *Nerium odorum* Aiton., *Pedilanthus tithymaloides* Poit. and *Bougainvillea spectabilis* Willd. respectively, also showed higher values of foliar dust deposition on the field. On the basis of good dust capturing capacities, frequent appearance, and performance under the dust chamber, four plant species viz; *Bougainvillea spectabilis* Willd., *Ficus benjamina L. var. nuda* (Miq.) M. F. Barrett, *Nerium odorum* Aiton. and *Pedilanthus tithymaloides* Poit. were short listed and chosen for further Active Phytomonitoring studies.

For understanding the reason behind better dust capturing capacity of some plants their anatomical study was conducted. Plants with good dust capturing capacities had trichomes or hairy outgrowths and many plants having smooth foliar surfaces were showing lesser amount of dust fall. *Nerium odorum* Aiton. was found to have high trichome density on its foliar surface. *Ficus hispida* L.f., *Ziziphus jujuba* Mill., *Hyptis suaveolens* (L.) Poit., *Nerium odorum* Aiton., *Calotropis gigantia* (L.) W. T. Aiton., *Malachra capitata* Linn., etc., were the plants having more trichomes on their leaves and good dust capturing capacities on the field and also still and simulated conditions. The plant species screened for dust fall within the dust chamber were examined for their DFI and DRI.

Tree species showing maximum DFI and DRI values in the study, *Pisoniaalba* Span. – 88.26 g/m^2, *Barringtonia asiatica* (L.) – 85.12 g/m^2 and *Ficus benjamina* L. var. *nuda* (Miq.) M. F. Barrett – 80.74 g/m^2, and *Ficus benghalensis* L. – 73.54 g/m^2, respectively.

Though *Pisonia alba* Span. and *Barringtonia asiatica* (L.) Kurz were showing maximum DFI values but could not retain the dust on the leaves in windy condition. Whereas *Muntingi acalabura* L., *Ficus benghalensis* L. and *Ficus benjamina* L. var. *nuda* (Miq.) M. F. Barrett were able to retain the foliar dust. Similarly, *Ziziphus jujuba* Mill. and *Mangifera indica* L. didn't show the high value of DFI but had high values of DRI. The Shrub species showing maximum DFI and DRI values in the study were *Lantana camara* L. – 86.37 g/m^2 *Pseuderanthemum carruthersii* (Seem.) Guillaumin – 85.7 g/m^2, *Ixora coccinea* L. Compacta – 81.86 g/m^2, *Nerium odorum* Aiton. – 81.08 g/m^2 and High values *Nerium odorum* Aiton. – 80.26 g/m^2, *Lantana camara* L. – 79.04 g/m^2, *Ixora coccinea* L. – 58.71 g/m^2. *Bougainvillea spectabilis* Willd., *Ficus benjamina L. var. nuda (Miq.) M. F. Barrett, Hyptis suaveolens (L.) Poit., Ixora coccinea L., Lantana Camara L., Nerium odorum Aiton., Pedilanthus tithymaloides Poit.* and *Tithonia rotundifolia* (Mill.) S.F. Blake, which requires low maintenance but are best performers can be used as phytomonitors and phytoremediation on road dividers on highways and main roads. Similarly for beautiful avenues, *Trema orientalis (L.) Blume, Ficus benghalensis L., Mangifera indica L.* and *Ziziphus jujuba Mill.* can be cultivated on either side of highways and main roads. And the plants like *Bryophyllum pinnatum (Lam.) Oken, Duranta plumieri Jacq., Stachytarpheta indica Vahl* and *Trachyspermum ammi Sprague* can be cultivated in residential areas because they are suitable for controlling dust pollution but require regular attention watering.

All the plant species showed more dust in still condition than simulated condition which justifies the role of wind and its speed in moving the particulates from place to place. In contrast to this, the rough foliar surfaces (containing hairs or trichomes) for example *Hyptis suaveolens (L.) Poit., Ixora coccinea L., Lantana Camara L., Nerium odorum Aiton., Muntingia calabura* L., *Tithonia rotundifolia* (Mill.) S.F. Blake, *Trema orientalis (L.) Blume, Ziziphus jujuba Mill.*, etc., have better dust holding capacity irrespective of wind action. *Hyptis suaveolens (L.) Poit.* and *Lantana Camara L.* have been more effectively used due to the presence of their trichomes (Joshi, 2014). Excellent dust capturing or/and retaining can be observed in all forms of vegetation, i.e., tree, shrub, herb, or climber have either higher DFI or/and DRI. Amongst 127 plant species, there were 95 excellent dust capturers with high DFI but only 11 were excellent dust retainers with high DRI. *Bryophyllum pinnatum* (Lam.) Oken, *Ficus benjamina* L. var. *nuda* (Miq.) M. F. Barrett, *Nerium odorum* Aiton., *Pedilanthus tithymaloides* Poit., *Pedilanthus tithymaloides* Poit. variegatus and *Tithonia rotundifolia* (Mill.) S.F. Blake was found to be the excellent dust retainers.

There were 52 trees, 17 shrubs, 20 herbs and 6 climber species showing excellence in capturing dust under still conditions. Whereas only 2 trees, 2 shrubs, 7 herbs and 0 climber species showed excellent dust retaining under simulated conditions. The members of Urticaceae, Asteraceae, Euphorbiaceae, Apocynaceae, and Crassulaceae families were found to be the excellent dust retainers by showing the DRI values from 71% to 90%. Hence it can be concluded that the above families are the better selection for foliar phytomonitoring and foliar phytoremediation of the city.

The study revealed that trees, shrubs, herbs, and climbers all play an important role in phytomonitoring and phytoremediation. The best tree species to retain dust on its foliar surface was *Ficus benjamina* L. var. *nuda* (Miq.) M. F. Barrett. Similarly, the best shrub species to retain dust on its foliar surface was *Nerium odorum* Aiton. Whereas the best dust retainer herb and climber were *Pedilanthus tithymaloides* Poit. and *Bougainvillea spectabilis* Willd., respectively. *Ficus benjamina* L. had been discovered as an active indicator of air pollution and found to be sensitive at residential, industrial, and heavy traffic areas as it showed APTI < 10 (Thambavani et al., 2012). Similarly, *Nerium species,* being tolerant to air pollution, along with many others have been suggested for greenbelt development in polluted areas (Choudhury et al., 2009). Also, *Bougainvillea species* had been reported as one of the tolerant species to air pollution (Radhapriya et al., 2009).

6.7 ROLE OF TREE CANOPIES IN RETAINING DUST BY URBAN PLANTS

Plants as a whole behave as active phytomonitor and control the environmental pollution. The morphological characters of plants viz. leaf surface, size, shape, number of trichomes, waxy deposition, etc., were mainly responsible for trapping the dust (Joshi, 1990). In order to control dust pollution; the nature of trees, type of canopy, branching pattern, surface of bark, phyllotaxy, venation, trichomes, etc., have significant role (Mancharkar, 2001). The canopies of alfalfa have been investigated for uptake of ordinary pollutants like SO_2 and NO_2 (Hill, 1971).

Trees planted in rows can provide several benefits to soil and crops. Such plantings are referred to as "windbreaks," "shelter belts" and occasionally "greenbelts." An impermeable shelter provides a greater degree of shelter immediately to the leeward but, because of eddying effect, it gives a comparatively shorter zone of effective shelter and hence an ideal shelterbelt should be (i) permeable to wind, (ii) of proper width and shape and (iii) of adequate height (Carbon, 1957). In the coastal areas exposed to sea spray flagging apparently results as the seaward side is preferentially tall, while buds on the leeward side continue to grow thus resulting after several growing periods in a leeward curvature of the stem (Boyce, 1954).

A canopy of vegetation can be defined as a more or less regular and continuous layer of vegetative crown or herbaceous foliage covering an area stratified fairly uniformly with respect to height. The term "Canopy" refers to an overhanging shelter and is therefore a specialized form of vegetative cover (Bennet and Hill, 1970). Shimwell (1972) has proposed three primary structural groupings to assess the general pattern of vegetation on a horizontal scale: the three categories are: (i) CLOSED CANOPY – when the vegetation forms a continuous cover laterally, including a stratified continuous and regular layer. (ii) OPEN CANOPY – layer of uniform vegetation where spaces less than twice the foliar diameters of the predominant organisms exist. (iii) SPARSE – where ground space dominates the aerial view.

Various types of canopy shapes have been described by Corner (1952); Thomas and Howard (1976); and Nayar (1985). The normal canopy shapes are often disturbed in an urban ecosystem. Due to the constant movement of vehicles, wind speeds are generated very regularly. Wind speed below 10 m above ground often has the greatest influence on plants whereby their growth is completely arrested at the surface (Nobel, 1983). Flag trees are commonly observed where branches are permanently swept to the leeward. This deformation indicates the direction of wind and with less fidelity the speed of the wind (Grace, 1947; Noguchi, 1979). Flagging occurs in response to night winds for valleys that have an uphill wind movement during the day and downhill cold air drainage at night (Holy-Royd, 1970).

Four basic shapes of crowns or canopies are considered as round, flat-topped conical and oblique (Joshi, 1990). A round crown shape represents a normal, undisturbed growth of branches; a conical shape represents undisturbed growth in crowded conditions. A flat-topped (including slightly convex ones) crown represents a disturbance of growth due to winds, including pollutants, while an Oblique crown is formed by unidirectional winds preventing growth on the windward side. It is true that different species of trees have normal crowns in one of the above categories, except in the case of oblique crowns.

Trees with spreading or round canopy capture dust more efficiently than the trees with oblong canopy (Mancharkar, 2001). A recent study revealed that microclimatic conditions, improper placement of trees, etc., led to deviation in their normal canopy shapes. Flattening of canopy in *Samanea saman*, *Peltophorum inerme* and *Erythrina indica* was very frequent and avenue trees showed a great variation in canopy shapes along the roads where they were surveyed. Though *Peltophorum pterocarpum* and *Thespesia populnea* generally bear a round canopy, some displayed oblique canopies too, and flat canopies were also present along a number of roads (Joshi and Joshi, 2014). The different canopies show different results. The amounts of dust calculated under various canopy on foliar surfaces of selected four ornamental plant species are expressed as follows *F. religiosa* > *F. benghalensis* > *T. catappa* > *D. regia* > *S. saman* > *P. pterocarpum* (Joshi, 2014).

6.7.1 PLANTS FOR MITIGATION OF AIR POLLUTION AND GREEN BELT DEVELOPMENT

Plant leaves function as a proficient gas exchange system. Their anatomical structure allows rapid diffusion of water-soluble gases. These characteristics allow the plant to respire and photosynthesize, and they can also remove pollutants from the air.

Air pollution levels are reduced when windblown particulates stick to the leaves and stems of plants Hosker and Lindberg (1982). Similarly, gaseous air pollutants, particularly carbon dioxide, can be dissolved or sequestered through stomata on plant leaves.

Akbari (2002) calculated that daytime temperature reduction due to trees would decrease the dependence on air conditioning and reduce emissions of NO_x from coal fired electricity plants resulting in an estimated 10% reduction in smog precursors or a reduction of 350 tons of Oxides of Nitrogen per day.

6.7.2 ROLE OF URBAN TREES

Trees act, as natural 'air conditioners,' a of the city (Rowntree, 1986) and can prevent solar radiation from heating the surrounding buildings, cooling the air by evapotranspiration, and reducing wind speed (Akbari and Taha, 2001).

Trees and parks influence air pollution by:

- Reducing air temperatures;
- Reducing air pollution;
- Reducing ultraviolet radiation;
- Reducing carbon dioxide (Nowak and Heisler, 2010).

Ministry of environment and forests gave some importance of plantations (2005):

- Trees are good SO_2 absorbers;
- Parks with trees have lower SO_2 level than urban streets;
- Roadside trees can reduce traffic generated airborne pollutants on back-side of plantation;
- Vegetation around industrial area help to purify industrial emissions;
- A 30 m belt of trees where even a single row of trees may bring about 25% reductions in airborne particulate;
- Plants may also be used as one of the ways to control pollution.

6.7.3 TREES IN AN INDUSTRIAL AREA AND ITS ROLE

Nowak and Heisler (2010) has stated various benefits of trees in an industrial area, viz, reducing air pollution, temperatures, etc.

6.7.3.1 BASIC ISSUES IN DEFINING GUIDELINES

There are four key steps in defining air quality guidelines:

- Selective sensitive species on which to base the guideline;
- Finding out the response parameter of concern and deciding what size of change in this parameter can be judged as adverse;
- The method of characterizing pollutant exposure; and
- Adequate experimental data, to determine the threshold pollutant exposure for the identified species and response parameter.

Difficulties arise from:

- Conceptual problems in defining what is meant by an 'adverse effect,' in all respects environmentally and socially;
- The adverse effects of air pollution may be clear in situations, while not so clear in somewhere the concentrations are high and mixtures

of pollutants are present. In such case, it becomes difficult to find out threshold values;

- Chronic exposures to air pollution on plants result in long-term accumulation of pollutants or long-term chemical changes in soils and vegetation.

It is difficult to test the consequences of these cumulative effects. Another approach that has been adopted to define critical loads of acidity to prevent long-term damage to ecosystems is to define a critical chemical concentration for biological effects in the relevant medium and then to develop mathematical models. The level of atmospheric deposition is thus found out, which would eventually lead to this chemical criterion being exceeded.

KEYWORDS

- **air pollution tolerance index**
- **air quality index**
- **air quality monitoring stations**
- **Maharashtra pollution control board**
- **suspended particulate matter**
- **volatile organic compounds**

REFERENCES

Abbasi, S. A., (1998). *Environmental Pollution and its Control* (pp. ix+445). Cogent International, Pondicherry.

Agarwal, S., & Tiwari, S. L., (1997). Susceptibility level of few plants on the basis of air pollution tolerance index. *Indian Forester, 123*, 319–322.

Akbari, H., (2002). *Heat Island Reduction: An Overview – Effects of Trees and Implementation Issues*. Presentation by Lawrence Berkeley Laboratory at the University of Pennsylvania, LAPR 760.

Akbari, H., Pomerantz, M., & Taha, H., (2001). Cool surface and shade trees to reduce energy use and improve air quality in urban areas. *Solar Energy, 70*(3), 259–310.

Alkama, G. F., Nitesh, C. J., & Ambika, N, J., (2014). Phytomonitoring of dust using *Pedilanthus Tithymaloides* poit. *International Journal of Research, 3*(2), 61–64.

Alkama, G. F., Nitesh, C. J., & Ambika, N. J., (2016). Statistical applications in environmental botany. *International J. Res., 3*(2), 61–64. ISSN: 2231 – 6124.

Barber, M. E., Brown, M. G., Lingenfelder, K. M., & Yonge, D. R., (2006). *Phase I: Preliminary Environmental Investigation of Heavy Metals in Highway Runoff.* Washington State Transportation Center (TRAC), Washington State University, Pullman, Washington.

Barrett, M. E., Malina, J. M., Charbeneau, R. J., & Ward, G. H., (1995a). *Characterization of Highway Runoff in the Austin, Texas Area.* CRWR 263. Center for research in water resources, Austin, Texas.

Beckerson, D. W., & Hofstra, G., (1980). Effects of Sulphur dioxide and ozone, singly and in combination, on membrane permeability. *Can. J. Bot., 58*, 451–457.

Bennett, J. H., & Hill, A. C., (1975). Interactions of air pollutants with canopies of vegetation. In: Mudd, J. B., & Kozlowski, T. T., (eds.), *Responses of Plants to Air Pollution* (pp. 273–306). New York, Academic Press.

Bhattacharya, T., Chakraborty, S., Fadadu, B., & Bhattacharya, P., (2011). Heavy metal concentration in street and leaf deposited dust in Anand city, India. *R. J. Chem. Sci., 1*(5), 61–66.

Central Pollution Control Board (CPCB), (2007). *Phytoremediation of Particulate Matter from Ambient Environment Through Dust Capturing Plant Species.* A report under central pollution control board, Ministry of Environment and Forests, Delhi.

Central Pollution Control Board (CPCB), (2011). "*Current Air Quality Issues in India.*" a paper presented by Central Pollution Control Board, Ministry of Environment and Forests, Delhi.

Chaphekar, S. B., (1972). Effect of atmospheric pollutants on plants in Bombay. *J. Biol. Science, 15*, 1–6.

Chaphekar, S. B., Boralkar, D. B., & Shetye, R. P., (1980). Plants for air monitoring in industrial area. In: Furtado, J. I., (ed.), *Tropical Ecology and Development* (pp. 669–675). I.S.T.E. Kuala Lampur.

Chaurasia, S., Karwaria, A., & Gupta, A. D., (2013). Cement dust pollution and morphological attributes of groundnut (*Arachis hypogaea*), Kodinar, Gujrat, India. *IOSR Journal of Environmental Science, Toxicology And Food Technology: 2319–2399, 4*(1), 20–25.

Choudhury, P., & Banerjee, D., (2009). Biomonitoring of air quality in the industrial town of Asansol using the air pollution index approach. *Research Journal of Chemistry and Environment, 13*(1), 46–51.

Chui, T., (1981). *Highway Runoff in the State of Washington: Model Validation and Statistical Analysis.* Master's thesis. University of Washington, Seattle, Washington.

Corner, E. J. H., (1952). *Wayside Trees of Malay.* Singapore.

Darley, E. F., (1966). Studies on the effect of cement kiln dust on vegetation. *J. Air. Poll. Contr. Assoc., 16*, 145–150.

Das, T. M., (1981). Plants and pollution. *Presidential Address, Section of Agricultural Sciences, 68th-Indian Sci. Congr.* (pp. 1–17). Varanasi.

Das, T. M., Bhaumik, A., Ghosh, A., & Chakraborty, A., (1981). Trees as dust filters. *Science Today*, 19–21.

Desai, Y., & Kapoor, M., (2013). Effect of building construction-dust on foliar micromorphology and biochemistry of Peltophorum pterocarpum (DC) Baker. *Bionano Frontier, 6*(1), 53–56.

Edmond, R. L., & Driver, C. H., (1974). Dispersion and deposition of spore fumes and fluorescent particles. *Phytopathology, 64*, 1313–1321.

Falla, J., Laval, G. P., Henryon, M., Morlot, D., & Ferard, J. F., (2000). Biological air quality monitoring: A review. *Environmental Monitoring and Assessment, 64*(3), 627–644.

Fritschen, J. A., Driver, C. H., Avery, C., Nuffo, J., Edmonds, R., Kinerson, R., & Schiess, P., (1970). *Dispersion of Air Traces into and Within a Forested Area (3).* Report No. OSD, College of Forest Resources, Washington Univ., Seattle, WA: 53.

Garg, S. S., Kumar, N., & Das, G., (2000). *Ind. J. Environ. Prot., 20*, 326–328.

Goodman, G. T., & Roberts, T. M., (1971). Plants and soils as indicators of metals in the air. *Nature, 231*, 287–292.

Gostin, I. N., (2009). Air pollution effects on the leaf structure of some Fabaceae species. *Not. Bot. Hort. Agrobot. Cluj., 37*, 57–63.

Grace, J., (1977). *Plant Responses to Wind.* Academic Press, London.

Grodzinska, K., (1978). Mosses as bioindicators of heavy metal pollution in polish national parks. *Water, Air and Soil Pollution, 9*(1), 83–97.

Gupta, R. B., Chaudhari, P. R., & Wate, S. R., (2008). Overview on attenuation of industrial air pollution by green belt development. *Jr. of Industrial Pollution Control, 24*(1), 1–8.

Hill, A. C., (1971). Vegetation: A sink for atmospheric pollutants. *J Air Pollut Cont Assoc., 21*, 341–346.

Holy-royd, E. W., (1970). Prevailing winds on white face mountains as indicated by flag trees. *For. Sci., 16*, 222–229.

Horaginamani, S. M., Ravichandran, M., & Kamdod, A. M., (2012). Air pollution tolerance of selected plant species considered for urban green belt development in Trichy. *World Journal of Environmental Biosciences, 1*(1), 51–54.

Hosker, R. P. Jr, & Lindberg, S. E., (1982). Review: Atmospheric deposition and plant assimilation of gases and particles. *Atmos. Environ., 16*(5), 889–910.

https://indianexpress.com/article/cities/mumbai/mumbai-pm-2-5-below-set-limits-at-11-stations-7111711 (accessed on 2 June 2022).

Joshi, N. C., & Joshi, A. N., (2014). Urban tree canopy analysis. *Pollution Research, 33*(2), 1–5.

Joshi, N. C., (1990). *Experiments in Phytomonitoring of Urban Atmosphere.* Ph.D. Thesis, University of Mumbai, Maharashtra, India.

Joshi, N., (2014). *Final Report to the UGC on Studies on Monitoring Suspended Particulate Matter Using Urban Plants and Understanding their Green Belt Potentials.* F.NO-40-320/2011.

Joshi, N., Bist, B., & Joshi, A., (2014). Screening of some common plants as potential dust monitors in Tarapur industrial area. *Int. J. Environmental Sciences, 3*(2), 101–107.

Joshi, P. C., & Chauhan, A., (2008). Performance of locally grown rice plants (*Oryza sativa* L.) exposed to air pollutants in a rapidly growing industrial area of district Haridwar, Uttarakhand, India. *Life Science Journal, 5*(3), 41–45.

Kayhanian, M., Singh, A., Suverkropp, C., & Borroum, S., (2003). Impact of annual average daily traffic on highway runoff pollutant concentrations. *Journal of Environmental Engineering 129*(11), 975–990.

Kearfott, P. J., Aff, M. P., Barrett, M. E., & Malina, J. F., (2005). *Stormwater Quality Documentation of Roadside Shoulders Borrow Ditches.* IDS-water white paper 179. Center for Research in Water Resources, University of Texas, Austin, Texas.

Keller, T., (1982). Physiological bioindicators of an effect of air pollution on plants. In: Steubing, L., & Jager, J., (eds). *Monitoring of Air Pollutants by Plants: Methods and Problems*. Dr. Junk Publishers, The Hague.

Kumar, K. S., Cini, C., & Valerie, S., (2012). Assessment of design wind speeds for metro cities of India. *The Seventh International Colloquium on Bluff Body Aerodynamics and Applications (BBAA7)*. Shanghai, China; September, 2–6.

Laisk, A., Pfanz, H., & Heber, U., (1988). Sulfur dioxide fluxes into different cellular compartments of leaves photosynthesizing in a polluted atmosphere. II. Consequences of SO_2 uptake as revealed by computer analyses. *Plama, 173*, 241–252.

Lancaster, C. D., (2005). *A Low Impact Development Method for Mitigating Highway Stormwater Runoff—Using Natural Roadside Environments for Metals Retention and Infiltration*. Master's thesis. Washington State University, Pullman, Washington.

Lee, J. A., & Tallis, J. H., (1973). Regional and historical aspects of lead pollution in Britain. *Nature, 245*(5422), 216–218.

Lone, P. M., Khan, A. A., & Shah, S. A., (2005). Study of dust pollution caused by traffic in Aligarh city. *Ind. J. of Env. Health, 47*(4), 33–36.

Maharashtra Pollution Control Board (MPCB), (2014). *Air Quality Status of Maharashtra*. (Compilation of Air Quality data recorded by MPCB 20132014).

Mancharkar, A. R., & Chaphekar, S. B., (2005). *Morphological Basis of Green Belt Efficiency*. A.W. College Otur.

Mancharkar, A. R., (2001). *Assessment of Characters of Plants to Optimize Green Belt Efficiency*. PhD Thesis University of Pune.

Manning, W. J., & Feder, W. A., (1980). *Biomonitoring Air Pollutants with Plants* (p. 142). Applied Science Publishers, London.

Meetham, (1950). Natural removal of pollution from the atmosphere. *Quart. J. Roy. Met. Soc., 76*, 359.

Meetham, (1954). Natural removal of atmospheric pollution during fog. *Quart. J. Roy. Met. Soc., 80*, 96.

Ministry of Environment and Forests (MOEF), (2005). *Manual on Norms and Standards for Environment Clearance of Large Construction Projects*. Government of India, New Delhi.

Nandi, P. K., Singh, M., & Rao, D. N., (1980). Effect of ozone, Sulphur dioxide and their mixture on germination of Phaseolus aureus seeds. *Indian Journal of Air Pollution Control, 3*(2), 50–55.

Nayar, M. P., (1985). Tree canopies. In: Subrahmanyam, G. V., Rao, D. N., Varshney, C. K., & Biswas, D. K., (eds), *Air Pollution and Plants: A State of the Art Report* (pp. 171–176). Publishers Dept of Environment, New Delhi.

Nimis, P. L., Lazzarin, G., Lazzarin, A., & Skert, N., (2000). Biomonitoring of trace elements with lichens in Veneto (NE Italy). *The Science of the Total Environment, 255*(1), 97–111.

Nobel, P. S., (1983). Wind as ecological factor. In: Lange, O. L., Nobel, P. S., Osmond, C. B., & Zieglar, H., (eds.) *Physiological Plant Ecology, IV: Ecosystem Process/Mineral Cycling, Productivity and Mass Influence* (pp. 472–500). Springer–Verlag, Berlin.

Noguchi, Y., (1979). Deformation of tree in relation to wind. *J. Ecology, 67*, 611–628.

Nowak, D. J., & Heisler, G. M., (2010). *Air Quality Effects of Urban Trees and Parks*. www.nrpa.org (accessed on 2 June 2022).

Pal, A., Kulshreshtha, K., Ahmad, K. J., & Yunus, M., (2000). Changes in leaf surface structures of two avenue tree species caused by auto exhaust pollution. *J. Environ. Biol., 21*, 15–21.

Pandey, A. K., Pandey, M., Mishra, A., Tiwary, S. M., & Tripathi, B. D., (2015). *Urban Forestry and Urban Greening, 14*, 866–871.

Posthumus, A. C., (1982). Biological indicators of air pollution. In: Unsworth, M. H., & Ormrod, D. P., (eds.), *Effects of Gaseous Air Pollution in Agriculture and Horticulture* (pp. 27–42). Butterworth Scientific, London.

Posthumus, A. C., (1983). Higher plants as indicator and accumulator of gaseous air pollution. *Environmental Monitoring and Assessment, 3*, 263–274.

Prajapati, S. K., & Tripathi, B. D., (2008). Seasonal variation of leaf dust accumulation and pigment content in plant species exposed to urban particulates pollution. *J. Environ. Qual., 37*, 865–870.

Preciado, H., & Li, L., (2006). Evaluation of metal loadings and bioavailability in air, water and soil along two highways of British Columbia, Canada. *Water, Air, and Soil Pollution, 172*, 81–108.

Radhapriya, P., Gopalkrishnan, A. N., Malini, P., & Ramachandran, A., (2009). *Journal of Environmental Biology, 33*, 635–641.

Rao, D. N., (1981). Phytomonitoring of air pollutants. *Proc. WHO Workshop Biol. Indicators Indices of Environmental Pollution* (pp. 1–8). Osmania University, Hyderabad, Cent. Bd. Prev. Cont. Water Poll.

Rao, D. N., (1985). Plant and air pollutant mixture. In: *Air Pollution and Plants: A State-of-the-Art Report* (p. 67). Dept. of Environment Ministry of Environment and Forests, Government of India, New Delhi.

Robinson, & Robins, (1970). Gaseous Sulphur pollutants form urban and natural sources. *J.A.P.C.A. 20*, 253–255.

Rühling, A., & Tyler, G., (1973). Heavy metal deposition in Scandinavia. *Water Air and Soil Pollution, 2*(4), 445–455.

Sahu, C., & Sahu, S. K., (2015). Air pollution tolerance index (APTI), anticipated performance index (API), carbon sequestration and dust collection potential of Indian tree species: A review. *International Journal of Emerging Research in Management &Technology, 4*(11), 37–40.

Schulz, H., Schulz, U., Huhn, G., & Schüürmann, G., (2000). Biomonitoring of airborne inorganic and organic pollutants by means of pine tree barks. I Deposition types and impact levels. *Journal of Applied Botany, 74*(5, 6), 248–253.

Shimwell, D. W., (1972). *The Description and Classification of Vegetation.* University of Washington Press, Seattle.

Singh, S. K., & Rao, D. N., (1983). Evaluation of plants for their tolerance to air pollution. In: *Proceedings Symposium on Air Pollution Control* (Vol. 1, pp. 218–224). Indian Association for Air Pollution Control, New Delhi, India.

Singh, S. N., & Rao, D. N., (1978). Possibilities of using chlorophyll and potassium contents in plants to detect cement dust pollution. *J. IPHE, 1*, 10–13. India.

Somashekar, R. K., Ravikumar, R., & Ramesh, A. M., (1999). *Pollut. Res., 18*, 445–451.

Steinnes, E., (1995). A critical evaluation of the use of naturally growing moss to monitor the deposition of atmospheric metals. *Science of the Total Environment, 160, 161*, 243–249.

Steubing, L., & Jager, H. J., (1982). Monitoring of air pollutants by plants-Methods and problems. *Proceedings of the International Workshop (1981).* Osnabruck (F.R.G.).

Steubing, L., (1982). Problems of Bioindication and the necessity of standardization. In: Steubing, L., & Jager, J., (eds.), *Monitoring of Air Pollutants by Plants* (pp. 19–24). Dr. Junk Publishers.

Steubing, L., Fangmier, A., & Both, R., (1989). Effects of SO_2, NO_2 and O_3 on population Development and morphological and physiological parameters of nature herb layer species in a beech forest. *Environmental Pollution, 58*, 281–302.

Sucharova, J., & Suchara, I., (1998). Atmospheric deposition levels of chosen elements in the Czech Republic determined in the framework of the International Bryomonitoring Program 1995. *The Science of the Total Environment, 223*, 37–52.
Sukumaran, D., (2012). *Effect of Particulate Pollution on various Tissue Systems of Tropical Plants*. Central Pollution Control Board (CPCB), Zonal Office, Kolkata, India.
Taylor, G. E., Pitelka, L. F., & Clegg, M. T., (1991). *Ecological Genetics and Air Pollution*. Springer-Verlag, New York.
Thomas, E., & Howard, S. I., (1976). *Urban Trees* (Vol. 25, p. 111). Scientific American.
Thomas, W., (1983). Using plants in spatial trace substance analysis for emission monitoring. *Staub–Reinhalt. Luft, 43*(4), 142–148.
Tingey, D. T., & Andersen, C. P., (1991). In: Taylor, G. E., & Pitelka, L. F., (eds.), *The Physiological Basis of Differential Plant Sensitivity to Changes in Atmospheric Quality.*
Tyagi, A., Singh, O. P., Singh, S. S., & Kumar, S., (2011). Climate of Jaipur, Meteorological Center, Jaipur India Meteorological Department, Ministry of Earth Sciences, Government of India.
Van, H. C., & Van, H. C. M., (2002). Bioindication: The community approach. In: Nimis, P. L., Scheidegger, C., & Wolseley, P. A., (eds.), *Book: Monitoring with Lichens–Monitoring Lichens, NATO Science Series IV* (pp. 39–64). Earth and Environmental Science, Kluwer Academic Publishers, Dordrecht.
Vergé, X., Chapuis, A., & Delpoux, M., (2002). Bioindicator reliability: The example of Bel W3 tobacco (*Nicotiana tabacum* L.). *Environmental Pollution, 118*, 337–349.
Verma, A., & Singh, S. N., (2006). Biochemical and ultrastructural changes in plant foliage exposed to auto-pollution. *Environmental Monitoring and Assessment, 120*(1), 585–602.
Vishaka, V., (2021). Ph.D thesis submitted to the University of Mumbai. *Studies in Urban Ecology in and Around Alibag.*
Warren, J. L., (1973). *Green Space for Air Pollution Control*. School of Forest Resources, Tech. Rep. No. 50, North Carolina State Univ., Raleigh, NC:118.
Wedding, J. B., Carlson, R. W., Stukel, J. J., & Bazzaz, F. A., (1975). Aerosl deposition on plant leaves. *Environ. Sci. Tech., 9*, 151–153.
Weinstock, (1969). Carbon monoxide residence time in the atmosphere. *Science, 166*, 224–225.
William, H. S., (1990). Interaction between air contaminants and forest ecology. In: *Air Pollution and Forest* (2nd edn.). Springer-Verlag, New York Resour. Tech. Rep.50. Raleigh. N. C. from Novak (1988).
Younis, U., Bokhari, T. Z., Malik, S. A., Ahmad, S., & Raja, R., (2013). Variations in leaf dust accumulation, foliage and pigment attributes in fruiting plant species exposed to particulate pollution from Multan. *International Journal of Agricultural Science and Research, 3*(3), 1–12.
Yunus, M., Dwivedi, A. K., Kulshreshtha, K., & Ahmad, K. J., (1985). Dust falling on some common plants near Lucknow city. *Env. Pollution, (Series B), 9*, 71–80.

CHAPTER 7

Air Pollution Induced Stress in Avenue Trees: A Case Study of Mysore City, Karnataka, India

C. HEMAVATHI[1] and J. SHOBHA[2]

[1]*Government First Grade College for Women, Vijayanagar, Mysore–570017, Karnataka, India*

[2]*Department of Studies in Botany, University of Mysore, Manasagangotri, Mysore–570006, Karnataka, India*

ABSTRACT

Changes in the environmental components are reflected in the growth pattern of plants. Air pollutants both gaseous and particulate bring about qualitative changes in the life support system. It may also affect the production capacity of plants. The effects of pollutants on vegetation are scientifically examined. In the present study, six existing avenue trees at all the 10 circles and control area were chosen for the assessment of *Polyalthia longifolia* Hook and Thomas., *Delonix regia* Rafin, *Peltophorum inerme* Naves, *Samanea saman* Merrill, *Azadirachta indica* A. Juss. and *Swietenia mahogany* Jacq to roadside automobile pollution. By determining a few physiological parameters, which include chlorophyll a, and b, total chlorophyll, carotenoids, ascorbic acid (AA), pH, and relative water content, the impact of automobile exhaust on these species were assessed. The mean, standard deviation and percent variation were calculated for the data collected. A significant change in all these parameters was found in the leaf samples collected from six road side trees exposed to automobile exhausts in comparison to the control area. Due to

Plant Metabolites under Environmental Stress: Mechanisms, Responses, and Adaptation Strategies.
Nivas M. Desai, Manasi Patil, Umesh R. Pawar (Eds.)

air pollution on roadside trees, there are qualitative and quantitative changes in secondary metabolite composition, withstand the pollution effects.

To evaluate the tolerance level of the selected trees to vehicular exhaust pollution, air pollution tolerance index (APTI) was assessed based on biochemical parameters like total chlorophyll, AA, pH of leaf extract, and relative water content. All the six trees in the present investigation showed an APTI value of less than 10 *A indica* value of 6.4 followed by *D. regia* (6); *P. longifolia* (5.9); *P. inerme* (5.7); *S. mahogany* (5.6) and *S. saman* (5.5). Since the APTI values were less than 16 in all the six trees, they may be categorized as 'sensitive' to air pollution. All these tree species could be conveniently employed as bio-indicators considering the relative advantages compared with conventional physicochemical monitors. These species are appropriate for biomonitoring of air pollution in urban air quality management.

7.1 INTRODUCTION

Physiological and biochemical changes are likely to occur much before the appearance of any visible injury symptoms in plants subjected to air pollutants. Though the plants appear green and normal, they generally function at a reduced efficiency. The invisible changes can be detected at the metabolic level by studying certain biochemical parameters such as chlorophyll, protein content, specific metabolites, enzymes, or a combination of one or both.

In photosynthesis and associated chemical processes there are numerous steps that could be blocked and interrupted by outside agents such as air pollutants. Not only chemistry in the plant cell but the sites and chemical reactions themselves can be destroyed (Treshow and Anderson, 1989). The air pollution tolerance index (APTI) is a new approach to improve green belt in and around industrial and urban environment and to reduce air pollution. The response of plants to air pollution at physiological and biochemical levels can be understood by analyzing the factors that determine resistance and susceptibility. The growth and development of plants is a function totally dependent on environmental factors such as solar radiation, air, water, and soil, which comprise life support system for all living beings. Changes in the environmental components are reflected in the growth pattern of plants and plants living in stressful habitats shown symmetrical and could be used as an 'index of habitat quality (Gostin, 2016).

It is well known that air pollutants both gaseous and particulate bring about qualitative changes in the life support system, which eventually may affect the production capacity of plants. The changes in the levels of plant production caused by different pollutants are primarily due to differences in their physicochemical properties. Higher levels of pollutants affect certain physiological and biochemical processes of plant (Williams and Banerjee, 1995).

The trees are complex organisms that possess numerous mechanisms such as buffering, filtering, detoxification, and repair, by which they are able to avoid or impede permanent damage to the environment. The dramatic rise in air pollution in most Indian metropolitan cities over the last two decades is a direct result of an inefficient state both in terms of lack of balancing responsibilities and precautionary activities. The antioxidants serve as a free oxyradical scavenging system in plants, which are exposed to gaseous air pollutants and protects membranes from peroxidation. One group of defense compounds synthesized by plants comprises phenolic compounds, which are located in the covering tissue and on the plant surfaces. The synthesis of phenol is a bio indicator of drastic changes in the environment. Some trees are tolerant and some are highly sensitive to the pollutants (Kousar et al., 1998).

The gaseous pollutants interact with protein and lipids in the cell wall and membranes leading to oxidation and lipid peroxidation (LPO). It initiates chain reactions giving rise to more free radicals followed by an increase in cell permeability. This can lead to severe metabolic disorder and cause damage to genetic material (Iqbal et al., 2000).

Each tree is a small air purifying and dust-absorbing force. At the same time, it functions as an oxygen factory. Trees act as natural filters as they remove pollutants from the atmosphere and thus improve air quality. The common metabolic antioxidant like ascorbic acid (AA) protects plants from injury, caused due to air pollution. It occurs in the cystol, chloroplast, mitochondria, and a pop last of the cell (Gavali and Gavali, 2003).

Chlorophyll pigments such as chlorophyll a, chlorophyll b, total chlorophyll and carotenoid are reduced by various types of air pollutants. The levels of injury are high in sensitive species and low in tolerant species. Singh and Rao (1983) used four-leaf parameters such as total chlorophyll, AA, pH, and relative water content to derive an empirical number indicating the APTI.

Generally, chlorophyll 'a' was found to be more sensitive than chlorophyll 'b.' (Ricks and Williams, 1975) and reduction in chlorophyll molecule is either due to reduced synthesis of chlorophyll molecule or conversion of chlorophyll to pheophytin under the influence of SO_2. Since the production potential of plants depends on the size of its canopy and the amount of chlorophyll available for photosynthesis, it will be pertinent to know the effects of these pollutants on the chlorophyll pigment. SO_2 is considered as the most important pollutant, which affects chlorophyll pigment.

Nitrogen oxides are the sum of the concentrations of NO and NO_2 of the many forms of nitrogen found in the atmosphere. These are the most significant pollutants, emitted by industrial and urban activities. Nitrogen dioxide causes visible injury in plants only at higher concentrations and growth reduction at lower concentration. Reduction in the pigment concentration is an indicator of leaf damage by air pollution.

Factors controlling atmospheric resistance are wind speed, leaf size, geometry, gas viscosity, and diffusivity. Stomata resistance is regulated by stomata aperture, which is influenced by water deficit, and carbon dioxide concentration. Mesophyll resistance is regulated by gas solubility in water, gas-liquid diffusion and leaf metabolism. The gaseous pollutants like HF, NOx, SO_2, and NH_3 after entering into sub-stomata cavity dissolve in extra-cellular fluid and induce acidification of subcellular compartments. Whereas other pollutants such as O_3, PAN, and NOx produce toxic oxygen radicals causing direct injury to membranes, enzymes, or DNA.

The presence of SO_2 POX increases the level of AA into dehydroascorbic acid. Since AA helps in chlorophyllide synthesis, its reduction inhibits this synthesis and causes decrease in chlorophyll pigment. The dose of the pollutant absorbed by plant tissue plays a key role in determining effects on the metabolism and physiology of plants. Since the waxy cuticle layer permits negligible diffusion of pollutants, essentially all gaseous exchange is carried out through the stomata openings. The rate of transfer of pollutants from the atmosphere into interior leaf cell is regulated by a series of resistance like atmospheric, stomatal, and mesophilic (Varshney, 1985).

SO_2 reduces the amount of total protein either due to disruption of protein-synthesizing enzyme or alternation of internal ionic media. Lower protein concentration was attributed to de-nova synthesis or an increase in amino acid degradation. Plants have a range of mechanisms to detoxify the toxic molecules. These include enzymes such as catalase, POX, superoxide dismutase (SOD), AA, α-tocopherol (Vitamin E), peptides (glutathione), carotenoids, polyamines, and malondialdehyde (Larson, 1988). They are used as early warning biochemical markers of oxidative stress in trees. An understanding of these markers, which provide protection against oxidative damage is essential for the development of plants more resistant to environmental stresses.

The growth pattern of plants and their response at physiological and biochemical levels depend totally on environmental factors such as solar radiation, air, water and soil.

Plants are used as bioindicators, which exhibits injury or stress symptoms when exposed to phytotoxic concentration of air pollutants or a combination of pollutants. Studies on the effect of air pollution on vegetation, therefore, provide an important basis, particularly to adopt preventive measures in pollution control.

Since the effect caused by air pollutants was not much on the morphological parameters to cause visible injury, the study was conducted on the effect of air pollutants on biochemical parameters to know the extent of damage caused by air pollutants, in particular, vehicular exhaust.

Mysore city is one of the historical centers of India which attracts a large number of tourists during the months of October to December (winter) and March to June (summer). During these periods there was an increase in the total number of vehicles in and around the city of Mysore. It is also well known for its educational institutions, which attracts a large number of students from all parts of the country. There has been a drastic increase in the number of registered vehicles in Mysore city over three decades. The vehicle numbers in the city have more than doubled in the past eight years. The total number of vehicles in Mysore city has increased from 6,333 in 1970 to 8.15 lakh in 2017. Consequently, this growth has led to the deterioration of air quality, especially in commercial areas.

7.2 MATERIALS AND METHODS

Leaf samples were collected from the selected six existing avenue trees at all the 10 circles was chosen for the study were *Polyalthia longifolia* Hook. f. and Thomas., *Delonix regia* Rafin, *Peltophorum inerme* Naves, *Samanea saman* Merrill, *Azadirachta indica* A. Juss. and *Swietenia mahogany* Jacq at all selected sites and control area (Mahadevapura) during all the seasons for two years. Leaf samples were washed and stored in polythene bags until further process. Fresh leaves were used for the estimation of chlorophyll a, chlorophyll b, total chlorophyll and carotenoids, total protein, total AA, phenol, relative water content and pH.

The chlorophyll content of leaves was estimated following the procedure described by Arnon (1949). Estimation of total protein was done using the method of Bradford (1976). The total AA content was estimated using the method of Behrens and Madere (1963). The total phenol content was estimated by using the method of Malik and Singh (1980). Relative water content of leaf sample was calculated by the method described by Singh and Rao (1968). The relative water content of leaf sample was estimated as percent moisture on oven dry weight basis. pH of aqueous leaf abstract was determined by digital pH meter. APTI was determined by calculating the AA contents, total chlorophyll content, pH of leaf extract and relative water content of leaf. APTI was calculated by the method described by Singh and Rao (1983). Mean, standard deviation and percent variation were calculated for the data collected and graphs were plotted. The data obtained were correlated with the pollutants.

7.3 RESULTS

In the present study all the biochemical parameters estimated in the leaves of the selected avenue trees from selected circles has showed a reduction when compared with that of trees growing at control area.

7.3.1 CHLOROPHYLL 'A'

There was not much significant reduction in the chlorophyll 'a 'content in the leaf samples from polluted areas. Significant three-fold reduction was observed in *P. longifolia* only at Milk dairy circle during rainy season. Two-fold reduction was observed at Canara Bank, Fountain, and Vijaya Bank circles during winter season, whereas no significant reduction was observed during summer season (Figures 7.1(a)–(f)).

7.3.2 CHLOROPHYLL 'B'

Compared to chlorophyll 'a,' chlorophyll 'b' showed a significant reduction in plants from polluted areas when compared to plants from the control area (Figures 7.2(a)–(f)).

7.3.3 TOTAL CHLOROPHYLL

The total chlorophyll content of the selected avenue trees at different circles and the control area is depicted in Figures 7.3(a)–(f).

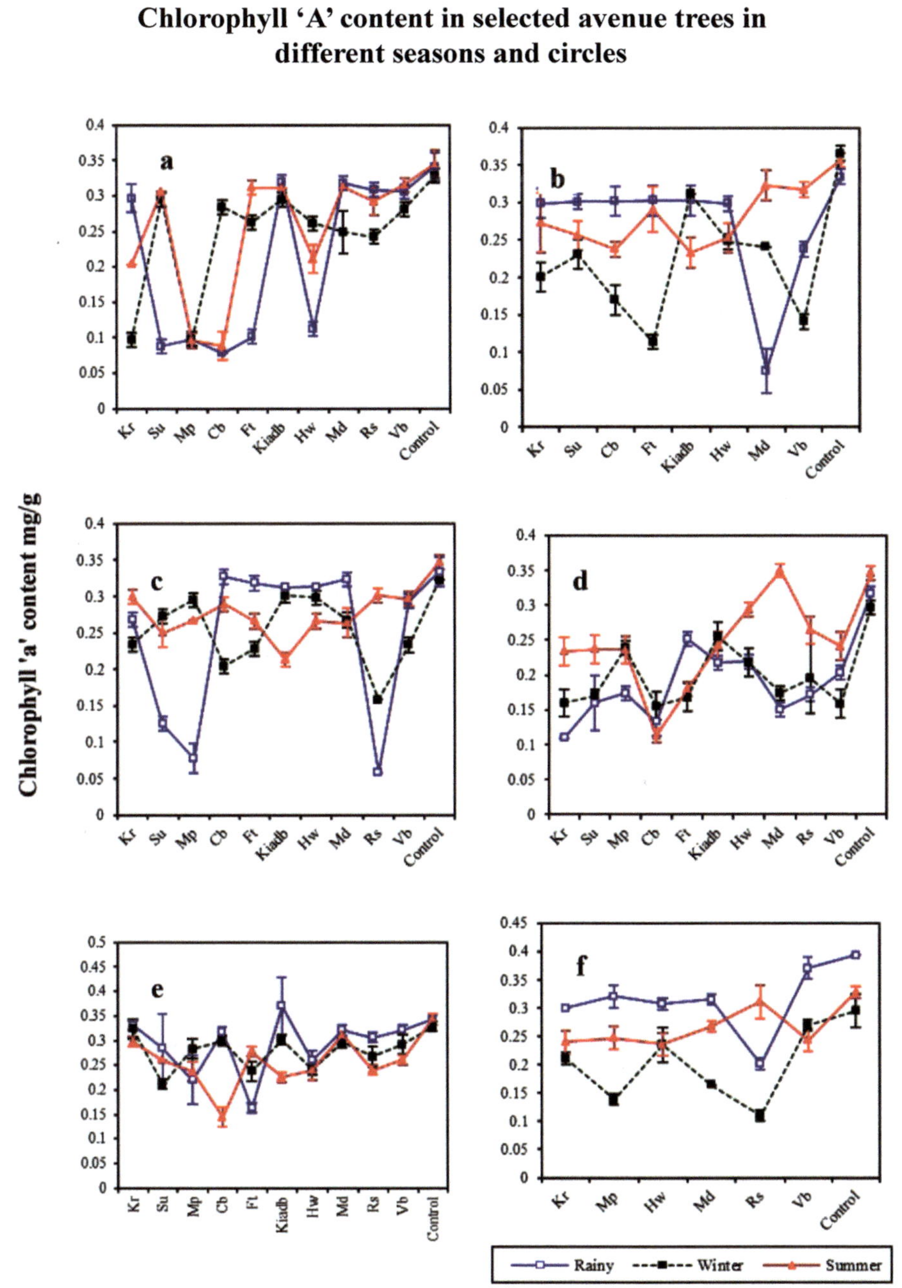

FIGURE 7.1 (a) *P. longifolia*; (b) *D. regia*; (c) *P. inerme*; (d) *S. saman*; (e) *A. indica*; (f) *S. mahogany.*

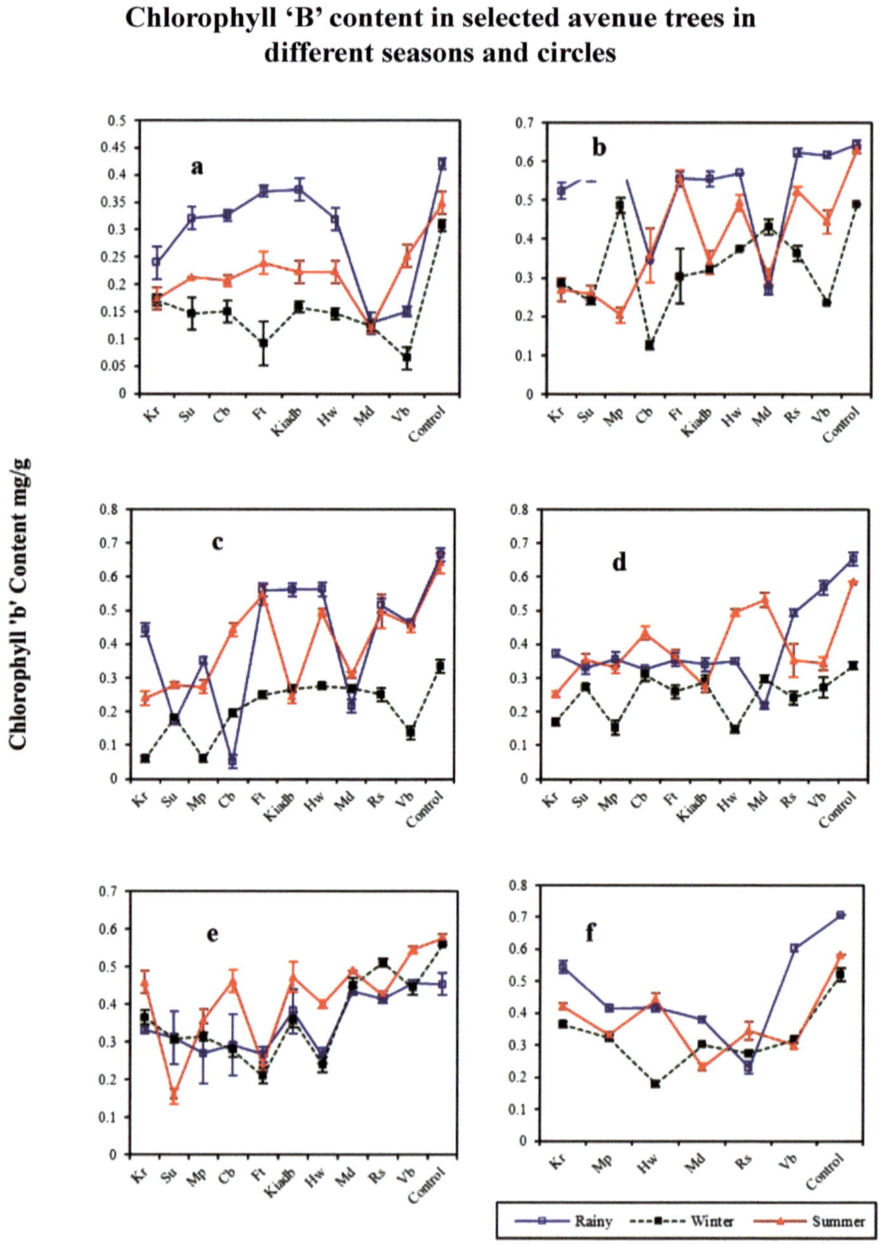

FIGURE 7.2 (a) *P. longifolia*; (b) *D. regia*; (c) *P. inerme*; (d) *S. saman*; (e) *A. indica*; (f) *S. mahogany.*

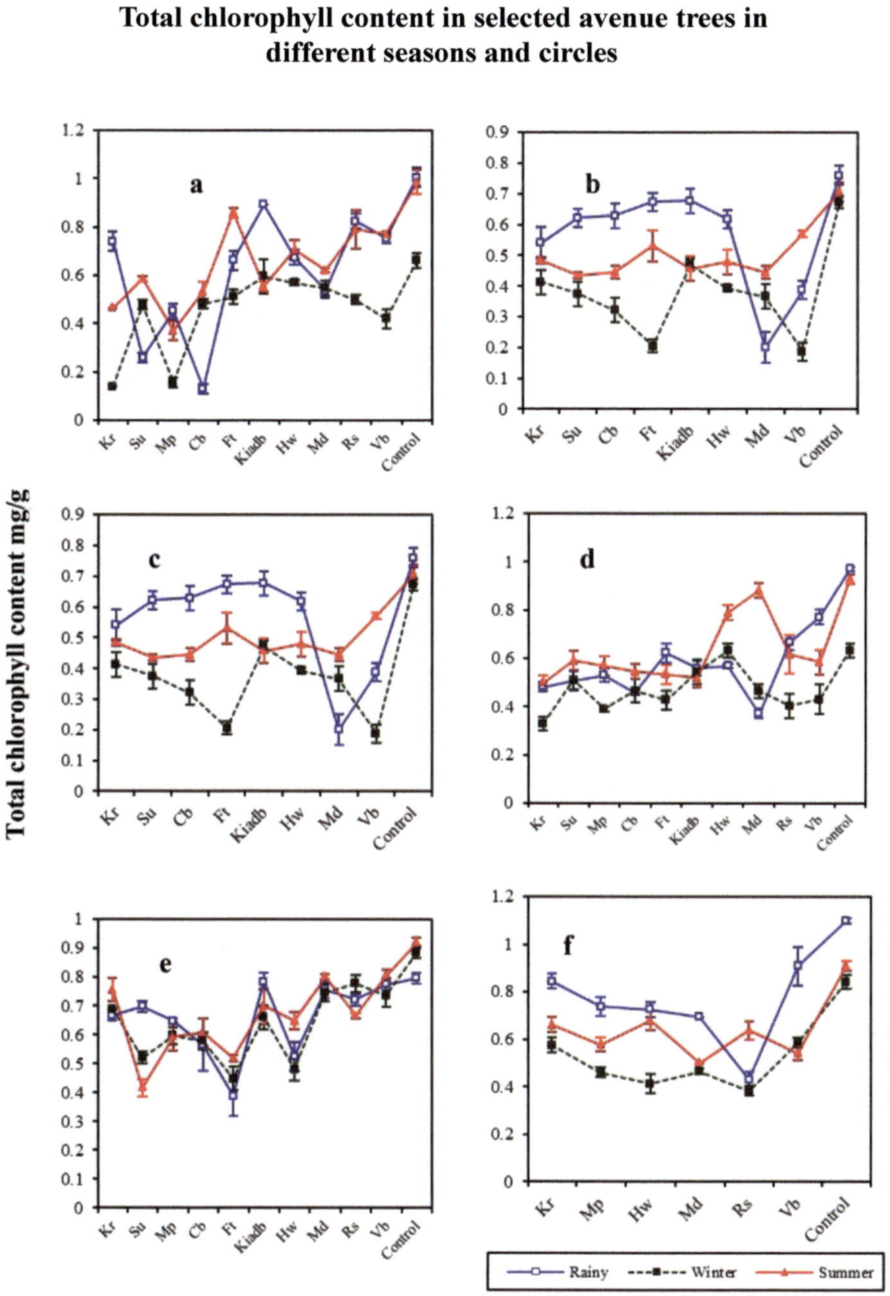

FIGURE 7.3 (a) *P. longifolia*; (b) *D. regia*; (c) *P. inerme*; (d) *S. saman*; (e) *A. indica*; (f) *S. mahogany.*

7.4 CAROTENOIDS

Reduction in carotenoid content was observed in different plants at different circles and seasons. *P. longifolia* showed no significant reduction during the rainy season. Whereas Suburban bus stand circle showed four-fold reduction during winter season, while the majority of the circles showed two-fold reduction. Two-fold reduction was observed in KR, Milk dairy and Highway circles during summer season (Figures 7.4(a)–(f)).

7.5 TOTAL PROTEIN

The total protein content in the leaf samples collected at various circles and the control area is depicted in Figures 7.5(a)–(f). Reduction in the total protein content was observed in the leaf samples of all the six avenue trees growing at selected circles when compared to the control area.

7.6 TOTAL ASCORBIC ACID (AA)

The total AA content in the leaf samples of selected avenue trees at different circles showed reduction when compared to the control area (Figures 7.6(a)–(f)).

7.7 TOTAL PHENOL

The leaf samples of the selected avenue trees growing at selected circles and control area showed a two-fold reduction in the total phenol content at all the circles, except at a few circles during different seasons. In *P. longifolia* two-fold reduction in the total phenol content was observed only at KR and Milk dairy circles during rainy season. Whereas during summer season, majority of the circles showed two-fold reduction, except Suburban bus stand circle. The majority of the circles showed two-fold reduction, during winter except KR, KIADB, highway, and milk dairy circles (Figure 7.7(a)).

7.8 RELATIVE WATER CONTENT

The leaf samples from different circles and control area showed no significant reduction in the relative water content during different seasons (Figures 7.8(a)–(f)).

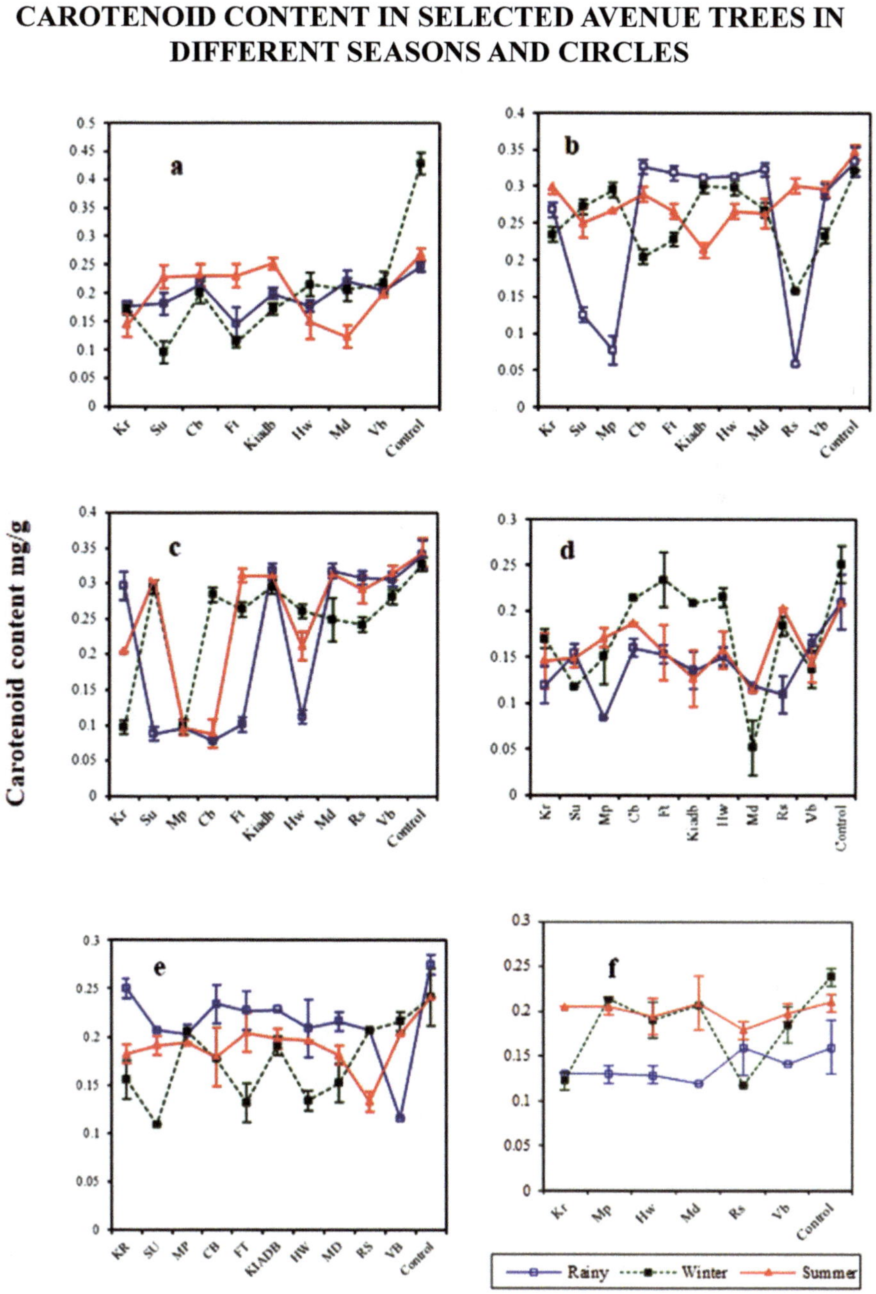

FIGURE 7.4 (a) *P. longifolia*; (b) *D. regia*; (c) *P. inerme*; (d) *S. saman*; (e) *A. indica*; (f) *S. mahogany.*

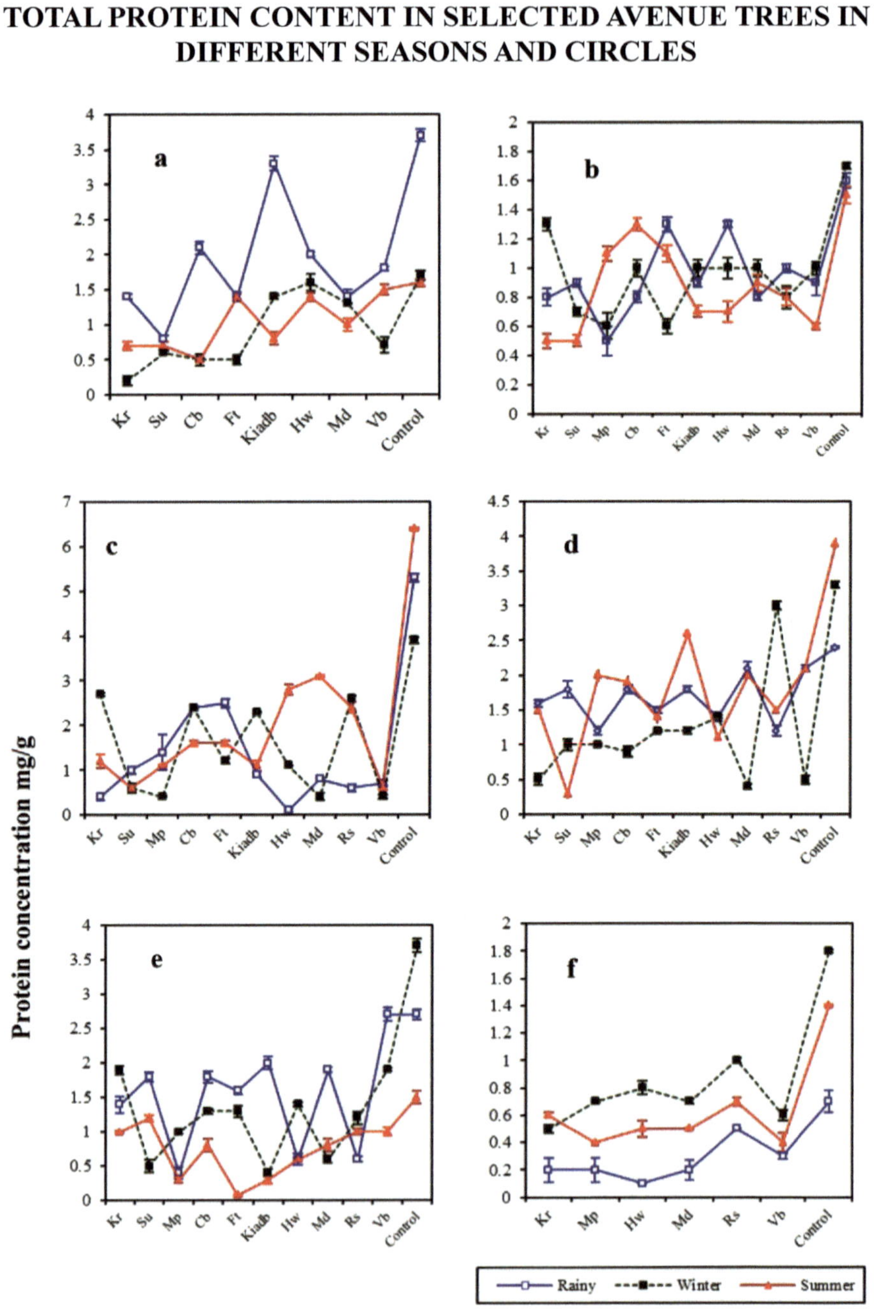

FIGURE 7.5 (a) *P. longifolia*; (b) *D. regia*; (c) *P. inerme*; (d) *S. saman*; (e) *A. indica*; (f) *S. mahogany.*

TOTAL ASCORBIC ACID CONTENT IN SELECTED AVENUE TREES IN DIFFERENT SEASONS AND CIRCLES

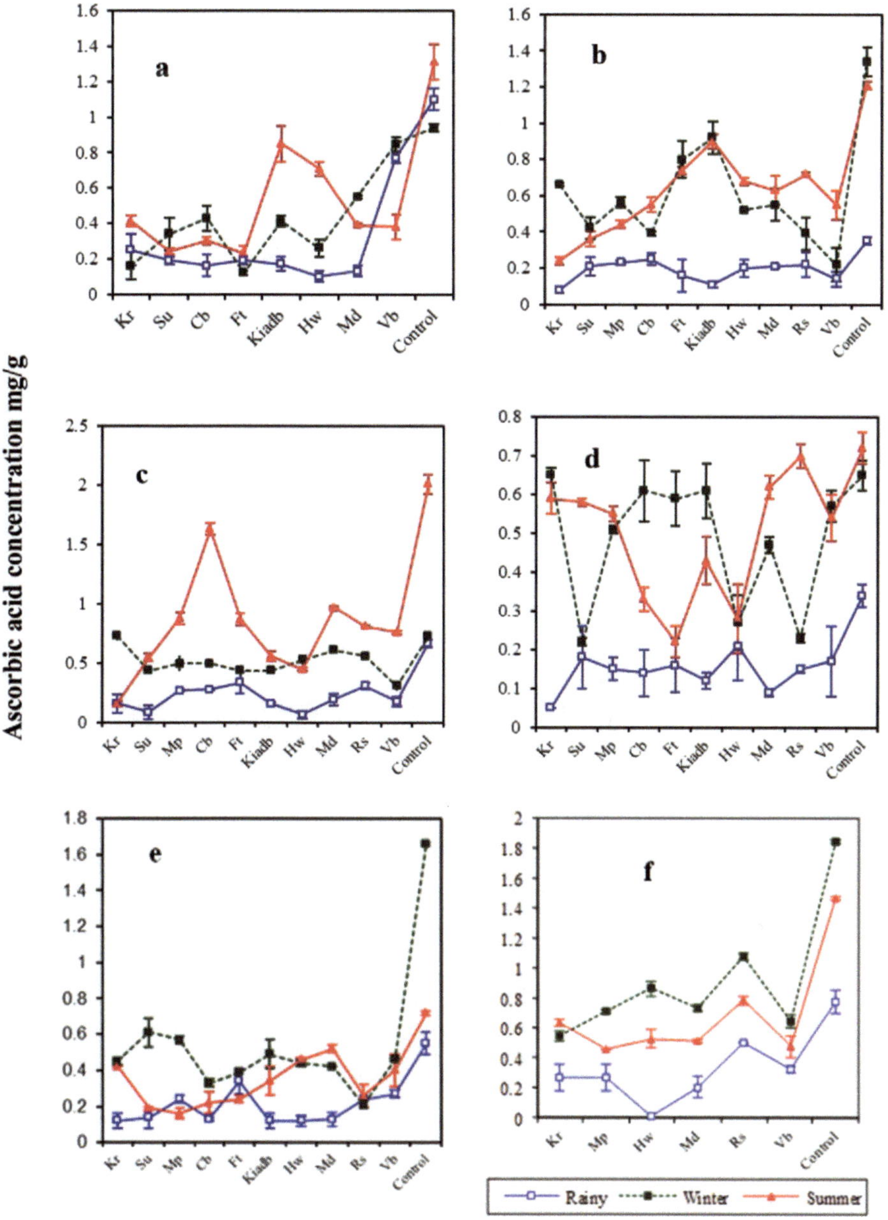

FIGURE 7.6 (a) *P. longifolia*; (b) *D. regia*; (c) *P. inerme*; (d) *S. saman*; (e) *A. indica*; (f) *S. mahogany.*

TOTAL PHENOL CONTENT IN SELECTED AVENUE TREES IN DIFFERENT SEASONS AND CIRCLES

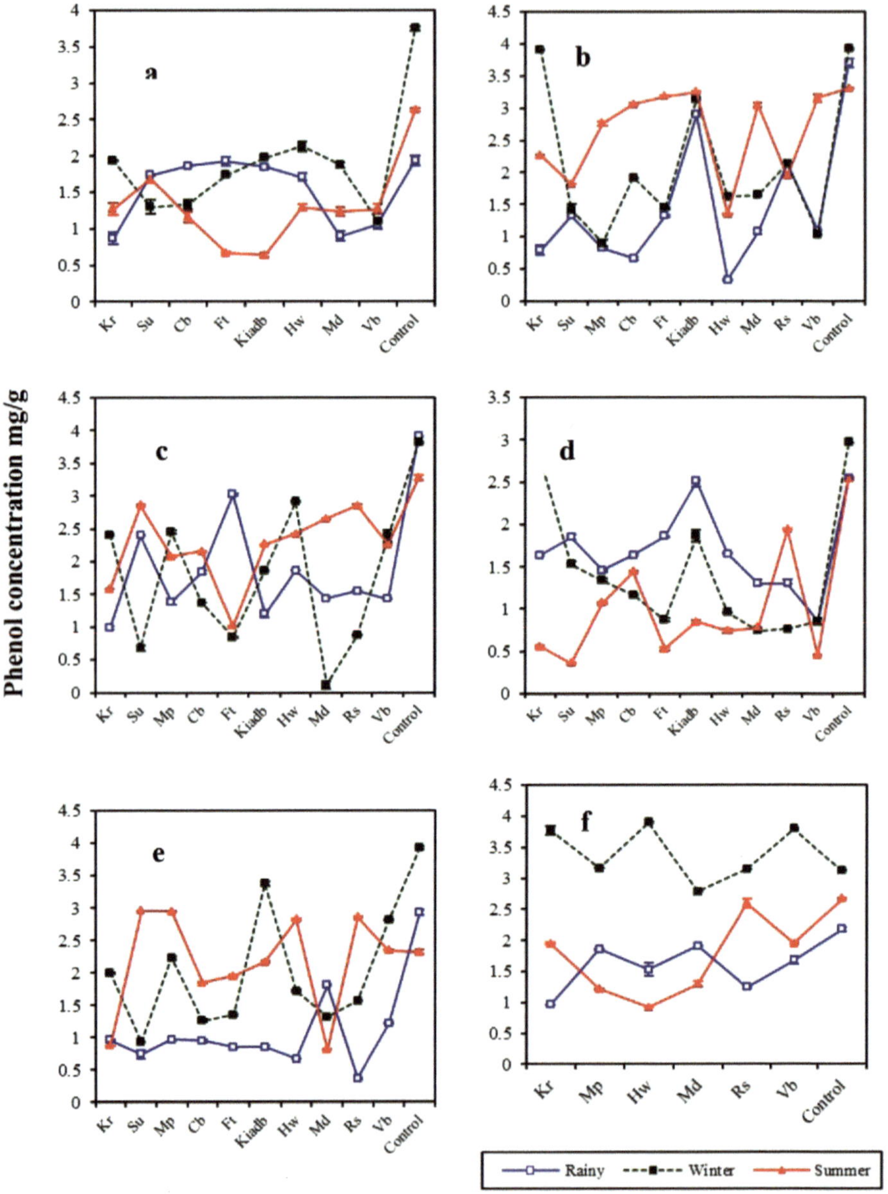

FIGURE 7.7 (a) *P. longifolia*; (b) *D. regia*; (c) *P. inerme*; (d) *S. saman*; (e) *A. indica*; (f) *S. mahogany.*

RELATIVE WATER CONTENT IN SELECTED AVENUE TREES IN DIFFERENT SEASONS AND CIRCLES

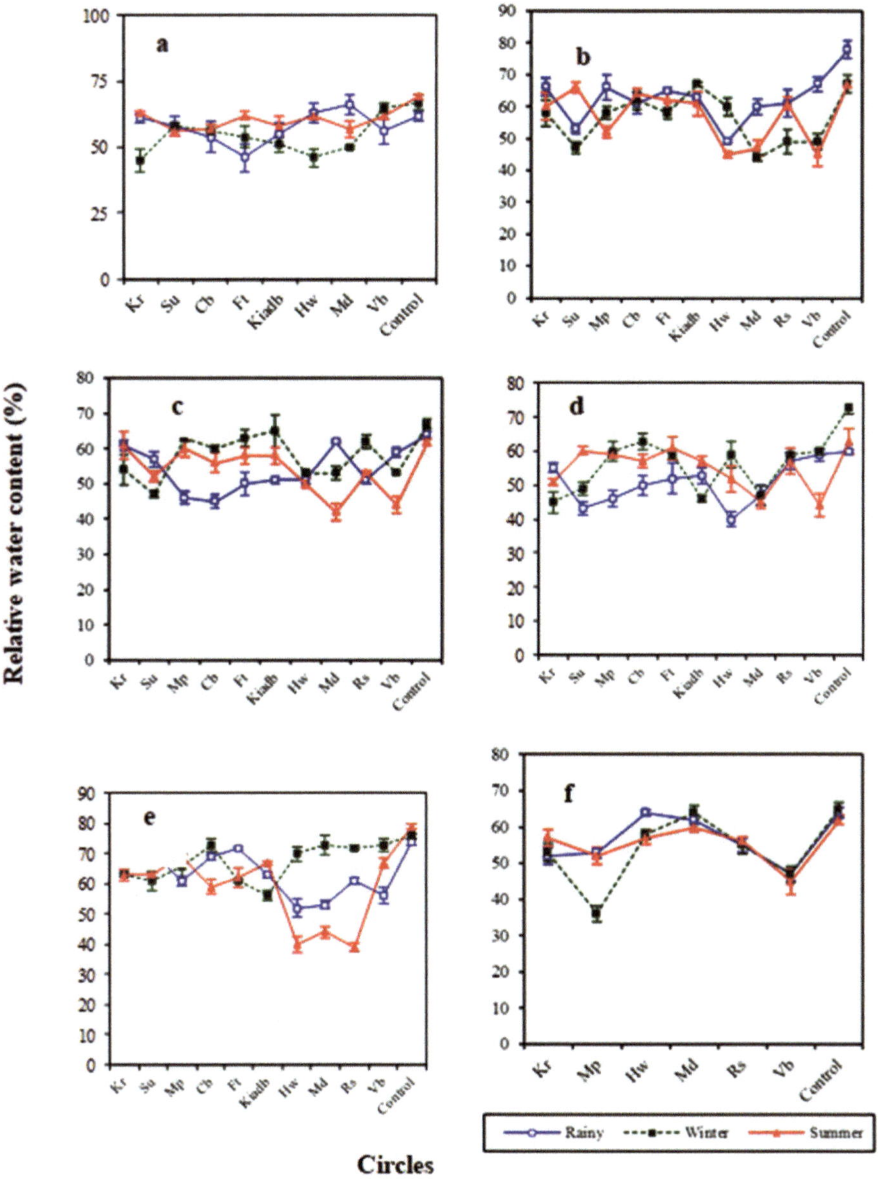

FIGURE 7.8 (a) *P. longifolia*; (b) *D. regia*; (c) *P. inerme*; (d) *S. saman*; (e) *A. indica*; (f) *S. mahogany.*

The percent variation of different biochemical parameters in the leaf samples of selected avenue trees, during different seasons and circles have been shown in Figures 7.9(a)–(c) and 7.10(a)–(c). There is a decrease in percent variation of total chlorophyll, total protein, total phenol, total AA, and relative water content in the leaf samples at all the selected circles when compared to control. The percent variation was less in case of total chlorophyll and relative water content, whereas, protein, AA, and phenol content, showed maximum percent variation.

When a total of all the biochemical parameters were considered percent variation in the plants follow the order of *P. inerme* > *P. longifolia* > *S. saman* > *A. indica* > *S. mahogany* > *D. regia*. Highest variation was observed in *P. inerme* and *P. longifolia* and lowest in *D. regia* (Figure 7.11). With respect to seasonal variation in all the biochemical parameters winter season showed relatively highest variation followed by rainy and summer season (Figure 7.12).

Of the circles selected for the present study maximum variation in micromorphological and biochemical parameters were observed in trees growing at Highway, Milk dairy, Suburban urban, Vijaya bank and Metropole circles in *P. longifolia, D. regia, P. inerme, A. indica* and *S. mahogany*, while *S. saman* showed maximum variation at Fountain and Highway circles (Table 7.1).

Variations in all biochemical parameters such as chlorophyll a, chlorophyll b, total chlorophyll, total protein, total AA, total phenol, relative water content, micromorphological and anatomical parameters were compared with the pollutants SPM, SO_2 and NOx levels in the ambient air (Figure 7.13). Decrease in percent variation of biochemical parameters was compared to SPM level both at selected circles and control area. The SPM showed a 13-fold increase in the selected circles when compared to the SPM level in the control area.

Although there was a 13-fold increase in the SPM level, approximately 15 to 70-fold decreases were observed in the various biochemical parameters studied. It varied in the total chlorophyll, total protein, total AA, total phenol, and relative water content indicating these are the susceptible molecules for the variable pollutant SPM. It was also observed that the percent variation of micromorphological parameters was correspondingly varied with that of SPM in the polluted area when compared to the control area. The gaseous pollutants SO_2 and NOx showed four- to five-fold increase than the control, which was very much negligible and found within the permissible limit of CPCB, in all the circles.

PERCENT VARIATION OF BIOCHEMICAL PARAMETERS OF SELECTED AVENUE TREES DURING DIFFERENT SEASONS

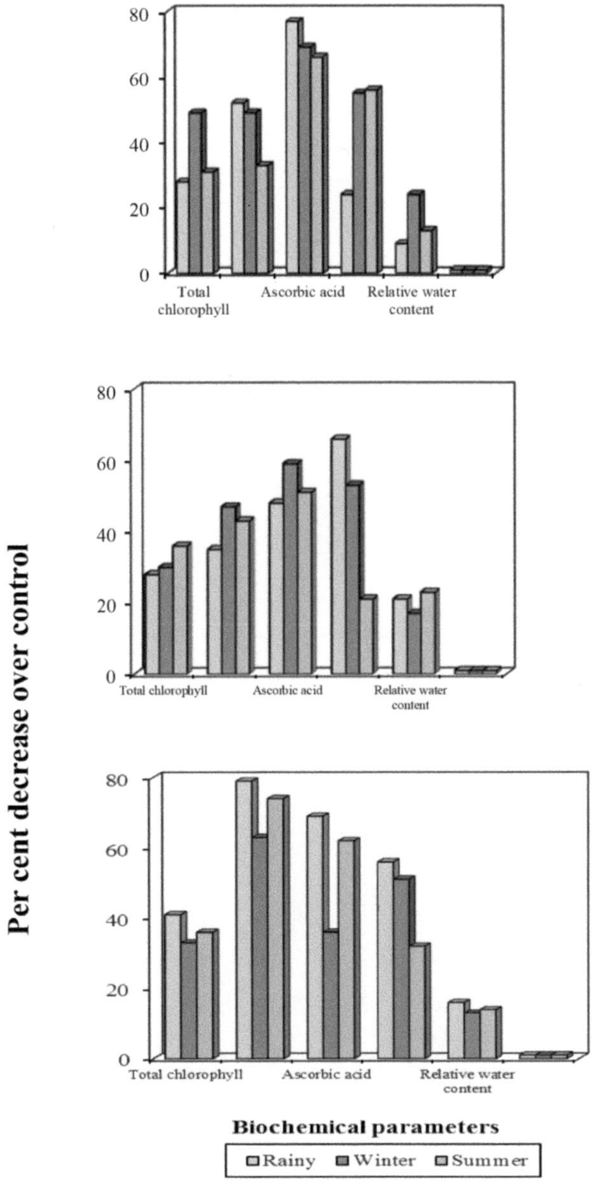

FIGURE 7.9 (a) *P. longifolia*; (b) *D. regia*; (c) *P. inerme*.

PERCENT VARIATION OF BIOCHEMICAL PARAMETERS OF SELECTED AVENUE TREES DURING DIFFERENT SEASONS

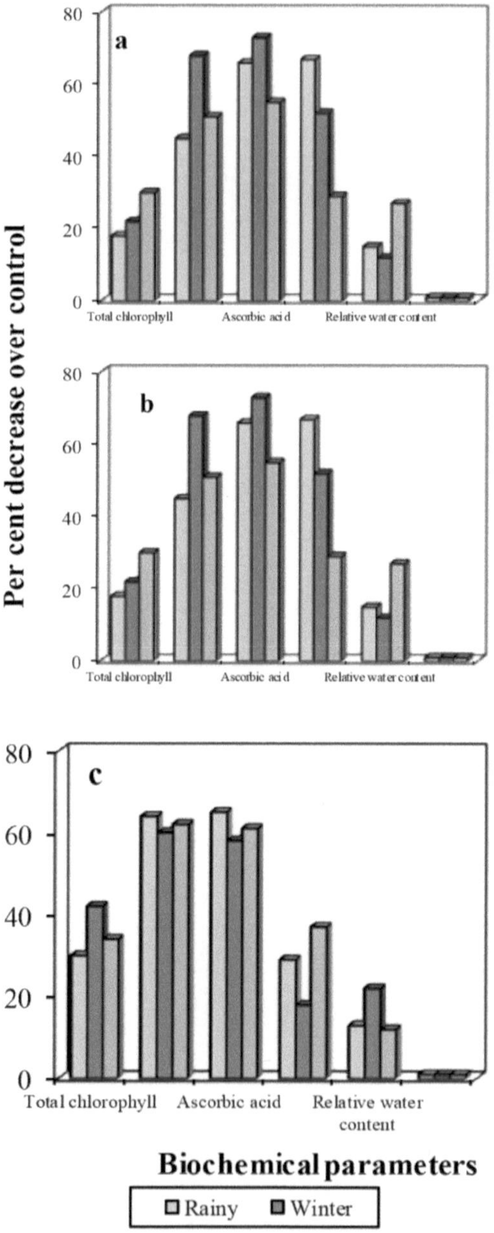

FIGURE 7.10 (a) *S. saman*; (b) *A. indica*; (c) *S. mahogany.*

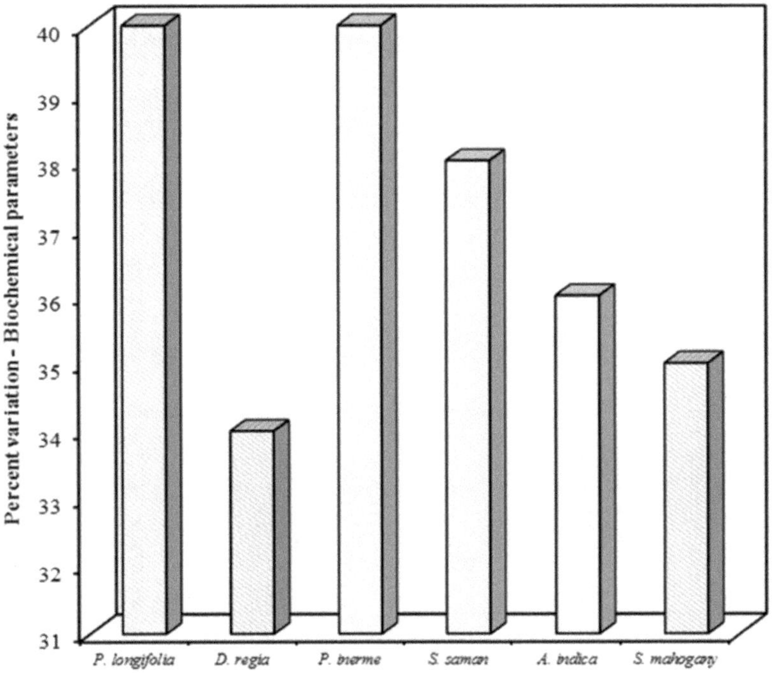

FIGURE 7.11 Percent variation of biochemical parameters of selected avenue trees from polluted areas.

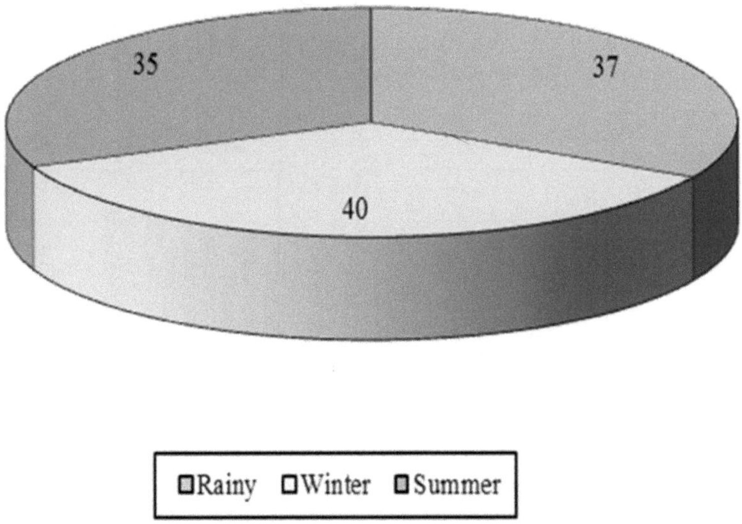

FIGURE 7.12 Seasonal variation of biochemical parameters.

VARIATION IN BIOCHEMICAL PARAMETERS AND COMPARISON WITH POLLUTANTS – SPM, SO_2, AND NOX

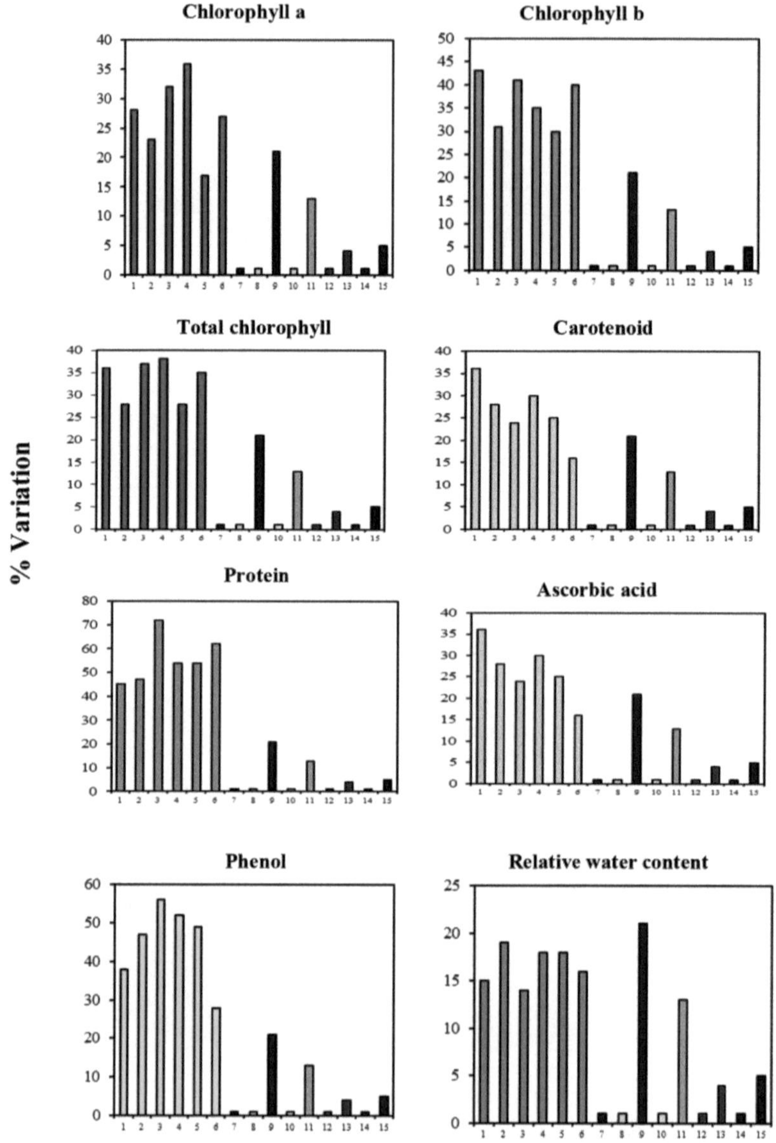

FIGURE 7.13 (1) *P. longifolia*; (2) *D. regia*; (3) *P. inerme*; (4) *S. saman*; (5) *A. indica*; (6) *S. mahogany*; (7) control (% decrease of biochemical parameters); (8) control-micromorphological % variation; (9) polluted-micromorphological % variation; (10) control-SPM; (11) polluted-SPM; (12) control-SO_2; (13) polluted-SO_2; (14) control-NOx; (15) polluted-NOx.

To evaluate the tolerance level of six tree species to vehicular pollution, APTI was analyzed on the basis of four biochemical parameters like AA, total chlorophyll, pH of leaf extract and relative water content. The results are tabulated in Table 7.1. *A. indica* showed an APTI value of 6.4 followed by *D. regia* (6.0); *P. longifolia* (5.9); *P. inerme* (5.7); *S. mahogany* (5.6); *S. saman* (5.5) since all the six species showed APTI value of less than 16, they are categorized as sensitive species.

TABLE 7.1 Air Pollution Tolerance Index of Selected Avenue Trees

Name of the Plant	Ascorbic Acid Content (mg/g)	Total Chlorophyll (mg/g)	pH	Relative Water Content (%)	APTI
P. longifolia	0.357	0.445	4.98	57	5.9
D. regia	0.435	0.649	6.17	57	6.0
P. inerme	0.492	0.552	6.15	54	5.7
S. saman	0.366	0.542	5.90	53	5.5
A. indica	0.314	0.642	6.26	62	6.4
S. mahogany	0.535	0.601	3.73	54	5.6

7.9 DISCUSSION

Photosynthetic activity is based on the quantity of chloroplast pigment and is directly proportional to the amount of photosynthetic area of the leaves. Reduction in the leaf size in all the tree species was associated with reduction in chlorophyll 'a,' chlorophyll 'b,' total chlorophyll and carotenoid pigments in the present study. However, increase in chlorophyll content was observed in *P. longifolia* due to accumulation of SO_2 and NOx present in the ambient air because of greater requirement of sulfur and nitrogen. Ziegler (1977) opined that reduction in chlorophyll was due to toxic ions formed by dissolution of SO_2 in water inside the leaf tissue, which gets preferentially incorporated in the thylakoid membrane.

Enhanced level of leaf sulfur preferentially degraded chlorophyll 'a' than chlorophyll 'b' due to increased acidic condition in leaf. Greater loss of chlorophyll 'a' than chlorophyll 'b' was ascribed to inactivation or denaturation of various enzymes associated with synthesis and action of chlorophyll.

Agrawal et al. (1985) have observed a reduction in the amount of chlorophyll 'a,' chlorophyll 'b,' total chlorophyll and carotenoids in O_3 exposed leaves of *Vicia faba*. It has been suggested that O_3 may either cause direct injury to chlorophyll structure or through oxidation (Nobel, 1974) or may impair chlorophyll synthesis through enzyme inactivation. O_2^- formed in O_3 exposed plants may cause destruction of chlorophyll and carotenoid pigments, following toxicity to physiological defense mechanism against free O_2^- radicals).

It is suggested that at pH 2.2 to 3.5, the free H^+ ions generated in the cell from splitting of $H_2SO_3^2$ into SO_3^{2-} and H^+, displace Mg^{++} from chlorophyll molecules to degrade them into pheophytin molecules. At pH above 3.5 SO_2 affect thylakoid membrane of chloroplast by oxidation of carotenoid through generation of superoxide radicals from HSO_3 (Pieser and Yang, 1978). This photosynthetically inactive brown pigment does not help in the continuation of the normal photosynthesis process thereby causing a decrease in total chlorophyll content in the polluted leaf samples.

Total chlorophyll and carotenoids showed significant seasonal variations at some of the circles in all the six trees species. Abnormal pattern of carotenoid pigment during senescence in leaves of *Quercus petracea* noticed at polluted site. In control plants, lutein was major component prior to leaf fall as β-carotenoid disappeared and Violaxanthin increased in relation to Neoxanthin Ricks and Williams (1975). These patterns were reversed in leaves at polluted regions, and changes in carotenoid pigments may themselves play a part in initiating early senescence. Increase in β-carotenoids were observed in plants exposed to auto exhaust (Madhumanjari and Mukherji, 1998).

They observed two-fold reduction in protein content in the leaves of the plants growing at various circles in the present study have also been reported in plants subjected to automobile exhaust (Banerjee et al., 1983). Enhanced level of amino acids has been attributed to an increase in the breakdown of protein in plants subjected to automobile exhaust. Reduction in protein content has been suggested as an indicator of air pollution. The protein content of a cell is reduced on exposure to SO_2 either due to disruption of protein-synthesizing enzymes or alteration of internal ionic media (Williams et al., 1996). The altered foliage metabolism followed by reduced translocation of metabolites to roots affected the development of nodules and nitrogen availability, which in turn restricted the protein synthesis in ozone exposed *Vicia faba* (Agrawal et al., 1985).

In the present study, *S. mahogany* showed two fold reductions in protein at all the circles and seasons. While *P. longifolia*, *A. indica* and *S. saman* showed reduction at all the circles only during winter and in *P. inerme* at all the circles during rainy and summer seasons. Seasonal variations in the protein content have also been observed earlier (Joshi et al., 1997).

The protein and nitrogen content was found to be higher during the rainy season, decreased during winter and increased during summer in several trees species (Swathi and Singh, 1996). According to them, a higher amount of protein was due to optimum conditions for growth, availability of water and absorption of nutrient from the soil. Reduction in protein content in winter was because of maximum dust fall on leaf.

In the present study, AA was reduced in all the selected tree species. Maximum reduction in AA was seen at majority of the circles in *P. longifolia*, *S. mahogany* and *A. indica* during all the seasons. *P. inerme* showed two fold reductions during rainy and summer seasons; *S. saman* during rainy and *D. regia* during winter season. This is in agreement with earlier findings in five tropical trees exposed to automobile emission. The reduction in AA varied widely in different species, and the quantum of changes was indicative of relative sensitivity of different species to urban air pollution. Similar variations in AA were also observed in *Cassia fistula* and *Cassia siamia* (Joshi et al., 1997).

Klumpp et al. (1994) reported reduction in AA content, increase in *gluthathion* and POX content in broad leaved species near industrial area of Cubato, Brazil. Reduction in AA content in pollution-exposed plants has been ascribed to enzyme toxicity and sulfonation of S-H groups (Williams et al., 1996).

Vijayan and Bedi (1988) studied the effect of SO_2 on *Syzygium cumini* and its amelioration by AA. They observed that reduction in chlorophyll content, carotenoids, protein, AA, carbohydrates was more in SO_2 exposed plants than plants treated with SO_2 and AA. Plants having high AA content were less susceptible to SO_2 pollution, i.e., the higher level of AA the greater was the tolerance (Varshney and Varshney, 1984).

Krishnayya and Date (1996) found that exogenous application of AA increased the internal AA content in SO_2 treated *Trigonella* plant thereby countering the SO_2 effect in a better way. Contrary to the findings of increased AA in plants, Wingsle and Hallgren (1993) reported no significant changes in antioxidants on exposure to SO_2 and NO_2. Antioxidants level increased only when the levels of pollutants were high. O_3 is known to limit the AA synthesis and reduced tolerance of plants to pollutants.

Plants exposed to SO_2, O_3, and $SO_2 + O_3$ pollutants showed a reduction in AA content, the extent of decrease being more in case of $SO_2 + O_3$ exposed plants (Agrawal et al., 1982). Decrease in AA, a scavenger of O_2^- in O_3 exposed plant suggested that O_3 related injury mechanisms were perhaps operational in *Vicia faba* (Agrawal et al., 1985).

The presence of AA in the chloroplast was found to maintain its integrity, thereby protecting the organelle from damage by SO_2, increasing the content of AA and photosynthetic rate. High reductant property of AA helps to reduce SO_2 to H_2S and tolerance of plants to SO_2 depends on the amount of H_2S emitted (Sekiya et al., 1982).

Phenolic compounds are the group of defense compounds synthesized by plants comprising only one group of secondary products that consists of thousands of compounds. They are localized in vacuoles, intercellular spaces, in the cuticle and epicuticular wax of plants. They are characteristic of the species, tissue-specific and their concentration can change depending on altered environmental conditions (Zobel, 1996).

Reduction in phenol content was observed in all six species, which confirms to the earlier findings. A decrease in phenolics in the extracellular matrix could be due to an increased oxidation of these compounds at cell wall level or due to the effect of air pollution on photosynthesis and carbohydrate metabolism. Reduction in phenol content indicates pronounced sensitivity of secondary metabolism to SO_2 pollution compared with primary metabolism and the reduced tolerance of birches to oxidative air pollutants. Reduction in some low molecular phenolic glycosides indicates that even short-time exposure of enhanced atmospheric SO_2 may cause moderate changes in the accumulation pattern of phenolics in birches (Lavola, 1998).

On the contrary, both decrease (*Nerium indicum*, *Boerhaavia diffusa* and *Tabernaemontana coronaria*) and increase (*Amaranthus* and *Cephalandra*) in phenolic compounds have been observed by Madhumanjari and Mukherji (1998). Increase in phenol content was observed in plants exposed to pollutants either singly or in combination (Agrawal et al., 1982).

Increased deposition of phenolic compounds was observed in mesophyll cells, directly under the stomata and disorganization of membranes in *Pinus nigra* and *P. resinosa* in response to acid rain (Zobel and Nighswander, 1991). Relative water content showed no significant reduction in the present study, when compared to other parameters. With respect to seasonal variation, all the six species showed maximum reduction in biochemical parameters during winter followed by rainy and summer seasons. *P. inerme* showed maximum reduction in biochemical parameters while *S. mahogany* showed minimum reduction.

All the six tree species evaluated for APTI in the present study showed APTI value of less than 10. Plants showing APTI value less than one were considered highly sensitive, and those in the range of 1 to 16 were considered as sensitive species to air pollution and plant showing a value of 17 to 21 were intermediately tolerant and between 30 and 100 as tolerant species. Similar, studies on APTI have been conducted by Karthiyayini et al. (2005) in Coimbatore – Ooty highway.

AA is a strong reductant that activates many physiological defense mechanisms and its reducing power is directly proportional to its concentration. It protects chloroplasts against SO_2 and protects the enzyme of the CO_2 fixation and chloroplast from inactivation. Plants having high level of AA under pollution conditions are considered to be tolerant to air pollutants (Varshney and Varshney, 1984). Total chlorophyll content was found to be decreased in all the six species studied. Plants show reduced efficiency of photosynthesis at low concentration of SO_2 under pollution stress. It is suggested that low pH triggers degradation of chlorophyll molecule into nonfunctional phaeophytin molecules, which does not help in normal photosynthesis leading to a decrease in chlorophyll content in leaf samples.

Relative water content was high in *D. regia* (67.78) and *A. indica* (63.96) while *S. mahogany* showed least value of 36.62. A higher level of relative water content helps to maintain the physiological balance in plants growing under conditions of stress due to air. Leaf pH extract in all the six tree species investigated was found to be acidic in nature. Scholz and Reck (1977) observed a drastic decline in pH in sensitive species due to the action of acidic pollutants, which decrease the efficiency of hexose sugar being converted to AA. However, the reducing activity of AA is dependent on pH. Hence higher pH provides tolerance to plants against pollutants.

7.10 BIOMONITORING

Pollutants present in the ambient air induce various kinds of physiological changes in the plants. Physiological and biochemical changes in plants are likely to occur much before the appearance of any visible injury symptoms on foliar surface subjected to various gaseous pollutants. The trees exposed to sub-lethal concentration of pollutants may appear green and normal, but their biochemical and physiological processes are subdued and generally functions at reduced efficiency. These invisible changes caused by exposure to various pollutants can, however, be detected at the metabolic level by examining certain biochemical parameters such as chlorophyll, protein, specific metabolic enzymes, or a combination of one or more parameters. The biochemical parameters studied included chlorophyll 'a,' chlorophyll 'b,' total chlorophyll content, carotenoids, protein content, AA, phenolic compound, and relative water content.

Air pollutants hamper the photosynthetic ability of plants by way of affecting chlorophyll pigments. Chlorophyll is the most sensitive pigment to air pollutants. Chlorophyll 'a,' chlorophyll 'b,' total chlorophyll of six species studied. Chlorophyll 'b' degradation was found to be more than the chlorophyll 'a.' Carotenoid pigment is considered to be an accessory pigment and protects chlorophyll from photo-oxidation, was also found to be degraded due to air pollution-induced stress. Protein content decreased in all the six tree species was investigated. *S. mahogany* showed significant reduction in protein content at all the circles and seasons while *P. longifolia, A. indica* and *S. saman* showed significant reduction only during winter season and *P. inerme* during rainy and summer seasons. Reduction in protein content has been and carotenoid pigment contents exhibited marked seasonal and site wise reduction in all the leaf samples suggested as an indicator of air pollution. Decrease in protein content may be attributed to decreased photosynthetic activity due to reduction in chlorophyll pigment.

AA has long been recognized as an important metabolite channelizing a number of important mechanisms such as carbon dioxide fixation besides activating defense mechanisms. Pollutants often inflict phytotoxicity by decreasing AA content in plants. All the tree species investigated showed a decrease in AA content, as compared to control plants. Maximum reduction in the AA content was seen in *P. longifolia, S. mahogany* and *A. indica* at the majority of the circles, during all the seasons. *D. regia* however showed significant reduction at the majority of the circles during winter season, *S. saman* during rainy and *P. inerme* during rainy and summer seasons. Phenolic content too showed a reduction in all the six tree species due to the effect of air pollutants. Significant reduction in phenol content was observed at the majority of the circles during winter and summer in *P. longifolia*; during rainy and summer in *D. regia*; during rainy and winter season in *A. indica* and only during rainy season in *P. inerme*. The relative water content is an important parameter for evaluation of the tolerance level of plants against air pollution stress. Maximum reduction was observed in relative water content in *P. longifolia, S. saman, S. mahogany*, during winter and in *A. indica* during summer season. With respect to seasonal variations, all the tree species showed maximum reduction in biochemical parameters during winter followed by rainy and summer season. *P. inerme* and *P. longifolia* showed maximum reduction in biochemical parameters and least in *D. regia*.

In the present study, APTI was calculated based on biochemical parameters like total chlorophyll, AA, pH of leaf extract and relative water content in the selected tree species. APTI value of all the six trees found to be less than 10.

A. indica showed an APTI value of 6.4 followed by *D. regia* (6); *P. longifolia* (5.9); *P. inerme* (5.7); *S. mahogany* (5.6) and *S. saman* (5.5). All the six trees were less than 16 in their APTI values, they may be categorized as "sensitive" to air pollution and could be appropriately used as bio-indicators based on relative advantages compared with conventional physicochemical monitors. These species are recommended to be appropriate for biomonitoring of air pollution in urban air quality management.

It is evident from the present analysis that vehicular air pollution in the city of Mysore is on the increase and has not yet reached significant levels to cause alarm. The pollutants monitored were well within the permissible limits except for suspended particulate matter (SPM), which had crossed the NAAQS limit at three out of the 10 circles monitored. SPM is of major concern as it is the cause for significant respiratory ailments. This has been found to clog the stomata openings of all the tree species in the present study. During the course of the study over 12% increase in traffic was registered. With no rapid transport options on the anvil, the vehicular population and associated pollution loads are expected to further deteriorate urban air quality at Mysore.

The vehicular pollution load, time, and duration of exposure cause air pollution. Traffic jams and congestion impede the flow of traffic, reduce the average speed, and cause more emissions. Vehicles utilize as much as 20% extra fuel due to reduced speed while in motion. Traffic movement can be streamlined by implementing efficient traffic management measures such as mass rapid transport system. This could greatly reduce the significant pollution loads. Declaring more roads as one-way, synchronized traffic signals, grade separators for fast and slow traffic could reduce accidents as well as delays. Prohibition of vehicles in congested market and residential areas and declaring them as pedestrian-only areas would greatly reduce the health risk. Introduction of certain healthy measures such as, bus-only lanes on peak routes, providing lanes for cyclists, carpooling, implementing parking restraints and high parking fees would improve the state of affairs. There are many other options in traffic management, such as staggering timing for vehicles either during part or whole of the day. This could be easily deployed for schools, hospitals, and factories. For economically highly productive professional business areas higher tax rates for very good transport services can be collected.

In addition to these, at individual and community levels, certain simple measures can be adopted which would go a long way in containing air quality deterioration. Maintenance of vehicles by regular servicing can reduce about

15% of the emissions. Further, environment friendly approaches like opting for bicycles, solar or battery driven vehicles, walking instead of using vehicles for short distances, etc., can contribute significantly toward reduction in air pollution. Popular eco-literary drives such as "you can prevent CO poisoning in fetus, heart patients and old aged people," could sensitize motor vehicular users. This could promote community pro-active responses in mitigating CO health risk at traffic signals.

As an implication for air quality management, the following measures can be used. Develop and display a systematic approach to measure ambient air quality with a source wise inventory and local concentrations using WHO guidelines. The information which was routinely collected, under national program such as Global Environmental Monitoring Systems can be standardized for urban centers and publicized regularly. This could be done by online display systems for ambient air quality, which also serves useful public education purpose. Identification of air pollutants such as ozone, benzene, formaldehyde, and hydrocarbons can be considered as additional parameters for ambient air quality monitoring. This could improve the scope of air quality monitoring. Biological indicators thus come as handy, cost-effective, and simple tools for evaluation of a sensitive zone in urban settlements. Monitoring the local impact of air pollution on health by conducting epidemiological studies and creating a comprehensive database for toxic effect for each of the emission gasses would be very useful.

7.11 CONCLUSION

Urban environmental quality monitoring policy matters need to be emphasized to mitigate air pollution and to prevent further irreversible damages on health, resources, and ecology. Biological monitoring of the environment is already a pioneering approach in monitoring water, air, and other natural resources. Evidence-based on the present study further justifies the need for a sustainable biological indicator in a comprehensive and integrated urban environmental quality monitoring. For over decades, WHO and UNEP have been using various parameters for urban air quality monitoring. It is high time that a better utilization of the present study and similar monitoring potential biological approach are integrated for improvement of the ongoing air quality management in Mysore and elsewhere. It is proposed as an outcome of this research that in choice of avenue trees short-listed, sensitive species can be planted for routine air quality monitoring in addition to conventional and advanced techniques. These could greatly substantiate ground truths for satellite and laser beam-based monitoring of urban air quality.

KEYWORDS

- **air pollution tolerance index**
- **ascorbic acid**
- ***Azadirachta indica***
- **chlorophyll**
- **photosynthesis**
- ***Polyalthia longifolia***

REFERENCES

Agarwal, M., Nandi, P. K., & Rao, D. N., (1982). Effect of ozone and Sulphur dioxide pollutants separately and mixture on chlorophyll and carotenoid pigments of *Oryza sativa. Water, Air and Soil Poll., 18*, 449–454.

Agrawal, M., Nandi, P. K., & Rao, D. N., (1985). Responses of *Vicia faba* plants to ozone pollution. *Indian J. Environ. Health, 27*(4), 318–329.

Arnon, D. I., (1949). Copper enzymes in isolated chloroplasts, poly phenoloxidase in beta vulgaris. *Plant Physiology, 24*(1), 1–15.

Banerjee, A. K., Sarkar, R. K., & Mukherji, S., (1983). Reduction in soluble proteins and chlorophyll content in a few plants as indicator of automobile exhaust pollution. *Intern. J. Environ. Studies, 20*, 239–243.

Behrens, W. A., & Madere, R., (1963). Ascorbic acid estimation by DNPH method. *Anal. Biochem., 92*, 510.

Bradford, M. M., (1976). A rapid and sensitive method for the quantification of microgram quantities of protein utilizing the principle of protein-dye binding. *Anal. Biochem., 72*, 248–254.

Gavali, J. G., & Gavali, D. S., (2003). Plants and air pollution. In: Hiremath, K. G., (ed.), *Recent Advances in Environmental Science* (pp. 169–192).

Gostin, I., (2016). *Air Pollution Stress and Plant Response*. 10.1007/978-981-10-1201-3_10.

Iqbal, M., Ali, T. S., & Mahmooduzzafar, (2000). Photosynthetic performance of some dicotyledonous tropical plants under degraded environment. In: Khan, M. A., & Farooq, S., (eds.), *Environment Biodiversity and Conservation* (pp. 409–427). APH publishing corporation, New Delhi.

Joshi, O. P., Wagela, D. K., & Pawar, K., (1997). Urban air pollution effects on two species of cassia. *Poll. Res., 16*(1), 1–3.

Karthiyayini, R., Ponnammal, N. R., & Joseph, R., (2005). Air pollution tolerance index of certain plants of Coimbatore–Ooty highway near ITI area Coimbatore, Tamilnadu. *Poll. Res., 24*(2), 363–365.

Klumpp, A., Klumpp, G., & Domingos, M., (1994). Plants as biomonitors of air pollution at the Serra Do Mar near the industrial complex of Cubato, Brazil. *Environ. Poll., 85,* 109–116.

Kousar, N., Seshikala, D., & Charya, S. M. A., (1998). Biomonitoring of air pollution by using plants in Warangal city. *IJEP, 19*(7), 488–492.

Krishnayya, N. S. R., & Date, M. V., (1996). The impact of SO_2 and SO^{2+} ascorbic acid treatments on growth and partitioning of dry matter in *Trigonella foenum graceum* L. *Environ. Poll., 9*(11), 121–125.

Larson, R. A., (1988). The antioxidants of higher plants. *Phytochem., 27*, 969–978.

Lavola, A., (1998). Soluble carbohydrates and secondary phytochemicals in Betula as affected by SO_2–pollution. *Water, Air and Soil Pollution., 197*, 25–34.

Madhumanjari, M., & Mukherji, S., (1998). Roles of ascorbic acid, β-carotene and phenolic compounds in protecting the plants exposed to automobile exhaust pollution. *Research Journal of Chemistry and Environment., 2*(2), 25–28.

Malik, C. P., & Singh, M. B., (1980). In: *Plant Enzymology and Histozymology* (p. 286). Kalyani Publishers, New Delhi.

Nobel, P. S., (1974). Ozone effects on chlorophyll A and B. *Natur. Wissenschaften, 61*, 80, 81.

Pieser, G. D., & Yang, S. F., (1978). Chlorophyll destruction in the presence of bisulfite and linoleic acid hydroperoxide. *Phytochemistry, 17*, 79–84.

Ricks, G. R., & Williams, R. J. H., (1975). Effects of atmospheric pollution on deciduous woodland. Part 3: Effects on photosynthetic pigments of leaves of *Quercus petraea* (Mattuschka) leibl. *Environ. Poll., 8*, 97–106.

Scholz, F., & Reck, S., (1977). Effects of acids on forest trees as measured by titration invitro inheritance of buffering capacity in *Picea abies*. *Water, Air and Soil Pollution, 8*, 41–45.

Sekiya, J., Wilson, G. L., & Filner, P., (1982). Resistance to injury by Sulphur dioxide, correlation with its reduction to an emission of hydrogen sulphide in Cucurbitaceae. *Plant Physiol., 70*, 432–437.

Singh, S. K., & Rao, D. N., (1983). Evaluation of plants for their tolerance to air pollution. In: *Proceedings of Symposium on Air Pollution Control* (Vol. I, pp. 218–224). Indian Association for Air Pollution Control, New Delhi.

Singh, S. N., & Rao, D. N., (1968). Effect of cement dust pollution on soil properties and on wheat plants. *Indian J. Environmental Health, 20*(3), 258–267.

Swati, S., & Singh, M. P., (1996). Effect of graphite factory emissions on the nitrogen and protein contents of some trees. In: Agarwal, S. K., Swarnlata, T., & Dubey, (eds.), *Bio-diversity and environment.*

Treshow, M., & Franklin, A. K., (1989). In: *Plant Stress from Air Pollution.* John Wiley and Sons Inc. New York, USA.

Varshney, C. K., (1985). Role of plants in indicating monitoring and mitigating air pollution. In: *Air Pollution and Plants* (pp. 146–170). A state of art report. Ministry of Environment and Forest.

Varshney, S. R. K., & Varshney, C. K., (1984). Effect of SO_2 on ascorbic acid in crop plants. *Environ. Poll., 35*, 285–291.

Vijayan, R., & Bedi, S. J., (1988). Effect of SO2 fumigation on *Syzygium cumini* Skeels and its amelioration by ascorbic acid. *Ind. J. Environ. Health., 30*, 155–162.

Williams, A. J., & Banerjee, S. K., (1995). Effect of power plant emissions on the metabolic activities of *Mangifera indica and Shorea robasta. Environ. and Ecol., 13*, 914–919.

Williams, A. J., Banerjee, S. K., & Gupta, B. N., (1996). Plant biochemical responses and biomonitoring of air pollution around thermal power plants at Korba, Madhya Pradesh. *Ecol. Env. & Conservation, 2*, 67–77.

Wingsle, G., & Hallgren, J. E., (1993). Influence of SO_2 and NO_2 exposure on glutathione; superoxide dismutase and glutathione reductase activities in scotpine needles. *J. Expt. Bot., 44*, 463–470.

Zobel, A. M., (1996). Phenolic compounds in defiance against air pollution. In: Mohammad, Y., & Mohammad, I., (eds.), *Plant Response to Air Pollution.* edited by Mohammad, Y., & Mohammad, I., (c). John Wiley & Sons Ltd.

Zobel, A., & Nighswander, J. E., (1991). Accumulation of phenolic compounds in necrotic areas of Australian and red pine needles after spraying with sulphuric acid: A possible bioindicator of air pollution. *New Phytol., 117*, 565–574.

CHAPTER 8

Plant Responses to Drought Stress: Morphological, Physiological, Molecular Approaches, and Drought Resistance

PRATAP VYANKATRAO NAIKWADE

Department of Botany, ASP College, Devrukh, Maharashtra, India, E-mail: naikwade.pratap@gmail.com

ABSTRACT

Water is the most vital component of all organisms for their existence and development. Though there are many abiotic factors, drought is the greatest imperative abiotic aspect responsible for various changes in plant metabolism and adversely affecting growth and yield. Drought independently causes more crop yield losses than all pathogens combined in a year. Plant responses to drought stress by various modifications in morphology, anatomy, physiology, biochemistry, etc., to enhance water usage proficiency by limiting the loss of water by transpiration and other ways. Plant responses to drought can be classified as short-term, mid-term, and long-term. Drought stress causes negative consequences on chlorophyll, photosystems which are important for photosynthesis, and it also disturbs electron transport mechanisms hence resulting in a decline in photosynthetic efficiency of plants. Drought resistance is a complex feature that endures through several mechanisms like escape, avoidance, and tolerance. It is estimated that in the future, water availability will be reduced due to various climate change factors, but demand for food is also increasing due to population explosion, so it is required to highlight how plants react to drought and various mechanisms involved in drought tolerance. The present chapter summarizes important findings of plant responses to drought stress, morphological, physiological, and molecular

Plant Metabolites under Environmental Stress: Mechanisms, Responses, and Adaptation Strategies.
Nivas M. Desai, Manasi Patil, Umesh R. Pawar (Eds.)

approaches, and highlights plant mechanisms of drought resistance which could be exploited to enhance the plant proficiency to survive under water deficit conditions. It is a need of the hour to comprehend various impacts of drought stress on plants for improved stress management strategies with logical implications.

8.1 INTRODUCTION

Stress is any adverse situation that negatively affects plant performance by changes in mechanisms of growth and development of plant. Abiotic stress can be caused by various factors like water, salinity, temperature, etc. Among these, water scarcity or drought stress is one of the most important adverse environmental factors, negatively affecting plant growth survival and productivity (Hui et al., 2018; Tanveer et al., 2019; Seleiman et al., 2021). Nevertheless, plant is affected by the availability of water in two ways, if excess water is accessible, i.e., waterlogged condition or scarcity of water, i.e., drought stress altering the plant's life by various ways (Vara Prasad and Rama Rao, 2019). Climatic factors and water present in soil available to roots affects water content in plant (Khorsand et al., 2021). The term drought implies a period with less rainfall and water deficit condition. Drought stress in plants occurs when environmental conditions result in the decline of soil moisture and relative humidity in air, but temperature is high, water is lost by transpiration or evaporation processes (Yadav et al., 2020). Drought stress initiates various plant responses, it adversely effects various processes of morphology, biochemistry, and physiology which are directly related to growth and yield of plant.

Plant responses to water shortage are mechanisms for survival of plants in drought stress. Drought stress has a negative impact on plant productivity (Naikwade, 2020a, b) causing about 70% yield loss in crops (Bray et al., 2000). In the last 10 years, nearly $30 billion are lost due to decrease in drought affected crop production in the world (Gupta et al., 2020). Water scarcity is a constraint to life that persists progressively in various parts across the globe due to global climate change as reported by IPCC (2007). Agriculture's water consumption could double by 2050, according to estimates, although freshwater availability is expected to drop by 50%, due to global warming and climate change (Gleick, 2000). Projected increases in temperatures, fluctuations in precipitation forms, occurrence of extreme weather events, and reduced water availability may result in reduced plant productivity (Naikwade, 2017, 2019). Increasing drought leads to severe water stress which has an impact on the survival of plants in natural ecosystems. The severity of drought is directly proportional to plant death rate (Liu and Feng, 2021).

As water is essential to plant for metabolism, water supply directly or indirectly impact every process in plants (Lobell et al., 2013). The impacts of water stress differ species-wise in plants. The extent of plant response to stress depends on the extent and duration of the water deficiency (Pugnaire et al., 1999). Severe drought condition causes a decrease in water pressure inside leaves and subsequent wilting of plant. Wilting is serious condition which will diminish growth of plant (Kaur and Gupta, 2005). Since water is crucial for the delivery of nutrients, a lack of it causes drought stress, leads to lose vigor in plants (Ashkavand et al., 2018).

As soon as the plant recognizes water stress, short-term responses of plant bring about closure of stomata (Zhou et al., 2021), declined water loss through transpiration and increase in CO_2 intake. Midterm responses of plant to water stress include fine-tuning of the osmotic potential by accumulation of solute, modifications in cell membrane elasticity and morphological variations. Long-term response is adaptation to drought by variation in gene responses, anatomical reforms of specific organs and procurement of modified physiological mechanisms. These responses also lead to Abscisic acid (ABA) accumulation and transportation which later inhibits shoot growth, cause abscission, etc. (Gai et al., 2020).

Plants must react swiftly in the event of a water shortage so entire biological processes are changed in plants in response to drought (Kogler and Soffker, 2017). Drought stress brings up multiple signal transduction pathway which follows second messengers formation which sequentially modulate intracellular Ca^{2+} level and activate kinases to start protein phosphorylation cascades. Drought stress causes stomatal and non-stomatal restrictions in the course of photosynthesis process (Izanloo et al., 2008; Carmo-Silva et al., 2010). The plant responses under drought stress are influenced by a multifaceted regulatory system of molecular mechanisms, including interaction between transcription factors (TFs), kinase cascades, manufacture of reactive oxygen species (ROS) along with contribution of heat-shock factors and small RNAs (Atkinson and Urwin, 2012). Drought stress limits water imbibition, reduces seedling vigor and by this way negatively impacts germination (Farooq et al., 2009). Various factors comprising period of drought, intensity as well as incidence of drought, properties of soil, growth circumstances and phases, and species of plant intensely impact the extent and duration of drought-related symptoms in plants (Zoghi et al., 2019).

Plant responses to drought stress can be classified into four types viz. morphological, physiological, biochemical in addition to molecular response as seen in Table 8.1. Drought stress persuades various physiological, morphological, and biochemical alterations in plants at different developmental stages (Tripathy et al., 2000). Plant growth is adversely affected by drought stress by impacting photosynthesis, respiration, sugar, and nutrient metabolism, translocation, phytohormones, ion uptake, etc. (Farooq et al., 2009; Prasad et al., 2011). Water stress, both moderate and severe, can cause changes in photosynthetic factors., for instance net photosynthetic rate, intercellular carbon concentration, stomatal conductance, and transpiration in plants (Zhao et al., 2020). Water stress has negative effects on chlorophyll content in leaves, photosynthetic elements, vegetation, and yield (Ghobadi et al., 2013; Chen and Hao, 2015; Ali et al., 2018). Desiccation is the most harmful effect of drought stress in which majority of protoplasmic water evaporates and the cell contains a modest amount of firmly bound water only.

TABLE 8.1 Plant Response Under Water Stress

Morphological Response	Reduced growth rate
	Reduced leaf area and stem length
Physiological Response	Reduced leaf water potential
	Loss of turgor and osmotic adjustment
	Decrease in stomatal conductance
	Decline in net photosynthesis
Biochemical Response	Increase in antioxidative enzyme, i.e., superoxide dismutase, catalase, peroxidase
	Transient decrease in photochemical efficiency
	Reduced reactive oxygen species accumulation
	Decrease efficiency of Rubisco
	Accumulation of stress metabolite pro, polyamine
Molecular Response	Increased expression in abscisic acid biosynthetic gene
	Synthesis of specific proteins, i.e., LEA, dehydrins
	Stress-responsive gene expression, i.e., DREB, WRKY, NAC

Drought affects three major periods of development in plants, including vegetative, pre-anthesis as well as terminal stage. Wilting and abscission of leaves, reduction in leaf surface area and architecture are responses of plants to decrease water loss through transpiration. Cell elongation is inhibited under serious water stress as water flow is disturbed within cells in higher plants (Shaheen et al., 2016). Plant height, area of leaf, growth, and development in plants are all affected by impaired mitosis cell division and expansion while facing drought situation (Cairns et al., 2012). Drought and abiotic stress also have negative impacts on photosystems (PS I and PS II), photosynthetic electron transport, and chlorophyll biosynthesis (Efeoglu and Terzioglu, 2009; Kalaji et al., 2016; Sharma et al., 2016). There is conventional water loss causing closing of stomata and disruption in cell structure along with plant metabolism (Yadav et al., 2019).

In response to drought stress, plants have developed plans to avoid loss of water, regulate optimum water supply to important plant organs, preserve water level in cells and endure through drought conditions. Plant response to drought vary from morphological to molecular level. Drought stress response aspects are crucial to comprehend because global climate change is producing more increasingly severe droughts (Kim et al., 2020). Identifying key physiological limitations to productivity under drought stress and mechanisms to face to water deficit stress will be important for improving the steadiness of yield in climate change situation. Hence this chapter summarizes the current understanding of plant responses to drought is explained along with the mechanisms to avoid drought.

8.2 ORIGIN OF DROUGHT STRESS

Restricted water supply to plant roots or increased rate of transpiration consequences in drought stress in plants. High temperatures helping for initiation of drought stress by working as an indirect driver of drought stress (Lobell et al., 2013). Roots are the site of water intake in plants so water deficit condition is first noticed by roots. Matric potential is the range of force required for water absorption from the soil. In water deficit situation, soil moisture is less available, so more energy is required by the plants to absorb water from the soil; hence the matric potential is higher (Bhattacharjee and Saha, 2014). In dried soil the matric potential is strong, plant experiences indications of water stress. This situation is known as the matric effect. The efficiency of water consumption is a broad perspective of photosynthetic and transpiration physiognomies of plants. As reported by Zhou et al. (2021) at the commencement of drought stress, the water use efficiency is significantly increased.

8.3 MORPHOLOGICAL RESPONSE TO DROUGHT STRESS

Plant development depends on complex exchanges concerning sources and sink restrictions of major organs of a plant, including root framework and the shoot system, setting up a balance. Drought stress enormously obstructs morphological aspects of vegetative parts of plants like reduced plant height, decreased number of tillers, leaf number and size, etc. (Akram et al., 2019). In response to drought stress, plants undergo morphological changes in root, shoot, leaf as explained in subsections.

8.3.1 CHANGES IN THE ROOT SYSTEM

Plant parts chiefly root and shoot systems are the most important morphological responses to drought for plant adaptation to water stress condition. As roots are water absorbing organ of plant, water deficit condition in soil is first noticed by roots (Kaashyap et al., 2018; Kim et al., 2018). In response to less soil moisture availability, plants increase root length in soil so that under drought stress, roots can absorb an adequate volume of water for various physiological processes. An increase in the length, density of roots, propagation, and size of roots are significant plant responses under drought stress condition (Kavar et al., 2007). In water-deficit region, the upright roots of tree species seedlings are 10 times more in height than above-ground plant height (Larcher, 2003). Jaleel et al. (2008) reported that drought stress increased root length of *Catharanthus roseus*. Changes in the root system, such as size, density, length, propagation, development rate is considered as central strategy to face drought and to achieve tolerance (Tzortzakis et al., 2020).

Some plants have vigorous ability to increase root growth in the initial stage of drought to overcome scarcity of water (Hu and Xiong, 2014). As reported by Mohamed et al. (2002) penetration ability of roots is correlated with drought resistance level in *Phaseolus* plant. In an experiment with *Reaumuria soongorica,* it was found that seedling root activity in this shrub was higher under moderate (12.89%) and severe (17.42%) water stress conditions when compared with control (Lishan et al., 2015). Water and nutrients were continued by roots to aerial portion of the plant so that growth will not decline even under drought stress. In drought stress root diameter is decreased so that water access is enhanced, it also enhance yield and production of plants (Wasson et al., 2012). However, in severe drought condition low soil water and increased soil resistance due to dryness decreases growth and development of roots (Comas et al., 2005). In *Populus* species dry biomass of roots was decreased under both mild as well as severe drought stress treatments (Wullschleger et al., 2005).

Root length densities in the range of 0.5 to 1 cm^{-3} are usually sufficient to achieve moisture requirement in plants generally (De Willigen et al., 2000), but in water deficit condition, soil is dry due to less moisture, hence to overcome hydraulic resistance in dry soil, greater root surface area density is necessary. While facing drought stress, plant shows adaptation and architecture of root system is changed by various ways (Den Herder et al., 2010). For escalation of surface area to increase water uptake, soil with low moisture encourages amplified distribution of the food assimilates to plant roots (Ryser, 2006). Increased carbon assimilation in the roots, leads to increased growth toward the water-sufficient soil layers in drought stress (Eapen et al., 2005). Furthermore, root hairs increase root surface area for absorption of water from soil, fine pores which are inaccessible to root apex (Brown et al., 2012). Amendment of organic manures to the soil also increase root and total biomass in plants (Naikwade et al., 2012; Ghadge et al., 2013; Naikwade, 2021) which may be beneficial to face water stress. The quantity of dividing cells in the root tip region and the characteristics of root caps are supplementary for root survival during water stress (Barlow, 2003).

The architecture of the root system is useful for the plants to develop drought resistance in plants (Foster et al., 2005). These adaptations are helpful to maintain higher water potential and regular rate of transpiration while facing drought circumstances, it is beneficial for the growth and development of plants. Well-watered plants can store water inside their cells and tissues however plants facing drought are unable to store water resulting in reduced inner mass of plants (Tandzi et al., 2019). Research also revealed that the response of roots to drought stress is mainly depend on the genotype of the plant as well as the period and intensity of stress (Xu et al., 2015).

8.3.2 CHANGES IN THE SHOOT SYSTEM

In normal conditions, plant shoot length is more due to adequate transpiration as well as translocation processes, while in plants facing water stress, shoot length is comparatively reduced as plants need to overcome scarcity of water and nutrients caused by drought conditions. Drought stress declines vegetative and reproductive growth and considerably reduce length, biomass, and weight of stem in plants under water stress (Shi et al., 2019). Once water deficit condition occurs, root growth is preferred over shoot growth. So, root:shoot ratio is increased in plants under drought stress situations (Fulda et al., 2011). Hence the root: shoot ratio is considered as important measure to describe drought resistance capacity of a particular plant (Pallardy, 2010). Many researchers found that drought stress reduces plant height, stem circumference, and fresh and dry biomass in various crops (Farooq et al., 2009; Zheng et al., 2016).

Plant facing drought stress exhibits decreased stem height as response to drought stress in the initial vegetative stage (Bhatt and Rao, 2005). Drought stress considerably decreased quantity of branches developed on soybean plant stem (Frederick et al., 2001). In water deficit situation length of shoot system was also reduced in chickling pea plant (Gheidary et al., 2017). The reason behind reduced stem length is limited uptake of nutrients by the roots as less soil water available in drought (Razmjoo et al., 2005). Semerci et al. (2017) described that water stress significantly decrease shoot height and biomass of plant due to decreased turgor pressure in plant.

8.3.3 CHANGES IN THE LEAF

The plants under drought stress, show morphological as well as physiological responses of leaves to control loss of water in various ways. Under drought stress, plants generally reduce the number of leaves as well as leaf area to decrease the water loss through leaves by transpiration, but it may result in yield loss. Drought stress also induces symptoms in plants like leaf rolling, yellowing leaves, leaf scorching (Corso et al., 2020). Poorter and Markesteijn (2008) reported that under severe water shortage condition, plum plants droop their leaves or roll as cell turgor pressure is lost. Leaf rolling is also observed in plants under drought stress to reduce water consumption and to increase water efficiency. Scarcity of water also initiate restrictions on the leaf surface area which is considered as the first line of protection while combating drought stress. In drought stress, growth, and development of cell is negatively affected, so leaf quantity and surface area is decreased. These modifications result in reduced water absorption from the soil along with declined transpiration from leaves of plants under drought stress (Fathi and Tari, 2016).

Leaves are arranged in such a way that incident solar radiation on leaves will be reduced, by this way, leaf surface temperature is decreased, and the plant is protected from excess water loss. It is common observation that plants with increased drought resistance have developed xeromorphic characters like smaller and thicker leaves, thicker cuticle above epidermis, smaller and denser stomata, thicker palisade tissue, more developed vascular bundle sheath, etc. The presence of trichomes on epidermis is also observed helpful to reflect light and to decline transpiration (Abclulrahaman and Oladele, 2011).

Lipids are accumulated in the epidermis to produce wax on the leaf surface, it increases sunlight reflection to reduce unnecessary transpiration and to control increased surface temperatures of leaves (Mohammadian et al., 2007). Palisade tissues and vascular bundles in leaves are helpful for transportation and retention of water and nutrients (Guha et al., 2010). It is supportive for plants to reduce excess water loss and increase water efficiency to check injury from drought. In an experiment, Shi et al. (2019) found that vegetative growth as well as leaf area was reduced under drought stress in rose plant.

In plants facing drought stress, leaf water potential is decreased so leaf growth is also negatively affected (Hsiao and Xu, 2000). Although in severe drought stress, flow of water in between xylem and other cells is interrupted, turgor pressure is reduced resulting in decreased cell expansion and leaf architecture (Taiz and Zeiger, 2006). Wilting of leaves and disturbance in occurrence of flowering has been reported by Bhatt and Rao (2005). Drought stress also increases the rate of premature leaf senescence under severe conditions (Farooq et al., 2009).

Leaf area was decreased in *Petroselinum crispum* L., *Stevia rabaudiana* and other plants under drought stress was also observed (Najla et al., 2012; Srivastava and Srivastava, 2014). Due to decreased soil water content, leaf water level is also decreased, which sequentially induces drop of turgor pressure of guard cells of stomata resulting in closing of stomata (Deka et al., 2018). The number of leaves decreased in *Andrographis paniculate* under water stress was also reported by Bhargavi et al. (2017). The rate of senescence in leaves is increased, some leaves are rolled, dropped, show brittleness, change color from green to yellow, and finally wilted (Khan et al., 2018; Ruehr et al., 2019).

8.4 GROWTH AND YIELD RESPONSE OF PLANT

Drought stress adversely affects the growth and yield of plant by influencing and altering various physiological and biochemical processes like photosynthesis, chlorophyll synthesis, ion uptake and translocation, nutrient metabolism, respiration, carbohydrates metabolism, etc. (Li et al., 2011). In response to drought stress, leaf area of plant reduces, subsequently photosynthesis is decreased, therefore crop growth and yield is reduced. As a response to drought, frequency of cell division, length of root, stem declines, water, and mineral nutrition association disturbs which leads to declined plant growth and yield (Kumawat et al., 2018).

Drought stress results in reduced turgor pressure in plants. As expansion of cell is affected by turgor potential, developing cells expand less and cell size is smaller in plants under water deficit conditions. Drought stress in plants causes reduction in turgor pressure, leaf water potential, and under development of cells (Farooq et al., 2009). In an experiment Sun et al. (2013) compared impact of water stress on growth of plant. *Nannong jingyan* plant species which is drought sensitive, upon exposure to water stress 7 days, resulted in reduced growth as compared to drought-tolerant species *Nannong xuefeng*. Biomass composition was drastically altered in *Miscanthus* under drought stress, with large reductions in cell wall and cellulose content, plant weight was also decreased by 45% (Weijde et al., 2017).

Drought stress adversely affects the yield of plants by reductions in yield traits which is the outcome of stress-initiated modifications in metabolisms and physiological mechanisms and subsequent negative consequences on plant reproductive organs with productivity. No doubt that drought stress has adverse effects on the growth and yield of plants but affects plant reproductive stage most as it is a more sensitive stage to drought as compared to pre- and post-reproductive stages (Hussain et al., 2018). Drought reduced the yield of various crops like maize (Kamara et al., 2003; Cakir et al., 2004) and wheat as reported by Barnabás et al. (2008). In case of sugarcane also yield is declined (Vasantha et al., 2005). Severe drought stress in rice can cause economic yield losses of 48–94% in the reproductive stage of rice (Lafitte et al., 2004) while in the grain-filling stage and losses are up to 60% (Basnayake et al., 2006). Decrease in yield and yield components due to drought stress has also been reported by many researchers on different crops such as sunflower, i.e., *Helianthus annuus* (Tahir et al., 2002), peanut, i.e., *Arachis hypogaea* (Furlan et al., 2012), cotton, i.e., *Gossypium barbadense*, etc. (Pettigrew, 2004). Severe drought stress reduced dry matter and grain yield traits including spike number per pot, grains per spike, grain weight and grain yield in wheat (Abid et al., 2018).

In legume plant, flower, and pod formation is reduced due to increased abortion thus quantity and quality of seed is reduced (Fang et al., 2010). As estimated by Kashiwagi et al. (2015) drought stress can cause a decrease in worldwide chickpea production by 33%. Drought stress leads to modification in canopy architecture, which leads to change in gaseous exchange, various water relations, plant growth as well as fruits or grains development in crops (Rahmati et al., 2018). Sometimes, plant shows symptoms like cracking of bark, branches, necrosis on leaves, inhibited growth and tree thinning are seen in plants under drought conditions (Toscano et al., 2019).

8.5 PHOTOSYNTHETIC RESPONSE TO DROUGHT STRESS

Drought stress has several effects on important physiological process which is very sensitive to drought stress, i.e., photosynthesis. Sun et al. (2013) reported that altering of photosynthesis mechanism due to water stress occurred at the beginning. Photosynthesis is decreased mainly by closing of stomata and injury to cell membrane due to drought conditions. Apart from that functioning of several enzymes associated with ATP synthesis is also changed under drought (Kumawat et al., 2018; Sharma et al., 2020). Figure 8.1 explains how drought stress influences photosynthetic machinery, disturbs different pathways resulting in declined rate of photosynthesis in plants.

Water deficit condition decreases water potential of the soil first and subsequently decrease water potential of the plant cells (Hartmann et al., 2013). Under drought stress, if leaf water potential is more than − 1 Mpa, process of photosynthetic efficiency is less affected, however further decrease in water potential, declines photosynthesis process (Wang et al., 2016). Drought also affects light reaction of photosynthesis by interruption of cyclic as well as non-cyclic electron transport (Obidiegwu et al., 2015). Hui et al. (2018) showed that minor water-deficit condition does not impact photosynthesis rate as compared to severe stress.

Plants reply to drought stress through lowering transpiration by closing of stomata. Stomatal closure is the major response of plants to drought stress caused by various factors, viz. root signal, reduced turgor pressure stomata guard cells, or rising vapor pressure gradient amongst leaves and air. Closing of stomata restricts water loss but also decreases CO_2 absorption and transportation of non-structural C (McDowell and Sevanto, 2010; Sevanto, 2014) which is an essential constituent of photosynthesis (Steppe et al., 2015). Stomatal closure may lead to a decrease of CO_2 concentration in the intercellular spaces, this is stomatal inhibition of photosynthesis, in due course photosynthesis is negatively affected (Ying et al., 2013; Rivas et al., 2016; Chen et al., 2017b). Drought stress leads to a reduction in photosynthesis and other physiological processes like transpiration, stomatal conductance in *Tectona grandis* (Galeano et al., 2019).

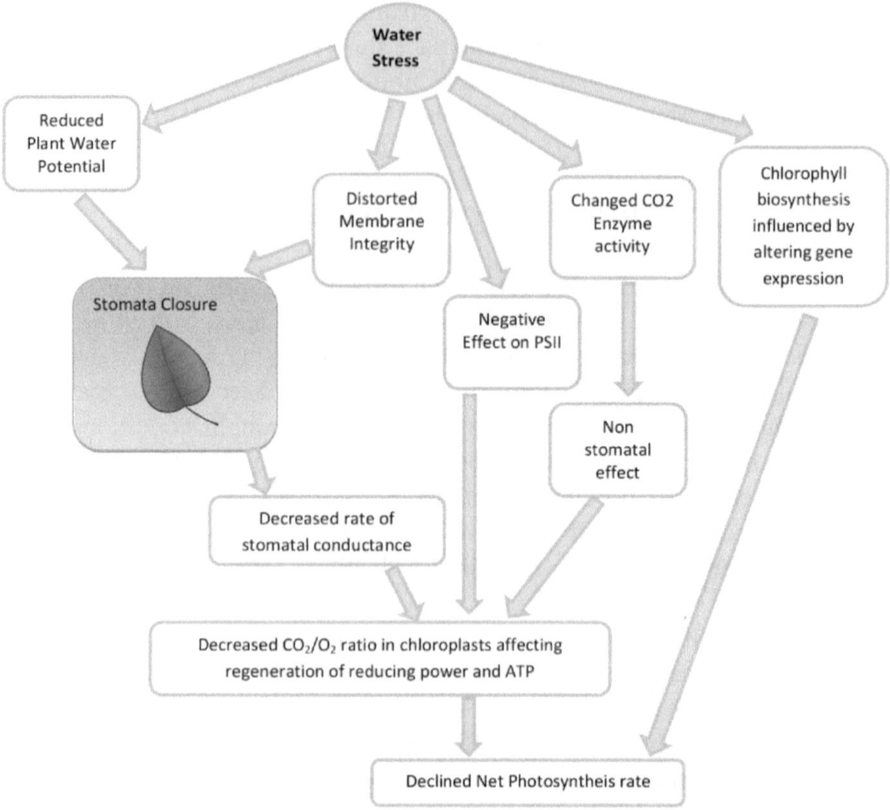

FIGURE 8.1 Water stress disturbs different pathways reducing photosynthesis rate.

Besides stomatal inhibition, other non-stomatal mechanisms are also responsible for decreased photosynthetic rate in plants facing drought stress (Hajiboland et al., 2017). Direct relation between stomatal conductance and CO_2 intake was proved by Osuna et al. (2015). As soon as the concentration of CO_2 decrease relative to O_2, alteration of C from the photosynthesis process to photorespiration is stimulated. Drought stress increases the photorespiratory pathway, primarily when RuBP oxygenation is high due to partial CO_2 fixation. About 70% of the total H_2O_2 (a type of ROS) is produced through photorespiration under drought stress (Noctor et al., 2002). In excessive-high light intensity, photo-inhibition of photosynthesis may occur leads to formation of free radicals usually active oxygen species in the chloroplasts which are injurious to protein environment of chloroplast reactions (Carvalho et al., 2008). Photosynthesis is also affected non-stomatal inhibition like photo-inhibition, chlorophyll degradation and the associated decrease in light-harvesting. During photosynthesis, electron transport-related with Photosystem II is also declined. Liu et al. (2018) described reduction in chlorophyll, chlorophyll fluorescence under drought stress.

Photo oxidation along with degradation of various chlorophylls can be seen in plants facing drought stress (Anjum et al., 2011). Chowdhury et al. (2017) also reported that in soybean leaves, drought stress causes alteration synthesis of chlorophyll and ratio of chlorophyll a:b also changed. One of the reasons for the decrease in chlorophyll is drought-promoted O_2^- and H_2O_2, which leads to lipid peroxidation (LPO) and eventually chlorophyll degradation (Karimpour, 2019). Gurumurthy et al. (2019) proved that drought stress changes the action of chlorophyll pigments and moieties of photosynthesis which subsequently decrease photosynthetic activities in black gram crop.

Decrease in CO_2 conductance via stomata, mesophyll limitations and reduced ability of stomatal movement result in declined photosynthesis in plants under drought stress (Marcinska et al., 2013; Singh and Thakur, 2018). This declined CO_2 intake alters activities of CO_2 fixing enzymes, membrane disruption leading to reduced ATP synthesis. It limits regeneration of RuBP, so Rubisco activity is declined, it also inhibits the task of sucrose phosphate synthase and nitrate reductase (NR) enzymes (Sicher and Barnaby, 2012). Severe drought stress cause protein degradation leading to inadequacy of Rubisco enzyme and ultimately strong inhibition of photosynthetic CO_2 assimilation and impairment of PS II reaction center (Reddy et al., 2004). Drought stress decreased quantity of various proteins required in Calvin cycle, comprising Rubisco in olive (Abdallah et al., 2018). In response to drought stress, PS II oxygen-evolving center is targeted, leading to a drop in the PS II reaction center as well as chlorophyll fluorescence is also reduced (Zhang et al., 2011). Draught stress also adversely influence PS II by decreasing quantum yield of PS II in plants, consequently photosynthesis rate is decreased (Albert et al., 2011; Tattini et al., 2014).

Drought stress may directly inhibit photosynthetic apparatus by reducing turgor pressure, it changes pH in chloroplast and ion concentrations. Rubisco and other enzymes used in Calvin cycle are sensitive to pH, so change in chloroplast pH, reduces activity of these enzymes ultimately resulting in reduced rate of photosynthesis. Drought stress causes degradation of chloroplasts thylakoid membrane, which negatively affects the activity of photosynthetic pigment (Bertioli et al., 2016). Drought stress resulted in damage to Rubisco and other enzymes associated with photosynthesis and loss of photosynthetic pigment content (Brito et al., 2019). Hu et al. (2010) opined that stomatal conductance as well as metabolic impairment result in declined photosynthesis. Du et al. (2020) found that in an experiment on soybean, drought stress suppress manufacturing of photosynthetic products and thus carbon balance is disturbed.

8.6 MOLECULAR RESPONSE TO DROUGHT STRESS

Plants facing drought stress initiates molecular responses of plants under comprise greatly regulated genes along with signal transduction mechanisms (Osakabe et al., 2014; Kapoor et al., 2020). Drought-responsive genes can be categorized into two types viz. ABA-dependent and ABA-independent genes (Kim, 2014).

Drought tolerance is achieved in plants by regulations of manifold genes expression prompted by drought (Lang and Bui, 2008). Plant assimilation is controlled by the transcription process in drought stress, DREB (dehydration-responsive element-binding proteins) and AREB (abscisic acid-responsive element-binding proteins) are main genes involved under water scarcity (Agarwal et al., 2006; Seki et al., 2007). Ali et al. (2017) also described the important role of water stress induced TFs such as DREB2A and DREB2B in the regulation of numerous genes which are useful for plants to face drought stress. Kaur and Asthir (2017) stated that responses of AREB at a transcriptional stage in addition to post-transcriptional stage of drought decide drought tolerance ability of plants. Drought stress prompts the expression of histone H_1 (Bhargava and Sawant, 2013), which is useful in the proper functioning of stress-responsive gene apart from it useful for closing of stomata. In experiments on many crops, various stress-related genes have been isolated and identified (Prabha et al., 2011; Singh et al., 2015).

Drought-responsive genes produce proteins used in signaling mechanisms and transcription processes like protein kinase, protein phosphatase, etc., as well as functional proteins associated with protection of cellular membranes, transportation of water, aquaporins, sugar transporters, etc. (Nakashima et al., 2014). C-repeat/dehydration-responsive element binding factors, mitogen-activated protein, cup-shaped cotyledon, zinc-finger proteins are important moieties to face drought stress in plants (Cong et al., 2008).

Drought stress activates ABA manufacture in roots which is further transported to shoot system causing closing of stomata, changes in root architecture and ultimately restricting growth in plants (Mittler and Blumwald, 2015). Moreover, ABA is produced in leaves and then transported to the plant organs. ABA concentration is regulated by key enzymes like 9-cis-epoxy carotenoid dioxygenase and d-arabino-1,4-lactone oxidase to face drought stress (Bao et al., 2016). Increased volume of ABA triggers signaling pathway in stomata guard cells which causes outflow of K^+ and decreased turgor pressure and finally closing of stomata (Lim et al., 2015; Salazar et al., 2015). Sugars translocated in the xylem can be helpful for opening and closing of stomata due to ABA in plants under drought stress (Le Gall et al., 2015).

ABA stimulates miscellaneous signaling pathways in plants under drought stress (Bucker-Neto et al., 2015). ABA-induced genes encoding drought-related proteins like ROS-detoxifying enzymes, dehydrins, and phospholipid signaling enzymes which are useful for plant to tolerate drought stress (Fahad et al., 2019). ABA balances stomatal conductance and transpiration rate, thus excessive water loss is prevented in drought conditions (Guo et al., 2008). Increased abscisic acid (ABA) accumulation is important mechanisms for dealing with drought stress in many plants (Bano et al., 2012; Brodribb and McAdam, 2013). Chen et al. (2017a) reported that increased levels of ABA resulted in enhanced drought stress tolerance in *Vaccinium corymbosum*.

Signal transduction in response to drought stress induce various important compounds like phytohormones to face drought condition. Concentration of cytokinin is reduced and ABA increased by drought stress which initiates signal pathways between root, stem, leaves, etc. Cytokinin is secreted by plants, it's increased volume in xylem sap induce opening of stomata by weakening its sensitivity to ABA (Bielach et al., 2017). Jasmonic acid is also synthesized which is helpful in increasing ROS scavenging, stimulating stomata closure and root growth, to increase drought tolerance in plants (Shi et al., 2014; Fang et al., 2016). Auxin is also produced by plants under stress which controls the growth of roots and ROS metabolism (Ullah et al., 2018). Ethylene is also helpful for plants as it controls ROS formation and thereby regulates stomata closing in plants facing drought stress (Watkins et al., 2014).

Galeano et al. (2019) reported that osmotic modification mechanism occurs in roots of plants under drought stress in which genes associated with lipid, carbohydrate, and various proteins are also playing a vital role. The proteins which are helpful in facing drought stress are chaperones, chaperonins, heat-shock proteins (HSPs), etc. Drought stress protein types can be seen in Figure 8.2. These proteins are further divided as regulatory and functional proteins. Aquaporin proteins are also important to regulate plant-water relations (Afzal et al., 2016). It also regulates membrane networks and transportation of glycerol, urea, formamide, and carbon dioxide too (Maurel et al., 2008). Proline is common well suited osmolytes, helps plants to maintain their homeostasis to face drought situations (Krasensky and Jonak, 2012). In plants under stress free proline content as well basic water use efficiency is enhanced (Galeano et al., 2019).

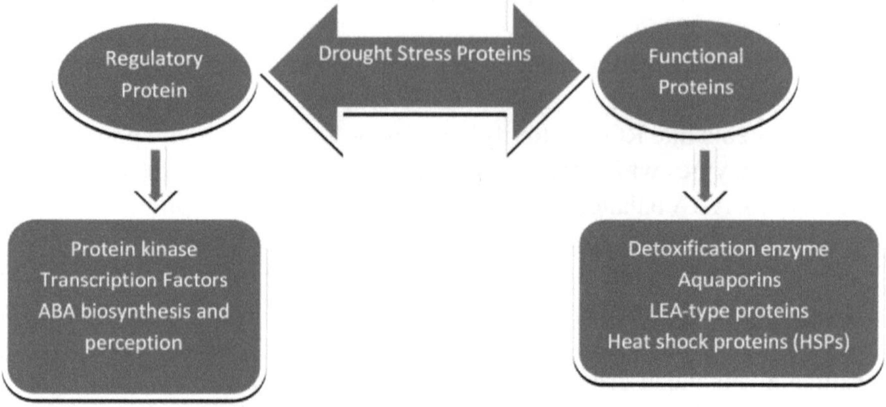

FIGURE 8.2 Drought-inducible proteins in stress tolerance.

Drought stress increases expression of genes important for osmolyte synthesis, aquaporins, and TFs, late embryogenesis abundant (LEA) proteins (Bhargava and Sawant, 2013). Water stress changed the expression of dehydrin/LEA genes and molecular chaperones, which are involved in proper protein folding and avoid protein denaturation in plant cell (Mahajan and Tuteja, 2005). LEA proteins would be useful in the prevention of aggregation of other proteins induced by drought stress (Veeranagamallaiah et al., 2011). LEA proteins are responsible for increased drought tolerance in plants (Juszczak and Bartels, 2017).

HSPs and LEA are main stress-induced proteins under drought stress which protect macromolecules, for example enzymes, lipids, and mRNAs from dehydration. HSPs or molecular chaperones regulates protein folding and assembly, as functions of protein depends on its structure, they also remove proteins which are nonfunctional due to its disturbed structure (Wang et al., 2004). Lee et al. (1995) proved that HSPs prevent thermal aggregation of proteins, so cell functions are regulated.

Fractional reduction of atmospheric O_2 results in the formation of ROS also known as active reactive oxygen intermediates (ROI) (Khan et al., 2018). ROS production is induced by various components of cells like nucleus, chloroplast, mitochondria, plasma membrane, cell wall, etc. (Gill and Tuteja, 2010). Drought and other abiotic stresses result in the production of ROS, which causes oxidation of lipids, proteins, DNA, and hence cell senescence and cell death is fasten (Akula and Ravishankar, 2011; Fang and Xiong, 2015). Plants have complex scavenging mechanisms and checking pathways to identify ROS redox homeostasis to neglect extra ROS in cells. In response to various stresses, plants accumulate antioxidant compounds like flavonoids, alkaloids, terpenoids, and brassinosteroids for avoiding cell death (Nakabayashi and Saito, 2015; Choudhury et al., 2017).

8.7 DROUGHT RESISTANCE RESPONSE

The capability of a plant to identify the water scarcity signal and initiate surviving strategies in response is called drought resistance. Plants have developed multiple strategies to respond and to face different stages of drought stress involving biochemical, physiological, and morphological modifications (Maximiliano et al., 2020). Drought resistance is a complex feature that endures through several mechanisms (Figure 8.3), such as drought escape, i.e., speeding up plant reproductive phase before harmful effects of stress on survival, drought avoidance, i.e., persistence with amplified water level and inhibition of cell and tissue damage, and tolerance, i.e., survival with less water level, sustaining growth over drought stress condition (Basu et al., 2016). Plants can employ these mechanisms while facing drought stress consecutively or simultaneously.

8.7.1 DROUGHT ESCAPE

Drought escape is strategy of completing life cycle of plant before onset of severe water deficit (Izanloo et al., 2008). Short-lived desert plants, also called ephemerals, escape drought by quick vegetative and reproductive growth, and dormant seeds are produced before the commencement of the dry season. These plants are also called drought escapers as they escape drought condition by germinating, growing, and flowering very quickly after rains. Escape strategies depend on early flowering (Tekle and Alemu, 2016), successful reproduction, better division of assimilates to developing fruits and seeds before drought. Certain plants store food material in plant organs like stems and roots and to mobilize them for fruit production, it has been reported in cereals (Bruce et al., 2002) as well as some leguminous plants (Chaves et al., 2002). During evolution, plants have developed adaptation to adjust growth duration to avoid water stress (Manavalan et al., 2009; Alvarez et al., 2018).

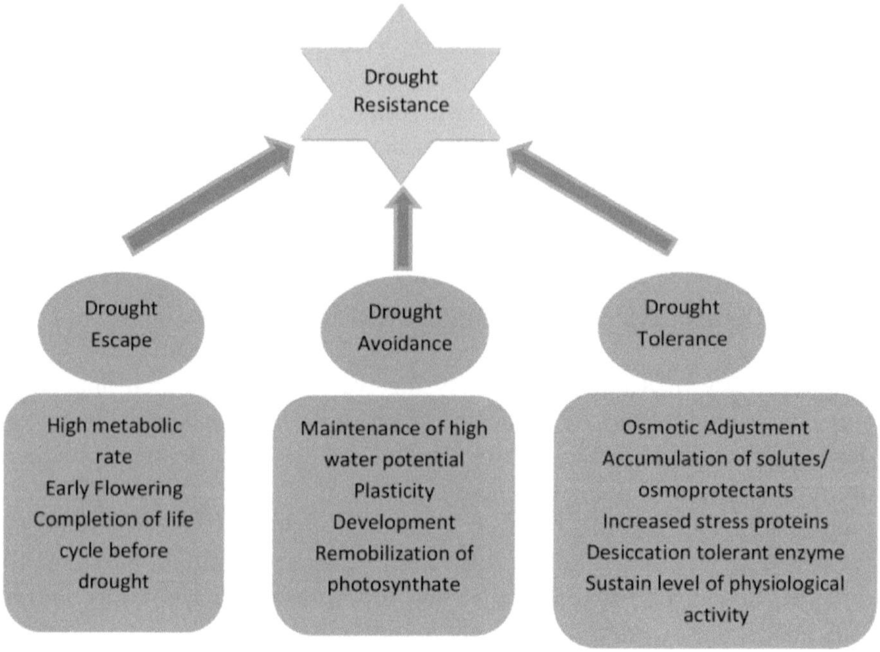

FIGURE 8.3 Drought resistance response mechanism in plants.

For drought escape very high metabolic rate is maintained, ensuring advanced cell expansion and rapid cell division. Stomata are kept open for high gas exchange rates leads to effective photosynthesis and photo-respiration and fast growth and development of plants (Bodner et al., 2015; Kooyers, 2015). Drought escape was designated as a standard tool for ephemeral local perennial plants with bulbs and shrubs during evolution (Dolferus, 2014; Kooyers, 2015) and in some annual grasses. In some plants, length of growing period and plant productivity is affected (Blum, 2011). Various different species of plant may accelerate post-anthesis development in harmful drought stress (Kottmann et al., 2016). Drought escape is an important strategy beneficial for yield stability and improvement of crops like wheat when drought is expected (Khan et al., 2018).

8.7.2 DROUGHT AVOIDANCE

Drought avoidance is supporting important physiological mechanisms such as closing of stomata while facing slight drought. It is achieved by enhanced water-conserving mechanism via stomatal closure (Dobra et al., 2020) and reduction of leaf area (Yu et al., 2017). Some plants avoid tissue dehydration, so avoid drought by regulating water uptake or minimizing water loss. Dehydration avoidance is common to annuals and perennials plants and accompanying with many adaptive traits. For increasing water absorption, root architecture and density is increased so that can penetrate deeper in soil (Kumawat and Sharma, 2018). In succulents, water is stored in plant tissue to face water scarcity condition hence drought avoidance is also known as 'succulent strategy' (Fang and Xiong, 2015). It is mainly based on policies of avoiding low water level, confirming continuous water absorption, and reducing water loss by various processes (Maximiliano et al., 2020). Drought avoidance involves producing barriers to reduces water evaporation and transpiration (Aharoni et al., 2004), and reductions on cell division and elongation to decrease leaf area of plants exposed to dry environment (Marron et al., 2003) and closing of stomatal to check transpiration through leaves (Choat et al., 2018).

Drought stress affects shoot conductivity by bringing embolism in xylem vessels (Sperry, 2003) or by decreasing vessel diameter (Lovisolo and Schubert, 1998). Xylem vessels with contracted diameters are less vulnerable to embolism. Plants facing drought stress increase or maintain the level of protection against cavitation and transport water more efficiently through xylem, revealing a balance between low vulnerability to cavitation and large hydraulic conductivity (Bacelar et al., 2007). Drought avoidance also includes reversible physiological changes like declining light absorption by reduced leaf area by rolling leaves and compact leaf architecture, transportation of nutrients from mature leaves, etc. (Chaves et al., 2003). In some plants, xeromorphic characteristics like the presence of hairy leaves and cuticles help to check water loss and maintain high water potentials in tissues (Boulard et al., 2017). Nevertheless, overdevelopment of these structures may reduce plant productivity and plant size (Wasaya et al., 2018). Decrease in leaf area is a drought avoidance strategy as it leads to decreased water loss and by maintaining less leaf area by reduced rate of cell division, ultimately resulting in loss of turgidity in cell (Deka et al., 2018).

8.7.3 DROUGHT TOLERANCE

Drought tolerance is the capability of plants to overcome severe dehydration through osmotic adjustment, osmoprotectants, and increased cell wall elasticity (Luo et al., 2010; Yi et al., 2016). If mitigation of drought stress is not possible with drought avoidance mechanisms, then drought tolerance mechanisms are activated by plants to protect tissues against damage. It is done by initiation of components to protect cell components from dehydration, processes to promote osmotic adjustment and detoxification of ROS (Du et al., 2009; Claeys and Inzé, 2013). Drought tolerance is associated with various biochemical, morpho-physiological, and molecular processes (Khan et al., 2018).

Drought tolerance is a vital strategy to face drought stress (Connor, 2005) with adaptations to face drought stress, devoid of fall in plant performance. It include mechanisms such as defense mechanism against antioxidant, accumulation of vital solutes, regulation of stomatal closure and enhanced root shoot ratio to tolerate negative effect caused due to drought condition (Lopez-Galiano et al., 2019). Plants under drought stress produce trichomes on leaves to tolerate water scarcity by reducing evaporation of water from the leaf surface (Zhang et al., 2019). Such structures reduce the temperature of leaf by enhancing light reflection from leaves and act as another extra layer to reduce water loss (Tiwari et al., 2020).

Osmotic adjustment can include an increase in the production of solutes like sugars, organic acids, and ions, mainly K^+ which are stored in mainly in cell vacuoles. Compatible osmolytes are organic compounds that don't interfere with enzyme functions but protect the cell membrane and components of metabolism processes in dry condition (Chaves et al., 2003). In plants under stress, osmotic compounds are synthesized viz. glutamic and aspartic acid, amino acids, e.g., proline (Bacelar et al., 2009), as well as quaternary ammonium compounds, etc. (Rathinasabapathi et al., 2001).

The level of drought tolerance ability of a plant is correlated with effectiveness of drought-adaptation processes as per genome (Tang et al., 2010). Some genes help the plants to achieve drought tolerance such as TaNAC69 in wheat (Budak et al., 2013), NF-YB2 in Maize (Vandenbroucke and Metzlaff, 2013), SodEFF3 in Sugarcane (Trujillo et al., 2008) and OSNAC10 in Rice (Oh et al., 2013). In rice crop ethylene response factor JERF1 is helpful to develop drought resistance (Zhang et al., 2010). Genotypes accumulating higher levels of glycine betaine (Virmani and Ilyas, 2007), mannitol, and other non-structural carbohydrates (Bardgett and van der, 2014) have more drought tolerance. Similarly, trehalose under drought stress helps to stabilize macromolecules like lipids, protein, etc., to enhance photosynthetic functioning, thus conferring drought tolerance to plants (Rasmussen et al., 2019). Increased antioxidant defenses are helpful to enhance drought tolerance and provide protection against oxidative stress caused as a result of water deficit conditions (Kim et al., 2020).

8.8 CONCLUSION

In recent climatic changes, drought stress is most severe hazard for worldwide food security and sustainability of production in plants. Drought stress affects different phases in the plant life cycle from embryo phase to reproductive and maturity phases. Drought stress alters specific physiological, biochemical as well as metabolic mechanisms in plants thus vegetative and reproductive growth and development, yield of plants is adversely affected. Morphological responses to drought stress include reduced plant growth rate, decreased shoot length, increased root length, decreased number of tillers, decreased leaf area and number, etc. Physiological responses of plants include decreased leaf water potential, inhibition of osmotic adjustment, decrease in stomatal conductance, decline in net photosynthesis, etc. Biochemical responses exhibited by plants under drought stress include an increase in antioxidative enzyme, transient decrease in photochemical efficiency, reduced ROS accumulation, decreased efficiency of Rubisco, accumulation of stress metabolite, etc. Molecular responses to drought stress consist of amplified expression in ABA biosynthetic gene, Synthesis of specific proteins, Stress-responsive gene expression, etc. Drought stress causes stress-initiated alterations in metabolic, biochemical, and physiological processes leading to stunted vegetative growth, underdeveloped plant reproductive organs and plant productivity.

Drought stress influences photosynthetic machinery, disturbs different pathways resulting in declined rate of photosynthesis in plants. Drought stress causes stomatal closure, reduces transpiration and photosynthetic rates consequently resulting in earlier crop maturity and poor yield. To curtail adverse impacts of drought stress, plants show various signaling pathways changing growth pattern, etc. Plants also show up-regulation of antioxidants, accretion of compatible solutes, and production of various stress induced proteins and molecular chaperones. The adverse effects are also tackled by adopting certain mechanisms which increase drought resistance in plants. Plants can adapt drought resistance mechanisms like drought escape, drought avoidance and drought tolerance in response to drought stress separately or concurrently. The drought tolerance strategies accepted by plants comprise numerous mechanisms at various levels like cell, tissue, organ, and whole plant at various stages of plant growth. Great effort has to be made to reveal the stress response pathways, comprising of interpretation of role and characterization of various genes leading to drought stress tolerance. Improved understanding of the molecular basis of plant responses to drought stress and their associated growth adjustments will be helpful to increase plant sustenance and productivity under drought stress.

Certain potential promising strategies to acquire higher drought tolerance plants include selection of plant species with increased water use efficiency, better antioxidant system, and production of vital osmolytes, secondary metabolites, etc. Furthermore, exogenous supply of compounds promoting drought tolerance in plants can be exploited in water stress conditions. In the future, the performance of plants under drought stress needs to be studied by using conditional promoter driving gene expression, specifically targeting developmental stages. It will be helpful in producing possible generation of transgenic crops with the ability to grow under water scarcity conditions without major losses in yield. It is essential to incorporate various branches like plant physiology, genomics, and plant breeding tactics for improved understanding of multifarious characters, molecular mechanisms of drought tolerance to develop the plant varieties which can survive in severe drought and in global changing climate.

KEYWORDS

- **abscisic acid**
- **heat-shock proteins**
- **late embryogenesis abundant**
- **reactive oxygen intermediates**
- **reactive oxygen species**
- **transcription factors**

REFERENCES

Abclulrahaman, A., & Oladele, E., (2011). Response of trichomes to water stress in two species of *Jatropha. Insight Bot., 1*, 15–21.

Abdallah, M. B., Trupiano, D., Polzella, A., De Zio, E., Sassi, M., Scaloni, A., Zarrouk, M., Youssef, N. B., & Scippa, G. S., (2018). Unraveling physiological, biochemical and molecular mechanisms involved in olive (*Olea europaea* L. cv. Chétoui) tolerance to drought and salt stresses. *J. Plant Physiol., 220*, 83–95.

Abid, M., Ali, S., Qi, L. K., et al., (2018). Physiological and biochemical changes during drought and recovery periods at tillering and jointing stages in wheat (*Triticum aestivum* L.). *Sci Rep., 8*, 4615.

Afzal, Z., Howton, T. C., Sun, Y., & Mukhtar, M. S., (2016). The roles of aquaporins in plant stress responses. *J. Dev. Biol., 9*, 1–22.

Agarwal, P. K., Agarwal, P., Reddy, M. K., & Sopory, S. K., (2006). Role of DREB transcription factors in abiotic and biotic stress tolerance in plants. *Plant Cell Rep., 25*, 1263–1274.

Aharoni, A., Dixit, S., Jetter, R., Thoenes, E., Van, A. G., & Pereira, A., (2004). The SHINE clade of AP2 domain transcription factors activates wax biosynthesis, alters cuticle properties, and confers drought tolerance when overexpressed in *Arabidopsis*. *The Plant Cell, 16*, 2463–2480.

Akram, R., Fahad, S., Masood, N., Rasool, A., Ijaz, M., Ihsan, M. Z., & Nasim, W., (2019). Plant growth and morphological changes in rice under abiotic stress. *Advances in Rice Research for Abiotic Stress Tolerance*, 69–85.

Akula, R., & Ravishankar, G. A., (2011). Influence of abiotic stress signals on secondary metabolites in plants. *Plant Signaling & Behavior, 6*, 1720–1731.

Albert, K. R., Mikkelsen, T. N., Michelsen, A., Ro-Poulsen, H., & Van, D. L. L., (2011). Interactive effects of drought, elevated CO_2 and warming on photosynthetic capacity and photosystem performance in temperate heath plants. *J Plant Physiol., 168*(13), 1550–1561.

Ali, F., Bano, A., & Fazal, A., (2017). Recent methods of drought stress tolerance in plants. *Plant Growth Regul., 82*, 363–375.

Ali, S., Xu, Y., Jia, Q., Ahmad, I., Wei, T., Ren, X., Zhang, P., Din, R., Cai, T., & Jia, Z., (2018). Cultivation techniques combined with deficit irrigation improves winter wheat photosynthetic characteristics, dry matter translocation and water use efficiency under simulated rainfall conditions. *Agric. Water Manag., 201*, 207–218.

Alvarez, S., Rodríguez, P., Broetto, F., & Sánchez-Blanco, M. J., (2018). Long term responses and adaptive strategies of *Pistacia lentiscus* under moderate and severe deficit irrigation and salinity: Osmotic and elastic adjustment, growth, ion uptake and photosynthetic activity. *Agric. Water Manag., 202*, 253–262.

Anjum, S. A., Xie, X. Y., Wang, L. C., Saleem, M. F., Man, C., & Lei, W., (2011). Morphological, physiological and biochemical responses of plants to drought stress. *Afr. J. Agric. Res., 6*, 2026–2032.

Ashkavand, P., Zarafshar, M., Tabari, M., Mirzaie, J., Nikpour, A., Kazem, B. S., et al., (2018). Application of SIO_2 nanoparticles AS pretreatment alleviates the impact of drought on the physiological performance of *Prunus mahaleb* (*Rosaceae*). *Boletin de la Sociedad Argentina de Botanica, 53*, 207.

Atkinson, N. J., & Urwin, P. E., (2012). The interaction of plant biotic and abiotic stresses: From genes to the field. *J. Exp. Bot., 63*(10), 3523–3543.

Bacelar, E. A., Moutinho-Pereira, J. M., Goncalves, B. C., Lopes, J. I., & Correia, C. M., (2009). Physiological responses of different olive genotypes to drought conditions. *Acta Physiol. Plant, 70831*(3), 611–621.

Bacelar, E. A., Santos, D. L., Moutinho-Pereira, J. M., Lopes, J. I., Gonçalves, B. C., Ferreira, T. C., & Correia, C. M., (2007). Physiological behavior, oxidative damage and antioxidative protection of olive trees grown under different irrigation regimes. *Plant Soil, 292*, 1–12.

Bano, A., Ullah, F., & Nosheen, A., (2012). Role of abscisic acid and drought stress on the activities of antioxidant enzymes in wheat. *Plant Soil Environ., 58*(4), 181–185.

Bao, G., Zhuo, C., Qian, C., Xiao, T., Guo, Z., & Lu, S., (2016). Co-expression of NCED and ALO improves vitamin C level and tolerance to drought and chilling in transgenic tobacco and stylo plants. *Plant Biotechnol. J., 14*(1), 206–214.

Bardgett, R. D., & Van, D. P. W. H., (2014). Belowground biodiversity and ecosystem functioning. *Nature, 515*, 505–511.

Barlow, P. W., (2003). The root cap: Cell dynamics, cell differentiation and cap function. *J. Plant Growth Regul., 21*, 261–286.

Barnabás, B., Jäger, K., & Fehér, A., (2008). The effect of drought and heat stress on reproductive processes in cereals. *Plant Cell Environ., 31*, 11–38.

Basnayake, J., Fukai, S., & Ouk, M., (2006). Contribution of potential yield, drought tolerance and escape to adaptation of 15 rice varieties in rainfed lowlands in Cambodia. In: *Proceedings of the Australian Agronomy Conference* (pp. 10–16). Australian Society of Agronomy: Brisbane, Australia.

Basu, S., Ramegowda, V., Kumar, A., Pereira, A., (2016). Plant Adaptation to Drought Stress. F1000Research, *5*, 1–10.

Bertioli, D. J., Cannon, S. B., Froenicke, L., Huang, G., Farmer, A. D., & Ozias-Akins, P., (2016). The genome sequences of *Arachis duranensis* and *Arachis ipaensis*, the diploid ancestors of cultivated peanut. *Nat. Genet., 48*(4), 438–446.

Bhargava, S., & Sawant, K., (2013). Drought stress adaptation: Metabolic adjustment and regulation of gene expression. *Plant Breed., 132*, 21–32.

Bhargavi, B., Kalpana, K., & Reddy, J. K., (2017). Influence of water stress on morphological and physiological changes in *Andrographis paniculata. Int. J. Pure Appl. Biosci., 5*, 1550–1556.

Bhatt, R. M., & Rao, N. K. S., (2005). Influence of pod load on response of okra to water stress. *Indian J. Plant Physiol., 10*, 54–59.

Bhattacharjee, S., & Saha, A. K., (2014). Plant water-stress response mechanisms. In: Gaur, R. K., & Sharma, P., (eds.), *Approaches to Plant Stress and their Management* (pp. 149–172). Springer India.

Bielach, A., Hrtyan, M., & Tognetti, V. B., (2017). Plants under stress: Involvement of auxin and cytokinin. *Int. J. Mol. Sci., 18*, 1427.

Blum, A., (2011). Plant water relations, plant stress and plant production. In: *Plant Breeding for Water-Limited Environments* (pp. 11–52). Springer: Berlin/Heidelberg, Germany.

Bodner, G., Nakhforoosh, A., & Kaul, H. P., (2015). Management of crop water under drought: A review. *Agron. Sustain. Dev., 35*, 401–442.

Boulard, T., Roy, J. C., Pouillard, J. B., Fatnassi, H., & Grisey, A., (2017). Modelling of micrometeorology, canopy transpiration and photosynthesis in a closed greenhouse using computational fluid dynamics. *Biosyst. Eng., 158*, 110–133.

Bray, E. A., Bailey-Serres, J., & Weretilnyk, E., (2000). Responses to abiotic stresses. In: Gruissem, W., Buchannan, B., & Jones, R., (eds.), *Biochemistry and Molecular Biology of Plants* (pp. 1158–1249). American Society of Plant Physiologists, Rockville.

Brito, C., Dinis, L. T., Moutinho-Pereira, J., & Correia, C. M., (2019). Drought stress effects and olive tree acclimation under a changing climate. *Plants, 8*, 232.

Brodribb, T. J., & McAdam, S. A. M., (2013). Abscisic acid mediates a divergence in the drought response of two conifers. *Plant Physiol., 162*, 1370–1377.

Brown, L. K., George, T. S., Thompson, J. A.,Wright, G., Lyon, J., Dupuy, L., Hubbard, S. F., & White, P. J., (2012). What are the implications of variation in root hair length on tolerance to phosphorus deficiency in combination with water stress in barley (*Hordeum vulgare*)? *Ann. Bot., 110*, 319–328.

Bruce, W. B., Edmeades, G. O., & Barker, T. C., (2002). Molecular and physiological approaches to maize improvement for drought tolerance. *J. Exp. Bot., 53*, 13–25.

Bucker-Neto, L., Paiva, A. L. S., Machado, R. D., Arenhart, R. A., & Margis-Pinheiro, M., (2017). Interactions between plant hormones and heavy metals responses. *Genet. Mol. Biol., 40*, 373–386.

Budak, H., Kantar, M., & Yucebilgili, K. K., (2013). Drought tolerance in modern and wild wheat. *Sci. World. J., 4*, 66–75.

Cairns, J. E., Sonder, K., Zaidi, P. H., Verhulst, N., Mahuku, G., Babu, R., et al., (2012). Maize production in a changing climate. In: *Advances in Agronomy* (pp. 1–58). Elsevier.

Cakir, R., (2004). Effect of water stress at different development stages on vegetative and reproductive growth of corn. *Field Crop Res., 89*, 1–16.

Carmo-Silva, A. E., Keys, A. J., Andralojc, P. J., Powers, S. J., Arrabaca, M. C., & Parry, M. A., (2010). Rubisco activities, properties, and regulation in three different C4 grasses under drought. *J. Exp. Bot., 61*, 2355–2366.

Carvalho, M. H. C. D., (2008). Drought stress and reactive oxygen species. *Plant Signal Behav., 3*, 156–165.

Chaves, M. M., Maroco, J. P., & Pereira, J. S., (2003). Understanding plant responses to drought—From genes to the whole plant. *Functional Plant Biology, 30*(3), 239.

Chen, X., & Hao, M. D., (2015). Low contribution of photosynthesis and water-use efficiency to improvement of grain yield in Chinese wheat. *Photosynthetica, 53*, 519–526.

Chen, X., Qiu, L., Guo, H., Wang, Y., Yuan, H., Yan, D., & Zheng, B., (2017a). Spermidine induces physiological and biochemical changes in southern high bush blueberry under drought stress. *Braz. J. Bot., 40*(4), 841–851.

Chen, Y., Zhang, C., Su, Y., Ma, J., Zhang, Z., Yuan, M., Zhang, H., & Yuan, S., (2017b). Responses of photosystem II and antioxidative systems to high light and high temperature co-stress in wheat. *Environ Exp. Bot., 135*, 45–55.

Choat, B., Brodribb, T. J., Brodersen, C. R., Duursma, R. A., López, R., & Medlyn, B. E., (2018). Triggers of tree mortality under drought. *Nature, 558*, 531–539.

Choudhury, F. K., Rivero, R. M., Blumwald, E., & Mittler, R., (2017). Reactive oxygen species, abiotic stress and stress combination. *The Plant Journal, 90*, 856–867.

Chowdhury, J., Karim, M., Khaliq, Q., & Ahmed, A., (2017). Effect of drought stress on biochemical change and cell membrane stability of soybean genotypes. *Bangladesh J. Agric. Res., 42*, 475–485.

Claeys, H., & Inzé, D., (2013). The agony of choice: How plants balance growth and survival under water-limiting conditions. *Plant Physiology, 162*, 1768–1779.

Comas, L. H., Anderson, L. J., Dunst, R. M., Lakso, A. N., & Eissenstat, D. M., (2005). Canopy and environmental control of root dynamics in a long-term study of concord grape. *New Phytol., 167*, 829–840.

Cong, L., Chai, T. Y., & Zhang, Y. X., (2008). Characterization of the novel gene BjDREB1B encoding a DRE-binding transcription factor from *Brassica juncea* L. *Biochem. Biophy. Res. Commun., 371*, 702–706.

Connor, D. J., (2005). Adaptation of olive (*Olea europaea* L.) to water-limited environments. *Aust. Agric. Res., 56*, 1181–1189.

Corso, D., Delzon, S., Lamarque, L. J., Cochard, H., Torres-Ruiz, J. M., King, A., & Brodribb, T., (2020). Neither xylem collapse, cavitation, or changing leaf conductance drive stomatal closure in wheat. *Plant Cell Env., 43*, 854–865.

De Willigen, P., Nielsen, N., Claassen, N., & Castrignanò, A., (2000). Modelling water and nutrient uptake. In: *Root Methods* (pp. 509–543). Springer: Morris, TX, USA.

Deka, D., Singh, A. K., & Singh, A. K., (2018). Effect of drought stress on crop plants with special reference to drought avoidance and tolerance mechanisms: A review. *Int. J. Curr. Microbiol. App. Sci., 7*, 2703–2721.

Den Herder, G., Van, I. G., Beeckman, T., & De Smet, I., (2010). The roots of a new green revolution. *Trends Plant Sci., 15*, 600–607.

Dobra, J., Motyka, V., Dobrev, P., Malbeck, J., Prasil, I. T., Haisel, D., Gaudinova, A., et al., (2010). Comparison of hormonal responses to heat, drought, and combined stress in tobacco plants with elevated proline content. *J. Plant Physiol., 167*, 1360–1370.

Dolferus, R., (2014). To grow or not to grow: A stressful decision for plants. *Plant Sci., 229*, 247–261.

Du, F. K., Petit, R. J., & Liu, J. Q., (2009). More introgression with less gene flow: Chloroplast vs. mitochondrial DNA in the *Picea asperata* complex in China, and comparison with other conifers. *Molecular Ecology, 18*, 1396–1407.

Du, Y., Zhao, Q., Chen, L., Yao, X., Zhang, W., Zhang, B., & Xie, F., (2020). Effect of drought stress on sugar metabolism in leaves and roots of soybean seedlings. *Plant Physioly Biochem., 146*, 1–12.

Eapen, D., Barroso, M. L., Ponce, G., Campos, M. E., & Cassab, G. I., (2005). Hydrotropism: Root growth responses to water. *Trends Plant Sci., 10*, 44–50.

Efeoglu, B., & Terzioglu, S., (2009). Photosynthetic responses of two wheat varieties to high temperature. *Eur. Asia J. BioSci., 3*, 97–106.

Fahad, S., Ullah, A., Ali, U., Ali, E., Saud, S., Hakeem, K., Alharby, H., Sabagh, A., Barutcular, C., Kamran, M., et al., (2019). Drought tolerance in plants role of phytohormones and scavenging system of ROS. In: Hasanuzzaman, M., Fujita, M., Oku, H., & Tofazzal, I. M.,(eds.), *Plant Tolerance to Environmental Stress Role of Phytoprotectants* (pp. 1–12). CRC Press: Boca Raton, FL, USA.

Fang, L., Su, L., Sun, X., Li, X., Sun, M., Karungo, S. K., Fang, S., et al., (2016). Expression of *Vitis amurensis* NAC26 in *Arabidopsis* enhances drought tolerance by modulating jasmonic acid synthesis. *J. Exp. Bot., 67*, 2829–2845.

Fang, X., Turner, N. C., Yan, G., Li, F., & Siddique, K. H. M., (2010). Flower numbers, pod production, pollen viability, and pistil function are reduced and flower and pod abortion increased in chickpea (*Cicer arietinum* L.) under terminal drought. *J. Exp. Bot., 61*, 335–345.

Fang, Y., & Xiong, L., (2015). General mechanisms of drought response and their application in drought resistance improvement in plants. *Cell. Mol. Life Sci., 72*, 673–689.

Farooq, M., Wahid, A., Kobayashi, N., Fujita, D., & Basra, S. M. A., (2009). Plant drought stress: Effects, mechanisms and management. *Agron. Sustain. Dev., 29*, 185–212.

Fathi, A., & Tari, D. B., (2016). Effect of drought stress and its mechanism in plants. *Int. J. of Life Sci., 10*(1), 1–6.

Foster, B., Thomas, W., & Chloupek, O., (2005). Genetic controls of barley root syste4ms and their associations with plant performance. *Aspects Appl Biol., 73*, 199–204.

Frederick, J. R., Camp, C. R., & Bauer, P. J., (2001). Drought stress effects on branch and main stem seed yield and yield components of determinate soybean. *Crop Sci., 41*, 759–763.

Fulda, S., Mikkat, S., Stegmann, H., & Horn, R., (2011). Physiology and proteomics of drought stress acclimation in sunflower (*Helianthus annuus* L.). *Plant Biol (stuttg)., 13*, 632–642.

Furlan, A., Llanes, A., Luna, V., & Castro, S., (2012). Physiological and biochemical responses to drought stress and subsequent rehydration in the symbiotic association peanut *Bradyrhizobium* sp. *ISRN Agron.,* 1–8.

Gai, Z., Wang, Y., Ding, Y., Qian, W., et al., (2020). Exogenous abscisic acid induces the lipid and flavonoid metabolism of tea plants under drought stress. *Sci. Rep., 10*, 12275.

Galeano, E., Vasconcelos, T. S., Novais De, O. P., & Carrer, H., (2019). Physiological and molecular responses to drought stress in teak (*Tectona grandis* L.F.). *PLoS One, 14*(9), 0221571.

Ghadge, S. A., Naikwade, P. V., & Jadhav, B. B., (2013). Utilization of problematic weed for improved yield of fenugreek, *Ind. Str. Res. J., 3*(4), 21–28.

Gheidary, S., Akhzari, D., & Pessarakli, M., (2017). Effects of salinity, drought, and priming treatments on seed germination and growth parameters of *Lathyrus sativus* L. *J. Plant Nutr., 40*, 1507–1514.

Ghobadi, M., Taherabadi, S., Ghobadi, M. E., Mohammadi, G. R., & Jalali-Honarmand, S., (2013). Antioxidant capacity, photosynthetic characteristics and water relations of sunflower (*Helianthus annuus* L.) cultivars in response to drought stress. *Ind. Crop. Prod., 50*, 29–38.

Gill, S. S., & Tuteja, N., (2010). Reactive oxygen species and antioxidant machinery in abiotic stress tolerance in crop plants. *Plant Physiol Biochem., 48*, 909–930.

Gleick, P. H., (2000). In: *The Worlds Water 2000–2001: The Biennial Report on Freshwater Resources* (p. 53–86). Island Press.

Guha, A., Sengupta, D., Kumar, & Ramachandra, R. A., (2010). An integrated diagnostic approach to understand drought tolerance in mulberry (*Morus indica* L.). *Flora Morphol. Distrib. Funct. Ecol. Plants., 205*, 144–151.

Guo, X. L., Ma, Y. Y., Liu, Z. H., & Liu, B. H., (2008). Effects of exterior abscisic acid on calcium distribution of mesophyll cells and calcium concentration of guard cells in maize seedlings. *Agric Sci China, 7*, 438–446.

Gupta, A., Rico-Medina, A., & Cano-Delgado, A. I., (2020). The physiology of plant responses to drought. *Science, 368*, 266–269.

Gurumurthy, S., Sarkar, B., Vanaja, M., Lakshmi, J., Yadav, S., & Maheswari, M., (2019). Morpho-physiological and biochemical changes in black gram (*Vigna mungo* L. Hepper) genotypes under drought stress at flowering stage. *Acta Physiol. Plant, 41*, 42.

Hajiboland, R., Cheraghvareh, L., & Poschenrieder, C., (2017). Improvement of drought tolerance in tobacco (*Nicotiana rustica* L.) plants by silicon. *J. Plant Nutr., 40*(12), 1661–1676.

Hartmann, H., Ziegler, W., Kolle, O., & Trumbore, S., (2013). Thirst beats hunger-declining hydration during drought prevents carbon starvation in Norway spruce saplings. *New Phytol., 200*(2), 340–349.

Hsiao, T. C., & Xu, L. K., (2000). Sensitivity of growth of root versus leaves to water stress; biophysical analysis and relation to water transport. *J. Exp. Bot., 51*, 1595–1616.

Hu, H., & Xiong, L., (2014). Genetic engineering and breeding of drought-resistant crops. *Annu. Rev. Plant Biol., 65*, 715–741.

Hu, L., Wang, Z., & Huang, B., (2010). Diffusion limitations and metabolic factors associated with inhibition and recovery of photosynthesis from drought stress in a C3 perennial grass species. *Physiol. Planta, 139*(1), 93–106.

Hui, D., Yu, C. L., Deng, Q., Dzantor, E. K., Zhou, S., Dennis, S., Sauve, R., et al., (2018). Effects of precipitation changes on switchgrass photosynthesis, growth, and biomass: A mesocosm experiment. *PLoS One, 13*(2), 0192555.

Hussain, H. A., Hussain, S., Khaliq, A., Ashraf, U., Anjum, S. A., Men, S., & Wang, L., (2018). Chilling and drought stresses in crop plants: Implications, cross talk, and potential management opportunities. *Front. Plant Sci., 9*, 393.

Intergovernmental Panel on Climate Change (IPCC), Climate change 2007: Mitigation, (2007). *Contribution of Working Group III to the Fourth Assessment Report of the Intergovernmental Panel on Climate Change*. Cambridge University Press, Cambridge.

Izanloo, A., Condon, A. G., Langridge, P., Tester, M., & Schnurbusch, T., (2008). Different mechanisms of adaptation to cyclic water stress in two South Australian bread wheat cultivars. *J. Exp. Bot., 59*, 3327–3346.

Jaleel, C. A., Gopi, R., Sankar, B., Gomathinayagam, M., & Panneerselvam, R., (2008). Differential responses in water use efficiency in two varieties of *Catharanthus roseus* under drought stress. *Comp. Rend. Biol., 331*, 42–47.

Juszczak, I., & Bartels, D., (2017). LEA gene expression, RNA stability and pigment accumulation in three closely related *Linderniaceae* species differing in desiccation tolerance. *Plant Sci., 255*, 59–71.

Kaashyap, M., Ford, R., Kudapa, H., Jain, M., Edwards, D., Varshney, R., & Mantri, N., (2018). Differential regulation of genes involved in root morphogenesis and cell wall modification is associated with salinity tolerance in chickpea. *Sci. Rep., 8*, 4855.

Kalaji, H. M., Jajoo, A., Oukarroum, A., Brestic, M., Zivcak, M., Samborska, I. A., Cetner, M. D., et al., (2016). Chlorophyll a fluorescence as a tool to monitor the physiological status of plants under abiotic stress conditions. *Acta Physiol Planta., 38*(4), 102.

Kamara, A. Y., Menkir, A., Badu–Apraku, B., & Ibikunle, O., (2003). The influence of drought stress on growth, yield and yield components of selected maize genotypes. *J. Agric. Sci., 141*, 43–50.

Kapoor, D., Bhardwaj, S., Landi, M., Sharma, A., Ramakrishnan, M., & Sharma, A., (2020). The impact of drought in plant metabolism: How to exploit tolerance mechanisms to increase crop production. *Appl. Sci., 10*, 1–19.

Karimpour, M., (2019). Effect of drought stress on RWC and chlorophyll content on wheat (*Triticum durum* L.) genotypes. *World. Ess. J., 7*, 52–56.

Kashiwagi, J., Krishnamurthy, L., Purushothaman, R., Upadhyaya, H. D., Gaur, P. M., Gowda, C. L. L., et al., (2015). Scope for improvement of yield under drought through the root traits in chickpea (*Cicer arietinum* L.). *Field Crop Res., 170*, 47–54.

Kaur, G., & Asthir, B., (2017). Molecular responses to drought stress in plants. *Biol. Plant., 61*, 201–209.

Kaur, N., & Gupta, A. K., (2005). Signal transduction pathways under abiotic stresses in plants. *Curr Sci., 88*(11), 1771–1780.

Kavar, T., Maras, M., Kidric, M., Sustar-Vozlic, J., & Meglic, V., (2007). Identification of genes involved in the response of leave of *Phaseolus vulgaris* to drought stress, *Mol. Breed., 21*, 159–172.

Khan, A., Pan, X., Najeeb, U., Tan, D. K., Fahad, S., Zahoor, R., & Luo, H., (2018). Coping with drought: Stress and adaptive mechanisms, and management through cultural and molecular alternatives in cotton as vital constituents for plant stress resilience and fitness. *Biol. Res., 51*, 47.

Khorsand, A., Rezaverdinejad, V., Hossein, A., Abolfazl, M. H., Amir, R., Sina, B., & Sadraddini, A. A., (2021). Linking plant and soil indices for water stress management in black gram. *Scientific Reports., 11*, 1–19.

Kim, T. H., (2014). Mechanism of ABA signal transduction: Agricultural highlights for improving drought tolerance. *J. Plant Biol., 57*, 1–8.

Kim, Y., Chung, Y. S., Lee, E., Tripathi, P., Heo, S., & Kim, K. H., (2020). Root response to drought stress in rice (*Oryza sativa* L.). *Int. J. Mol. Sci., 21*, 1–22.

Kim, Y., Seo, C. W., Khan, A. L., Mun, B. G., Shahzad, R., Ko, J. W., Yun, B. W., Park, S. K., & Lee, I. J., (2018). Exo-ethylene application mitigates waterlogging stress in soybean (*Glycine max* L.). *BMC Plant Biol., 18*, 254.

Kogler, F., & Soffker, D., (2017). Water (stress) models and deficit irrigation: System-theoretical description and causality mapping. *Ecol. Model., 361*, 135–156.

Kooyers, N. J., (2015). The evolution of drought escape and avoidance in natural herbaceous populations. *Plant Sci., 234*, 155–162.

Kottmann, L., Wilde, P., & Schittenhelm, S., (2016). How do timing, duration, and intensity of drought stress affect the agronomic performance of winter rye? *Eur. J. Agric., 75*, 25–32.

Krasensky, J., & Jonak, C., (2012). Drought, salt, and temperature stress-induced metabolic rearrangements and regulatory networks. *J. Exp. Bot., 63*, 1593–1608.

Kumar, S., Sachdeva, S., Bhat, K. V., & Vats, S., (2018). Plant responses to drought stress: Physiological, biochemical and molecular basis. In: Vats, S., (ed.), *Biotic and Abiotic Stress Tolerance in Plants*. Springer Nature Pte Ltd., Singapore.

Kumawat, K. R., & Sharma, N. K., (2018). Effect of drought stress on plants growth. *Popular Kheti, 6*, 239–241.

Lafitte, H., Ismail, A., & Bennett, J., (2004). Abiotic stress tolerance in rice for Asia: Progress and the future. *Proceedings of 4th International Crop Sci. Congress*. Brisbane, Australia.

Lang, N. T., & Bui, B. C., (2008). Fine mapping for drought tolerance in rice (*Oryza sativa* L.). *Omonrice, 16*, 9–15.

Larcher, W., (2003). *Physiological Plant Ecology: Ecophysiology and Stress Physiology of Functional Groups.* Springer, Netherlands.

Le Gall, H., Philippe, F., Domon, J. M., Françoise, G., Jérôme, P., & Rayon, C., (2015). Cell wall metabolism in response to abiotic stress. *Plants, 4*, 112–166.

Lee, G. J., Pokala, N., & Vierling, E., (1995). Structure and in vitro molecular chaperone activity of cytosolic small HSP from pea. *J. Biol. Chem., 270*, 10432–10438.

Li, C., Jiang, D., Wollenweber, B., Li, Y., Dai, T., & Cao, W., (2011). Waterlogging pretreatment during vegetative growth improves tolerance to waterlogging after anthesis in wheat. *Plant Sci., 180*, 672–678.

Lim, C. W., Baek, W., Jung, J., Kim, J. H., & Lee, S. C., (2015). Function of ABA in stomatal defense against biotic and drought stresses. *Intern. J. Mol. Sci., 16*, 15251–15270.

Lishan, S., Caihong, Y., Yi, L., Yanan, D., Dongmei, G., Zhenyin, L., Zhang, R., & Duan, G., (2015). Effects of drought stress on root physiological traits and root biomass allocation of *Reaumuria soongorica, Acta Ecologica Sinica, 35*(5), 155–159.

Liu, S., & Qin, F., (2021). Genetic dissection of maize drought tolerance for trait improvement. *Mol. Breeding, 41*(8), 1–13.

Liu, X., Li, L., Li, M., Su, L., Lian, S., Zhang, B., Li, X., et al., (2018). AhGLK1 affects chlorophyll biosynthesis and photosynthesis in peanut leaves during recovery from drought. *Sci Rep., 8*(1), 2250.

Lobell, D. B., Hammer, G. L., McLean, G., et al., (2013). The critical role of extreme heat for maize production in the United States. *Nat. Climate Change, 3*, 497–501.

Lopez-Galiano, M. J., García-Robles, I., González-Hernández, A. I., Camañes, G., Vicedo, B., Real, M. D., & Rausell, C., (2019). Expression of miR159 is altered in tomato plants undergoing drought stress. *Plants, 8*, 201.

Lovisolo, C., & Schubert, A., (1998). Effects of water stress on vessel size and xylem hydraulic conductivity in *Vitis vinifera* L. *J. Exp. Bot., 49*, 693–700.

Luo, L. J., (2010). Breeding for water-saving and drought-resistance rice (WDR) in China. *J. Exp. Bot., 61*, 3509–3517.

Mahajan, S., & Tuteja, N., (2005). Cold, salinity and drought stresses: An overview. *Arch. Biochem. Biophys., 444*, 139–158.

Manavalan, L. P., Guttikonda, S. K., Tran, L. S. P., & Nguyen, H. T., (2009). Physiological and molecular approaches to improve drought resistance in soybean. *Plant Cell Physiol., 50*, 1260–1276.

Marcinska, I., Czyczylo-Mysza, I., Skrzypek, E., Filek, M., Grzesiak, S., Grzesiak, M. T., Janowiak, F., et al., (2013). Impact of osmotic stress on physiological and biochemical characteristics in drought-susceptible and drought-resistant wheat genotypes. *Acta Physiol. Plant., 35*, 451–461.

Marron, N., Dreyer, E., Boudouresque, E., Delay, D., Petit, J. M., Delmotte, F. M., & Brignolas, F., (2003). Impact of successive drought and re-watering cycles on growth and specific leaf area of two *Populus × canadensis* (Moench) clones, 'Dorskamp' and 'Luisa_Avanzo'. *Tree Physiol., 23*, 1225–1235.

Maurel, C., Verdoucq, L., Luu, D. T., & Santoni, V., (2008). Plant aquaporins: Membrane channels with multiple integrated functions. *Annu. Rev. Plant Biol., 59*, 595–624.

Maximiliano, E. B., Mattera, M. G., Soliani, C., Bellora, N., Opgenoorth, L., Heer, K., & Arana, M. V., (2020). Molecular bases of responses to abiotic stress in trees. *J. of Exp. Bot., 71*(13), 3765–3779.

McDowell, N. G., & Sevanto, S., (2010). The mechanisms of carbon starvation: How, when, or does it even occur at all? *New Phytol., 186*(2), 264–266.

Mittler, R., & Blumwald, E., (2015). The roles of ROS and ABA in systemic acquired acclimation. *Plant Cell, 27*, 64–70.

Mohamed, M. F., Keutgen, N., Tawfika, A. A., & Noga, G., (2002). Dehydration-avoidance responses of tepary bean lines differing in drought resistance. *J. Plant Physiol., 159*, 31–38.

Mohammadian, M. A., Watling, J. R., & Hill, R. S., (2007). The impact of epicuticular wax on gas-exchange and photonhibition in *Leucadendron lanigerium* (Proteaceae). *Acta Oecol., 31*, 93–101.

Naikwade, P. V., (2017). Impact of climate change on agricultural production in India: Effect on rice productivity. *Biosci. Discov., 8*(4), 897–914.

Naikwade, P. V., (2019). Soil organic carbon sequestration by long-term application of manures prepared from *Trianthema portulacastrurm* Linn. *Comm.in Soil Science and Plant Analysis, 50*(20), 2579–2592.

Naikwade, P. V., (2020a). Effect of climate change on growth and productivity of cotton: Global scenario. *The Int. J. of Ana. and Exper. M. Ana., 12*(4), 64–80.

Naikwade, P. V., (2020b). Impact of climate change on potato production and mitigation measures. *O.H. Journ., 68*(38), 1296–1305.

Naikwade, P. V., (2021). Influence of *Celosia argentea* L. weed manure on the yield of spinach. *Bioinfolet., 18*(1A), 64–68.

Naikwade, P. V., Sankpal, S. T., & Jadhav, B. B., (2012). Management of waste by composting, vermicomposting and its use for improvement of growth, yield and quality of fodder maize, *ARPN J. of Sci. and Tech., 2*(5), 184–194.

Najla, S., Sanoubar, R., & Murshed, R., (2012). Morphological and biochemical changes in two parsley varieties upon water stress. *Physiol. Mol. Biol. Plants, 18*, 133–139.

Nakabayashi, R., & Saito, K., (2015). Integrated metabolomics for abiotic stress responses in plants. *Current Opinion in Plant Biology, 24*, 10–16.

Nakashima, K., Yamaguchi-Shinozaki, K., & Shinozaki, K., (2014). The transcriptional regulatory network in the drought response and its crosstalk in abiotic stress responses including drought, cold, and heat. *Front. Plant Sci., 5*, 1–7.

Noctor, G., Veljovic-Jovanovic, S., Driscoll, S., Novitskaya, L., & Foyer, C. H., (2002). Drought and oxidative load in the leaves of C3 plants: A predominant role for photorespiration? *Ann Bot., 89*, 841–850.

Obidiegwu, J. E., Bryan, G. J., Jones, H. G., & Prashar, A., (2015). Coping with drought: Stress and adaptive responses in potato and perspectives for improvement. *Front. Plant Sci., 6*, 542.

Oh, S. J., Kim, Y. S., Kwon, C. W., Park, H. K., Jeong, J. S., & Kim, J. K., (2013). Overexpression of the transcription factor AP37 in rice improves grain yield under drought conditions. *Plant Physiol., 150*, 1368–1379.

Osakabe, Y., Osakabe, K., Shinozaki, K., & Tran, L. P., (2014). Response of plants to water stress. *Plant Physiol., 5*, 1–8.

Osuna, J. L., Baldocchi, D. D., Kobayashi, H., & Dawson, T. E., (2015). Seasonal trends in photosynthesis and electron transport during the Mediterranean summer drought in leaves of deciduous oaks. *Tree Physiol., 35*(5), 485–500.

Pallardy, S. G., (2010). *Physiology of Woody Plants*. Academic Press, London.

Pettigrew, W. T., (2004). Physiological consequences of moisture deficit stress in cotton. *Crop Sci., 44*, 1265–1272.

Poorter, L., & Markesteijn, L., (2008). Seeding traits determine drought tolerance of tropical tree species. *Biotropica., 40*, 321–331.

Prabha, R., Ghosh, I., & Singh, D. P., (2011). Plant stress gene database: A collection of plant genes responding to stress condition. *ARPN J Sci Technol, 1*, 1.

Prasad, P. V. V., Pisipati, S. R., Momčilović, I., & Ristic, Z., (2011). Independent and combined effects of high temperature and drought stress during grain filling on plant yield and chloroplast EF-Tu expression in spring wheat. *J. Agron. Crop Sci., 197*, 430–441.

Pugnaire, F. I., Serrano, L., & Pardos, J., (1999). Constraints by water stress on plant growth. In: Pessarakli, M., (ed.), *Hand Book of Plant and Crop Stress* (2nd edn., pp. 271–283) Marcel Dekker, Inc., New York.

Rahmati, M., Mirás-Avalos, J. M., Valsesia, P., Lescourret, F., Génard, M., Davarynejad, G. H., Bannayan, M.,Azizi, M., & Vercambre, G., (2018). Disentangling the effects of water stress on carbon acquisition, vegetative growth, and fruit quality of peach trees by means of the quali tree model. *Front. Plant Sci., 9*, 3.

Rasmussen, P. U., Bennett, A. E., Tack, A. J. M., & Singh, B., (2019). The impact of elevated temperature and drought on the ecology and evolution of plant-soil microbe interactions. *J. Ecol.*

Rathinasabapathi, B., Fouad, W. M., & Sigua, C. A., (2001). B-alanine betaine synthesis in the 1003 *Plumbaginaceae*: Purification and characterization of a trifunctional, S-adenosyl-L-methio-1004 nine-dependent N-methyltransferase from *Limonium latifolium* leaves. *Plant Physiol., 1005126*, 1241–1249.

Razmjoo, K., Heydarizadeh, P., & Sabzalian, M. R., (2008). Effect of salinity and drought stresses on growth parameters and essential oil content of *Matricaria chamomile*. *Int. J. Agric. Biol., 10*, 451–454.

Reddy, A. R., Chaitanya, K. V., & Vivekanandan, M., (2004). Drought-induced responses of photosynthesis and antioxidant metabolism in higher plants. *J. Plant Physiol., 161*(11), 1189–1202.

Rivas, R., Falcão, H. M., Ribeiro, R. V., Machado, E. C., Pimentel, C., & Santos, M. G., (2016). Drought tolerance in cowpea species is driven by less sensitivity of leaf gas exchange to water deficit and rapid recovery of photosynthesis after rehydration. *S. Afr. J. Bot., 103*, 101–107.

Ruehr, N. K., Grote, R., Mayr, S., & Arneth, A., (2019). Beyond the extreme: Recovery of carbon and water relations in woody plants following heat and drought stress. *Tree Physiol.,* 39, 1285–1299.

Ryser, P., (2006). The mysterious root length. *Plant Soil, 286*, 1–6.
Salazar, C., Hernández, C., & Pino, M. T., (2015). Plant water stress: Associations between ethylene and abscisic acid response. *Chil. J. Agric. Res., 75*, 71–79.
Seki, M., Umezawa, T., Urano, K., & Shinozaki, K., (2007). Regulatory metabolic networks in drought stress responses. *Curr. Opin. Plant Biol., 10*, 296–302.
Seleiman, M. F., Al-Suhaibani, N., Ali, N., Akmal, M., Alotaibi, M., Refay, Y., Dindaroglu, T., et al., (2021). Drought stress impacts on plants and different approaches to alleviate its adverse effects. *Plants, 10*, 259.
Semerci, A., Cicek, N., Karahan, F. A., Özyürek, E., Arici, Y. K., & Ekmekci, Y., (2017). Some growth and chlorophyll fluorescence parameters of black and hybrid poplar clones under water stress. *Turk J. Agric., 41*(5), 348–356.
Sevanto, S., (2014). Phloem transport and drought. *J. Exp. Bot., 65*(7), 1751–1759.
Shaheen, T., Rahman, M., Shahid, R. M., Zafar, Y., & Rahman, M., (2016). Soybean production and drought stress. In: *Abiotic and Biotic Stresses in Soybean Production* (pp. 177–196). Elsevier.
Sharma, A., Kumar, V., Shahzad, B., et al., (2020). Photosynthetic response of plants under different abiotic stresses: A review. *J. Plant Growth Regul., 39*, 509–531.
Sharma, A., Thakur, S., Kumar, V., Kanwar, M. K., Kesavan, A. K., Thukral, A. K., Bhardwaj, R., Alam, P., & Ahmad, P., (2016). Pre-sowing seed treatment with 24-epibrassinolide ameliorates pesticide stress in *Brassica juncea* L. through the modulation of stress markers. *Front. Plant Sci., 7*, 1569.
Shi, H., Chen, L., Ye, T., Liu, X., Ding, K., & Chan, Z., (2014). Modulation of auxin content in *Arabidopsis* confers improved drought stress resistance. *Plant Physiol. Biochem., 82*, 209–217.
Shi, L., Wang, Z., & Kim, W. S., (2019). Effect of drought stress on shoot growth and physiological response in the cut rose 'charming black' at different developmental stages. *Hortic. Environ. Biotechnol., 60*, 1–8.
Sicher, R. C., & Barnaby, J. Y., (2012). Impact of carbon dioxide enrichment on the responses of maize leaf transcripts and metabolites to water stress. *Physiol. Planta, 144*, 238–253.
Singh, B. P., Jayaswal, P. K., Singh, B., Singh, P. K., Kumar, V., Mishra, S., Singh, N., et al., (2015). Natural allelic diversity in OsDREB1F gene in the Indian wild rice germplasm led to ascertain its association with drought tolerance. *Plant Cell Rep., 34*, 993–1004.
Singh, J., & Thakur, J. K., (2018). Photosynthesis and abiotic stress in plants. In: *Biotic and Abiotic Stress Tolerance in Plants* (pp. 27–46). Springer: Singapore.
Sperry, J. S., (2003). Evolution of water transport and xylem structure. *Int. J. Plant Sci. (Suppl.), 1051164*, S115–S127.
Srivastava, S., & Srivastava, M., (2014). Morphological changes and antioxidant activity of *Stevia rebaudiana* under water stress. *Am. J. Plant Sci., 5*, 3417.
Steppe, K., Sterck, F., & Deslauriers, A., (2015). Diel growth dynamics in tree stems: Linking anatomy and ecophysiology. *Trends Plant Sci., 20*(6), 335–343.
Sun, J., Gu, J., Zeng, J., Han, S., Song, A., Chen, F., Fang, W., Jiang, J., & Chen, S., (2013). Changes in leaf morphology, antioxidant activity and photosynthesis capacity in two different drought-tolerant cultivars of *Chrysanthemum* during and after water stress. *Sci. Hortic., 161*, 249–258.
Tahir, M. H. N., Imran, M., & Hussain, M. K., (2002). Evaluation of sunflower (*Helianthus annuus* L.) inbred lines for drought tolerance. *Int. J. Agric. Biol., 3*, 398–400.

Taiz, L., & Zeiger, E., (2006). *Plant Physiology* (4th edn.). Sunderland, MA: Sinauer Associates Inc.

Tandzi, L. N., Bradley, G., & Mutengwa, C., (2019). Morphological responses of maize to drought, heat and combined stresses at seedling stage. *J. of Bio. Sci., 19*(1), 7–16.

Tang, H., Sezen, U., & Paterson, A. H., (2010). Domestication and plant genomes. *Curr. Opin. Plant Biol., 13*, 160–166.

Tanveer, M., Shahzad, B., Sharma, A., & Khan, E. A., (2019). 24-epibrassinolide application in plants: An implication for improving drought stress tolerance in plants. *Plant Physiol Biochem., 135*, 295–303.

Tattini, M., Velikova, V., Vickers, C., Brunetti, C., Di Ferdinando, M., Trivellini, A., Fineschi, S., et al., (2014). Isoprene production in transgenic tobacco alters isoprenoid, nonstructural carbohydrate and phenylpropanoid metabolism, and protects photosynthesis from drought stress. *Plant Cell Environ., 37*(8), 1950–1964.

Tekle, A. T., & Alemu, M. A., (2016). Drought tolerance mechanisms in field crops. *World J. Biol. Med. Sci., 3*, 15–39.

Tiwari, P., Srivastava, D., Chauhan, A. S., Indoliya, Y., Singh, P. K., Tiwari, S., Fatima, T., Mishra, S. K., Dwivedi, S., Agarwal, L., et al., (2020). Root system architecture, physiological analysis and dynamic transcriptomics unravel the drought-responsive traits in rice genotypes. *Ecotoxicol. Environ. Safety, 207*, 111252.

Toscano, S., Ferrante, A., & Romano, D., (2019). Response of Mediterranean ornamental plants to drought stress. *Horticulturae, 5*, 6.

Tripathy, J., Zhang, J., Robin, S., Nguyen, T. T., & Nguyen, H., (2000). QTLs for cell-membrane stability mapped in rice (*Oryza sativa* L.) under drought stress. *Theor. Appl. Genet., 100*, 1197–1202.

Trujillo, L. E., Sotolongo, M., Menendez, C., Ochogavia, M. E., Coll, Y., Hernandez, I., & Hernandez, L., (2008). SodERF3, a novel sugarcane ethylene-responsive factor (ERF), enhances salt and drought tolerance when over expressed in tobacco plants. *Plant Cell Physiol., 49*, 512–525.

Tzortzakis, N., Chrysargyris, A., & Aziz, A., (2020). Adaptive response of a native Mediterranean grapevine cultivar upon short-term exposure to drought and heat stress in the context of climate change. *Agronomy, 10*, 249.

Ullah, A., Manghwar, H., Shaban, M., Khan, A. H., Akbar, A., Ali, U., Ali, E., & Fahad, S., (2018). Phytohormones enhanced drought tolerance in plants: A coping strategy. *Environ. Sci. Pollut. Res. Int., 25*, 33103–33118.

Vandenbroucke, K., & Metzlaff, M., (2013). Abiotic stress-tolerant crops: Genes, pathways and bottlenecks. In: *Sustainable Food Production* (pp. 1–17). Springer: Berlin/Heidelberg, Germany.

Vara, P. K. V. G. K., & Rama, R. G., (2019). Various plant's responses and strategies to cope with the water deficit: A review. *J. of Pharmaco. and Phytochem., 8*(1), 159–168.

Vasantha, S., Alarmelu, S., Hemaprabha, G., & Shanthi, R. M., (2005). Evaluation of promising sugarcane genotypes for drought. *Sugar Technol., 7*, 82, 83.

Veeranagamallaiah, G., Prasanthi, J., Reddy, K. E., Pandurangaiah, M., Babu, O. S., & Sudhakar, C., (2011). Group 1 and 2 LEA protein expression correlates with a decrease in water stress induced protein aggregation in horse gram during germination and seedling growth. *J. Plant Physiol., 168*, 671–677.

Virmani, S. S., & Ilyas-Ahmed, M., (2007). Rice breeding for sustainable production. In: *Breeding Major Food Staples* (pp. 141–191). Blackwell Pub.: Hoboken, NJ, USA.

Wang, Q., Li, C., Zheng, R., & Que, X., (2016). Phytoremediation of chlorpyrifos in aqueous system by riverine macrophyte, *Acorus calamus*: Toxicity and removal rate. *Environ Sci Pollut. Res., 23*, 16241–16248.

Wang, W. X., Vincour, B., Shoseyov, O., & Altman, A., (2004). Role of plant heat-shock proteins and molecular chaperones in the abiotic stress response. *Trends Plant Sci., 9*, 244–252.

Wasaya, A., Zhang, X., Fang, Q., & Yan, Z., (2018). Root phenotyping for drought tolerance: A review. *Agronomy, 8*, 241.

Wasson, A. P., Richards, R. A., Chatrath, R., Misra, S. C., Prasad, S. V., Rebetzke, G. J., Kirkegaard, J. A., Christopher, J., & Watt, M., (2012). Traits and selection strategies to improve root systems and water uptake in water-limited wheat crops. *J. Exp. Bot., 63*, 3485–3498.

Watkins, J. M., Hechler, P. J., & Muday, G. K., (2014). Ethylene-induced flavonol accumulation in guard cells suppresses reactive oxygen species and moderates stomatal aperture. *Plant Physiol., 164*, 1707–1717.

Weijde, T. V. D., Huxley, L. M., Hawkins, S., Sembiring, E. H., & Trindade, L. M., (2017). Impact of drought stress on growth and quality of miscanthus for biofuel production, *Global Change Biology Bioenergy, 9*, 770–782.

Wullschleger, S. D., Yin, T. M., DiFazio, S. P., Tschaplinski, T. J., Gunter, L. E., Davis, M. F., et al., (2005). Phenotypic variation in growth and biomass distribution for two advanced-generation pedigrees of hybrid poplar. *Can. J. For. Res., 35*, 1779–1789.

Xu, W., Cui, K., Xu, A., Nie, L., Huang, J., & Peng, S., (2015). Drought stress condition increases root to shoot ratio via alteration of carbohydrate partitioning and enzymatic activity in rice seedlings. *Acta Physiol. Plant, 37*, 1760.

Yadav, S., Modi, P., Dave, A., Vijapura, A., Patel, D., & Patel, M., (2020). *Effect of Abiotic Stress on Crops* (p. 1–18). IntechOpen.

Yadav, S., Vijapura, A., Dave, A., Shah, S., & Memon, Z., (2019). Genetic diversity analysis of different wheat [*Triticum aestivum* (L.)] varieties using SSR markers. *Int. J. of Cur. Microbio. and Appl. Sci., 8*(02), 839–846.

Yi, X. P., et al., (2016). Rapid recovery of photosynthetic rate following soil water deficit and re-watering in cotton plants (*Gossypium herbaceum* L.) is related to the stability of the photosystems. *J. Plant Physiol., 193*, 23–34.

Ying, Y., Yue, Y., Huang, X., Wang, H., Mei, L., Yu, W., Zheng, B., & Wu, J., (2013). Salicylic acid induces physiological and biochemical changes in three red bayberry (*Myric rubra*) genotypes under water stress. *Plant Growth Regul., 71*, 181–189.

Yu, T., et al., (2017). Improved drought tolerance in wheat plants overexpressing a synthetic bacterial cold shock protein gene SeCspA. *Nat. Publ. Gr., 7*, 44050.

Zhang, F., Wang, P., Zou, Y. N., Wu, Q. S., & Kuˇca, K., (2019). Effects of mycorrhizal fungi on root-hair growth and hormone levels of taproot and lateral roots in trifoliate orange under drought stress. *Arch. Agron. Soil Sci., 65*, 1316–1330.

Zhang, L. T., Zhang, Z. S., Gao, H. Y., Xue, Z. C., Yang, C., Meng, X. L., & Meng, Q. W., (2011). The mitochondrial alternative oxidase pathway protects plants against photoinhibition by alleviating inhibition of the repair of photodamaged PSII through preventing the formation of reactive oxygen species in Rumex K-1 leaves. *Physiol. Planta, 143*(4), 396–407.

Zhang, Z., Li, F., Li, D., Zhang, H., & Huang, R., (2010). Expression of ethylene response factor JERF1 in rice improves tolerance to drought. *Planta, 232*, 765–774.

Zhao, W., Leizhen, L., Qiu, S., Yang, J., Han, X., Tian, F., & Wu, J., (2020). Effects of water stress on photosynthesis, yield, and water use efficiency in winter wheat. *Water, 12*, 2127.

Zheng, M., Tao, Y., Hussain, S., Jiang, Q., Peng, S., Huang, J., et al., (2016). Seed priming in dry direct-seeded rice: Consequences for emergence, seedling growth and associated metabolic events under drought stress. *Plant Growth Regul., 78*, 167–178.

Zhou, J., Chen, S., Shi, W., et al., (2021). Transcriptome profiling reveals the effects of drought tolerance in giant juncao. *BMC Plant Biol., 21*, 2.

Zoghi, Z., Hosseini, S. M., Kouchaksaraei, M. T., Kooch, Y., & Guidi, L., (2019). The effect of biochar amendment on the growth, morphology and physiology of *Quercus castaneifolia* seedlings under water-deficit stress. *Eur. J. For. Res., 138*, 967–979.

CHAPTER 9

Devastating Effect of Amphan Cyclone on the Biodiversity of the Sundarbans Mangroves and the Need of Its Rehabilitation

MADHUMITA MAJUMDER

Department of Botany, Raidighi College, Raidighi, P.O.+P.S.–Raidighi, South 24 Parganas, West Bengal–743383, India, E-mails: madhumita2363@rediffmail.com; madhumitamajumder1981@gmail.com

ABSTRACT

The Sundarbans is the largest UNESCO world lineage site and a crucial wetland region of concern. The Sundarbans is the greatest tidal forest in this planet, covering about 10,000 km^2 in the Northern Bay of Bengal. Before anything else, it can be considered that the complete tract of mangrove swamp forestland is a very important natural environment of plants and animals. Mangrove vegetation of Sundarbans shows high productiveness as well as diverse type of ecosystem. They also contribute so many ecosystem services to riverine communities. The Sundarbans encompasses a group of compact, small depressed islands in the Ganga-Brahmaputra Meghna River delta. Mangrove forests of Sundarbans have the capacity to reconcile with rising sea level, grab runoff from soil disintegration that leads a prime role for augmentation of coastal areas and are obviously remarkable 'blue carbon' sink. The tree mass along with its various species decreases the entry of water and builds a characteristic buffer zone against floods and storm surges. The magnitude and the health of Sundarbans mangroves have

Plant Metabolites under Environmental Stress: Mechanisms, Responses, and Adaptation Strategies.
Nivas M. Desai, Manasi Patil, Umesh R. Pawar (Eds.)

reduced due to destruction of forests, utilization of lands, anthropological infiltrations including tourism and farming. Experts correlate these changes to this Amphan's devastation. Actually, mangroves over India keep going to diminish due to the infrastructural establishments, urbanization, and conversion of agricultural land, the people of the country's coastal and offshore will only suffer even most severely. So, in order to facilitate the conservation of mangrove forest, the application of inventive, reliable, and impartial conservative approaches is the need of the hour.

9.1 INTRODUCTION

The word "mangrove" is related to the Spanish word "mangal," the English term "grove" and the Portuguese word "mangue." Thus, worldwide mangroves are known as "mangals." "Coastal woodlands," "tidal forests," etc. Mangroves consist of woody shrubs or trees forming reducibly diverse forests alongside complex network of food webs and unique dynamics of its ecosystem. Mangrove means formation of trees and shrubs which is derived from West India, inhabiting the top intertidal zones of saltwater (30–90 ppt), the coastline and the stream estuaries of tropical and sub-tropical seas within 30° of the equator. Mangroves predominates about three-quarters of tropical coastline covering 172,000 km^2 of the surface of the earth in the large delta zones, regions of estuaries as well as Barrier Island. They are specialized in relation to the range in soil salinity, tidal height, and aeration in the soil and water. Mangroves forests have got features which are very different from other forests. They prevent soil erosion, and decrease the concentration of carbon dioxide in the environment by absorbing and storing them. Therefore, mangrove forests help to reduce the accumulation of greenhouse gas in the atmosphere (Saenger, 2002). Mangroves store large quantities of carbon-di-oxide as biomass, thereby acting as carbon sink since ancient times (Hogarth, 2007). Mangrove forests are facing rapid deforestation, the cause of which includes land use for agriculture purpose, exploitation of forests, setting up of industries, buildings, dams, and rapid urbanization (Pendleton et al., 2012). These wetland soils bury Co_2 for many years and do not reach saturation. These stored CO_2 does not synthesize methane gas and possess limited nutrients (Kumara et al., 2010), thereby facilitating coastal habitat, and adaptation to sea-level rise (Krauss et al., 2014). Thus, riverine areas are a strategic way to arrest and store carbon (Kennedy et al., n.d.). Mangrove vegetation occurs in two different bio-geographical zones:

- The Indo-West Pacific region covering Australia, Asia, Oceania, and the African eastern coast; and
- Atlantic-Caribbean-East Pacific areas covering America and the African Western Coast.

But Indian Mangrove vegetation is abundant in the Sundarbans of the Ganga estuary and also in the Brahmaputra system, the Andaman Islands, and the Irrawaddy delta.

9.2 THE SUNDARBANS MANGROVE FOREST

The Sundarbans are the single largest stakeholder of mangrove-forest worldwide. UNESCO has announced Sundarbans as a world heritage in 1997. It boasts of three wildlife sanctuaries within its domain. It is nationally and internationally hailed for its role in biodiversity and environmental services (Iftekhar and Saenger, 2008), Gangetic delta's southern areas comprising Eastern India and Bangladesh, situated between the Hooghly River on the west and the Meghna River on the east represents the Sundarbans. The word Sundarbans means beautiful forest and the term came from the "Sundari" (*Heritiera fomes*) trees which are found in large numbers in Sundarbans. The name has been coined from a Bengal word "Samudraban" means 'sea forest.' The mangrove forests of Sundarbans form a complex ecosystem as one of the only three largest single traits of mangroves worldwide. The forest area is about 10.000 sq.kms of which 60% is in Bangladesh and 40% is the Indian part of Sundarbans (Figure 9.1). About 1,700 sq. km. of Sundarbans is filled by bodies of water in the form of canals, rivers, etc. The Sundarbans includes an interwoven system of many tiny islands, mudflats, and tidal waterways. These small islands are dominated by salt tolerant mangrove forests.

9.3 MANGROVE VEGETATION

As per Champion and Seth (1968) Sundarbans Forest is classified under the subgroup 4B Tidal Swamp Forest. The variation in the distribution of plant species implies a distinct shift in their species composition from east to west parts of Sundarbans. A study revealed the presence of 24 true mangrove species, 21 mangrove associated plants and 9 mesophytic bioinvasive plants in the north-east, while 17 true mangrove species, 19 mangrove associated plants and 10 mesophytic bio-invasive plants are predominant in South-west parts of Sundarbans. According to Chakraborty et al. (2009), the central part is occupied only by 16 true mangrove plants and 8 mangrove associated plants. These variations in species distribution are due to salinity; cations exchange capacity, water holding capacity and textural composition of soil (Chaudhuri and Choudhury, 1994). Species richness is greatest in monsoon and reaches its nadir in pre-monsoon (Chaudhuri and Choudhury, 1994; De, Mitra, and Choudhury, 1987). According to Smith (1992), local forest structure of mangroves is determined by various factors like: (1) procedures of building lands influencing mangroves' autogenic succession; (2) rise in sea-level; (3) Stress adaptation like soil-salt content in the intertidal-zone; (4) patterns of disposal of mangroves propagules according to the size as well as rooting-time; and (5) following the process of disposal determining the establishment of propagules, like competition and predation. Optimal requirement for growth of mangrove varies from species to species in salinity, nutrient abundance, and soil characteristics (Hutchings and Saenger, 1987), which thereby determine species recruitment, establishment, and dominance (Lovelock et al., 2005). Mangrove forests exhibit less bio-diversity, even though they possess a thick-covering and the undergrowth consists largely of mangrove tree-derived seedlings.

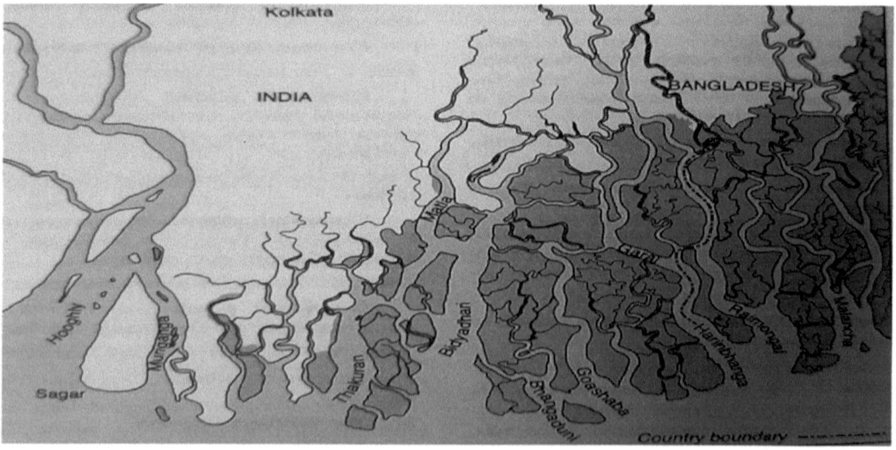

FIGURE 9.1 Sundarbans mangroves.

Mangroves can be divided into three categories, viz. True Mangrove, Back Mangrove and Mangrove Associates. Actually, true mangroves are true halophytes and mostly salt-tolerant, and occur near the ocean (Table 9.1).

TABLE 9.1 List of True Mangroves from Sundarbans Mangrove Forest

Species	Family	Local Name
Bruguiera gymnorrhiza	Rhizophoraceae	Kankra
B. cylindrica	-do-	Bakul kankra
B. parviflora	-do-	Bakul kankra
B. sexangula	-do-	Kankra
Rhizophora mucronata	-do-	Garjan
R. apiculata (Syn. R. conjugata)	-do-	Garjan
Ceriops decandra (Syn. C. roxburghiana)	-do-	Jhamti goran
C. tagal (Syn. C. candolleana)	-do-	Math goran
Kandelia candel	-do-	Garia
Sonneratia apetala	Sonneratiaceae	Keora
S. griffithii	-do-	Ora
S. caseolaris	-do-	Chak keora
Avicennia marina	Avicenniaceae	Pyara baen
A. officinalis	-do-	Jat baen
A. alba	-do-	Kalo baen
Xylocarpus granatum	Meliaceae	Dhundul
X. mekongensis	-do-	Passur/Pitamari
X. moluccensis	-do-	-do-
Aglaia cuculata	-do-	Golpati/Gulga
Lumnitzera racemosa	Combretaceae	Kripa
Excoecaria agallocha	Euphorbiaceae	Genwa
Aegialitis rotundifolia	Plumbaginaceae	Tora
Heritiera fomes	Sterculiaceae	Sundari
Aegiceras corniculatum	Myrsinaceae	Khalsi
Nypa fruticans	Arecaceae	Golpata
Phoenix paludosa	-do-	Hental

9.3.1 VALUATION OF ECOSYSTEM SERVICES OF SUNDARBANS MANGROVE FOREST

Why Mangrove ecosystem is one of the most prolific ecosystems in the world? In the course of high tide, substantial parts of the forest subsystem get submerged and receive major contributions from estuarine water as moisture that restores the sedimental deposition of the bottom soil, and supply of nutrients such as macro, micro as well as the trace elements for forest subsystem. During low tide, the receding water retreats large amount of mangroves litter contributed detritus to the adjoining aquatic subsystem.

The nutrients released from the remnants are utilized by phytoplankton along with abundance of sunlight and water available in the open water system. Mangrove being a perennial evergreen plant, produces a large quantity of leaf litters year-long. These, after falling on the moisture surface of the bottom-soil loaded with silt-clay, are decayed by benthic fauna such as gastropods, crabs, microarthropods, etc., into small fragments giving more opportunities for the microbes like bacteria, protozoa, and fungi to further decompose them to produce detritus via decomposition of the litter. The feeding activities of the deposit feeders, such as mollusks, crabs, nematodes, polychaetes, etc.), turn over the surface sediment layer, thereby showing new surfaces of litter to microbial activities. Heal and French (1974) outlined three major processes of decomposition: 1. Carbon release as gas by micro-flora and -fauna via respiration, 2. Leaching out of soluble materials and 3. Corrosion by physical factors as well as by the fauna (Heal and French, 1974). Therefore, two interlinked subsystems of the mangrove ecosystem switch biotic and abiotic components in between them and avail the benefits in respect to higher biological productivity. The forest subsystem's shallow water transmits nutrients and energy to deeper estuarine water and conjoining coastal shelf. The free moving planktons and nektons between the two fixed subsystems produce, convert, and transport energy and nutrients in relation to the diurnal tidal as well as seasonal periods (Odum, 1971). Mangrove ecosystem productivity can be dependent on four factors: (1) three units of primary production such as phytoplankton, benthic algae, and marsh vegetation (2) tidal action mediated ebb and flow of water-movements; (3) sufficient nutrient supplies and (4) fast regeneration and conservation of nutrients because of the action of microbes and filter feeders (Schelake and Odum, 1962). Various food chains are interwoven to each other to generate a substantially complex food web in mangrove estuaries. A vast range of interplays among different structural components (such as water, soil, flora, and fauna), the biological, chemical, and physical processes nurture a rich level of productivity in the ecosystem. The decaying plant litters are taken up by a verities of organisms, and they discharge a considerable amount of nutrients to the mangrove ecosystem for the purpose of recycling.

9.3.2 FUNCTION OF SUNDARBANS MANGROVES IN SAFEGUARDING GLOBAL WARMING AND BLUE CARBON STORAGE

Today's world is in front of prime challenges mediated by atmospheric CO_2, the primary cause of global warming (IPCC, 2013) which is the results of anthropogenic-activities-mediated emissions of greenhouse gasses (primarily CO_2) (IPCC, 2013). Fossil fuel combustion and destruction of forests are the two main reasons for the elevation of CO_2 level in the atmosphere (Detwiler and Hall, 1988; Woodwell et al., 1983). It has been presumed that, by the year 2050, atmospheric CO_2 will be around 467 ppm to 555 ppm, in 1,750 which was around 278 ppm and in 2011, 390 ppm and that the average global temperature will be elevated by 2 to 42°C by 2050 (IPCC, 2007, 2013; Anderson and Bows, 2011). The enormity of this global warming could remarkably reorient the climate of the earth, land use pattern, and main vegetation areas and perhaps more significantly the polar ice melting of possibly resulting in raising of sea level by 5 m killing many lives and devastating severe loss of land and properties (Woodwell et al., 1983; IUFRO, 2009) especially in the developing countries.

Forests play a crucial role in neutralizing the climate changes via sequestration of atmospheric carbon (Adame et al., 2013). Khan et al. (2007); and Donato et al. (2011) depicted mangroves as efficacious sinks, responsible for sequestration of four times carbon/unit area in comparison to the tropical terrestrial forests (Khan, Suwa, and Hagihara, 2007; Donato et al., 2011). However, mangrove deforestation was very abundant during the last few decades (Giri et al., 2007), may provide them huge amounts of atmospheric carbon (Donato et al., 2011). More studies have been done on carbon sequestration (Khan, Suwa, and Hagihara, 2007; Bouillon et al., 2008) and organic carbon dynamics (Machiwa and Hallberg, 2002; Kristensen et al., 2008) in mangrove forests. In mangrove ecosystem Carbon stock varies with species level (Laffoley and Grimsditch, 2009), vegetation type (Adame et al., 2013; Laffoley and Grimsditch, 2009; Cero´n-Breto´n et al., 2011; Mitra, Sengupta, and Banerjee, 2011; Sapit et al., 2011) and salinity (Adame et al., 2013).

Considering the immense importance of Sundarbans, assessing its above- and belowground carbon sequestration can be treated as an aid to the forest conservation and the benefit offered by many global initiatives such as carbon trading. The inherent heterogeneity of mangrove vegetation covering wide areas, mangrove types influenced by salinity zones and the functional parameters of vegetation such as canopy cover, mean tree height, basal area, etc., that might determine the carbon stock at both above and belowground levels (Wahid, Mukand, and Bhuiyan, 2007).

To restrain from the destructive tree among various vegetation types, Sundri dominated forest demonstrated the highest amount of ecosystem (both above- as well as belowground) carbon (360.0 Mg ha^{-1})/unit area followed by Sundri-Gewa, Sundri-Passure-Kankra, many others (Gewa Coppice, Sundri-Passur Passur-Kankra-Baen), Gewa, Gewa-Sundri, nontree vegetation, Goran-Gewa, Goran, and GewaGoran (Figure 9.1). The cause behind this disparity could be above the ground level vegetation height as the carbon stock was traced greatly associated with the tree size (diameter and height) and basal area. The trees with greater height were found to outcompete their neighbors in accumulating more carbon (Klimešova et al., 2008; Westoby et al., 2002; Westoby, 1998). In natural ecosystems, the tree-diameter is a determinant of the biomass aboveground (Chave et al., 2004).

Tree density provides a strong indicator of carbon stock of any ecosystem (Gross, Flores, and Schwendenmann, 2014). Therefore, higher canopy height of the vegetation, and larger diameter of trees with more basal areas, provide higher ecosystem carbon. Interestingly, Sundri is one of the tallest trees in Sundarbans, attaining a height of 15 m to 21 m (Karim, 1988) with a larger diameter than many other trees apart from Baen and Keora. There is low carbon stock in vegetations with less or no Sundri. Here, salinity depends significantly on the fluctuating freshwater volume from the upstream and literally the absence of freshwater flow makes the strong saline zones (Wahid, Mukand, and Bhuiyan, 2007). Therefore, the reasons for diversification of carbon stock in various STs in Sundarbans could be attributed to the spatial variety of freshwater nutrients and the unfavorable effect of growing salinity on the biomass productivity (Wahid, Mukand, and Bhuiyan, 2007). Generally, in mangrove ecosystem there is a significant tradeoff between salinity and the species distribution, growth, and productivity of mangrove vegetation (Twilley and Chen, 1998) and thus on carbon stock (Crooks et al., 2011). In mangrove forests, vegetation is highly prolific in lower STs which shows higher productivity implicated to both higher biomass as well as carbon stock (Crooks et al., 2011; Kathiresan and Bingham, 2001; Ball, 1998, 2002). The contribution of Soil carbon (up to 1 m) to total ecosystem carbon stock was almost identical to total carbon stock aboveground both among the STs and types of vegetation indicating the mangroves' ability to accumulate huge quantity of organic carbon into sediment (Khan, Suwa, and Hagihara, 2007; Donato et al., 2011; Bouillon et al., 2003). Moreover, like other mangroves, the contribution of belowground carbon to ecosystem carbon stock, was higher than that of aboveground carbon as soil and root carbon both were taken (Kauffman and Cole, 2010). This result was attributed to the higher root-shoot ratios found in the mangroves (Khan, Suwa, and Hagihara, 2007; Fujimoto, 2004; Fujimoto et al., 1999; Page et al., 2002; Komiyama et al., 2000; Komiyama, Ong, and Poungparn, 2008), that store considerable quantity of soil carbon (Bouillon et al., 2003). There exists a remarkable difference in carbon stock in various types of mangrove vegetation and various forest zones with different regimes of salinity. The attributes of vegetation like height and basal area of the dominating mangrove plant species in each vegetation type are the major index to determine the nature of the ecosystem's carbon stock.

9.3.3 REPERCUSSIONS ON CLIMATE FOLLOWING CHANGES IN MANGROVE ECOSYSTEM

The Mangroves in flora of the Sundarbans represent natural buffer, an embankment preventing erosion in the coastal regions and seawater enters into one of the most densely populated areas of the globe. The eco-sensitive zones of Sundarbans are not only unique in its function and diversity, but also extraordinarily fragile due to richness in its diverse ecology, and unique biota which has been constantly threatened by anthropological disturbances changing the local climate and disturbing the exquisite balance among sea, land, and air.

By assisting groups of shellfishes and fishes, Mangroves perform a significant role in food chains and coastal ecosystems. Mangroves lessen the damage produced by storms like cyclones by acting as a flood barrier. Soil erosion and damage make the coast gradually exposed and it becomes increasingly susceptible to such storms. The Ganges-Brahmaputra Delta being the densely populated coasts remain mostly vulnerable to violent storms. Scientists gave research that the undamaged and healthy mangroves in Orissa, a coastal Indian state of severe casualty in a 1999 cyclone emanating from the Bay of Bengal. Researchers have also argued that the 2004 Indian Ocean tsunami created surprisingly lesser damage in the mangrove protected regions. Simulations therefore suggest that a thick mangroves belt can remarkably reduce the high point pressure of tsunami. The mangrove in Sundarbans provides a prime defense in controlling the impacts of climate changes, protecting the coastal belts from cyclones and other storms. In the Bay of Bengal, more or less eight storms with speed greater than 63 km/hr. are formed every year and from it average of two become tropical cyclones. Tropical cyclones like Sidr (in 2007) and Aila (in 2009) produced significant damage, prior Amphan. Mangroves act as a biological shield securing coastal communities from the perilous damages of storm upsurge. Stimulating rise in sea level as a result of global warming may submerge the Sundarbans in the near future. This will terminate the safety they provide against progressively vigorous tropical storms. Scientists forecast that the gradual sea-level rise will attempt to submerge most of the Sundarbans in three decades, and may extinguish the region by the end of this century (Nicholls et al., 2007).

Therefore, Sundarbans mangroves serve as the buffer against intense and frequent storms without which life and property are at stake in those areas. The following measurements are needed to be taken in this regard:

- Integrated marine and coastal management;
- Keeping track on the consequences of climate change on vegetation, Royal-Bengal Tiger, Coral Reef, Sundari tree, and crocodiles;
- Assessing the vulnerability of coasts and the associated risks;
- Evaluation of the wealth of Sundarbans;
- Use of salt resistance plant species to prevent afforestation and facilitate reforestation Accelerating natural reproduction of native trees;
- Generating public awareness using various media resources;
- Facilitating forestation of the coasts using local communities;
- Preventing cutting and encroachment to protect existing mangroves;
- Implementing techniques to get the global carbon market benefit.

9.3.4 CHALLENGES IN MAINTAINING THE MANGROVE ECOSYSTEM

The Sundarbans ecosystem acts as an equilibrium between salty ocean and freshwater ecosystems. Freshwater flow via the Ganges and the Gorai rivers, especially at dry regimes, is needed in order to balance the intrusion of salinity. The total mangrove vegetation did not alter remarkably during the last few centuries. As per change detection analyses, natural dynamics of internal mangroves experienced a change with land loss and gain at a small scale by the effects of erosion and sediment accumulation within the tidal waterways. New sedimentary deposits along with small tidal channels within a couple of years are colonized by light dependent pioneer plant communities which are later reinstated by various successional communities because of land-rise, modify inundation and saline condition as shown by Change Detection Analysis (Iftekhar and Saenger, 2008). Because of the hydrological modifications and the fluctuating soil salinity and flooding and parameters, floristic changes can affect the whole food chain within the Sundarbans forests. Moreover, soil salinity is reduced because of the Himalayan runoffs, and, it causes more erosion, elevated siltation curtails transparency and thus reduces photosynthesis of the phytoplankton with harmful effects on food chains and oxygen generation (Raha et al., 2012; Banerjee, 2013; Mitra et al., 2009). Large-scale transformations in mangrove area as well as coastline are triggered by periodic cyclonic effects. Sundarbans in India and Bangladesh were ravaged by a tropical cyclone Aila which had a speed of 110 km/hr.

on 25th May, 2009. A million people were rendered homeless and more than 8,000 people had been missing in the both the countries. Almost 300 people had been killed in Sagar Island alone in the Indian Sundarbans only (Gupta, Basu, and Bhatta, 2009). Repercussions on mangrove forests range from fracturing of delicate root structures because of rough wind, defoliation which alter according to the sharpness of the storm. Mangrove forests typically own a high durability to natural disasters like tsunamis or tropical storms. Thus, saplings of mangroves must be propagated as a significant measure for substantial mangrove management, and thus, ecosystem alteration can be avoided (Harun-or-Rashid et al., 2009). Tidal bores and high tides are also equally devastating and have already started playing damage on the ecosystem. On 13–15 July, 2014 a tidal bore displaced more than 50,000 residents. Moreover, more than 10,000 dwelling places were totally destroyed and 5,000 more were partially affected. These highly frequented but low-intense calamities are generally energized by weather fluctuating interplaying with the local ecosystems. It directly converts into higher devastating capacities of the cyclones. The other significant and crucial alteration is the rise of sea level due to global warming, land subsidence, and human activities (Raha et al., 2012), which in course influences the mangrove vegetation as well as the vulnerable, coastal people. Sea level rise is a prime concern for recent loss in coastal lands (Cornforth et al., 2013) which has significant consequences on the forest health of the Sundarbans. The areas with reduced salinity will be significantly minimized, producing alterations in a huge number of species and the flora, with reduction of less salt-resistant species that includes the commercially important tree *Heritiera fomes* (Shams-Uddin et al., 2013). Two effects of enhanced soil-salinity and pollution on the Sundarbans forests is the current attention of a reduction in canopy closure and the extensiveness of a top dying disease-causing hazards to the formerly dominant *Heritiera fomes* (Rahman, 1990; Zaman et al., 2013). Such deterioration may have a significant adverse influence on the various interwoven parameters of the biodiversity of Sundarbans (Selvam, 2003). There are scattered records of a few interesting research, but a thorough report of the different economic, ecological, and biophysical attributes demonstrated by mangroves is still absent. In order to elevate the realization of the various benefits of mangroves, there is a necessity to organize more scientific research and concentrate more on broadening mangrove areas in participation with local residents and other prominent stakeholders. Unless there are reasonable reduction in pollution and emissions, vast areas of Sundarbans may be lost underwater, and the remaining could be in danger of incursion of saltwater. Few studies suggest altering salinity is the major cause of ecosystem destruction both via sea-level rise of the Bay of Bengal and decreased flow of freshwater. Sundari trees are being killed by the top-dying disease, which is known locally as 'agamora.' It is a very alarming fact that this tree species may be soon become extinct due to the fast progress of the disease. The disadvantageous outcome of enhanced salt levels on the Sundarbans ecosystem are manifested in the dying tops of Sundari trees, degeneration of forest types, slower growth of forest, and reduced productivity of forest sites. For nearby cities like Kolkata, it is likely alarming to come from higher tides and more destructive storms destroying the urban vegetation, as witnessed in Amphan – with storm surges unchecked by the rapidly vanishing Sundarbans mangroves (Sen, 2020; Biswas and Sen, 2020). It's high time to choose wisely to determine whether the rich and diverse Sundarbans stay on the map – influencing the bare survival of residents in Kolkata and the adjacent Ganges basin.

9.4 CONCLUSION

Mangroves are very distinct and immensely productive ecological alliance at the land-sea interface. The Sundarbans, the largest mangrove in the globe bestow an array of significant ecosystem services. For example, food, and water, habitat for millions of its species(including the protected Royal Bengal Tiger) and inhabitants; protection against the damaging effects of natural hazards, the capability to act as a colossal long-term carbon sink; the reservation of terrestrial sediments. The usefulness of the Sundarbans therefore broadens from the local to the global with diverse objectives of its stakeholders. Sundarbans mangrove ecosystem has been majorly affected by anthropogenic causes and extreme weather events for the last two centuries. Human activities have also increased the salinity of water, relative mean sea level and land destruction. Protecting mangroves is a complex, multi-scalar task because of the interaction of climatic threats, environmental governance, etc. Despite possessing a high resilience to natural disturbances, the impacts of anthropogenic degradation is often irremediable. This is why it is of prime importance to reconsider the existing plans by keeping in mind the local need and to deal the danger via a multicolor and polycentric strategy, instead of keeping climate change adaptation and conservation separate. The Sundarbans desperately need an intertwined collaboration between stakeholders such as natural and social scientists to make policies in relation to the protection of the environment, conservation of natural resources, and human and climate change solutions.

KEYWORDS

- **Amphan**
- **blue carbon storage**
- **climate change**
- **global warming**
- **mangrove ecosystem**
- **Sundarbans**

REFERENCES

Adame, M. F., Kauffman, J. B., Medina, I., Gamboa, J. N., Torres, O., Caamal, J. P., & Reza, M., (2013). Carbon stocks of tropical coastal wetlands within the karstic landscape of the Mexican Caribbean. *PLoS One, 8*(2), e56569. doi: 10.1371/ journal.pone.0056569.

Anderson, K., & Bows, A., (2011). Beyond 'dangerous' climate change: emission scenarios for a new world. *Philos. Trans. R Soc. A, 369*(1934), 20–44. doi: 10.1098/rsta.2010.0290.

Ball, M. C., (1998). Mangrove species richness in relation to salinity and waterlogging: A case study along the Adelaide River floodplain, northern Australia. *Glob. Ecol. Biogeogr. Lett., 7*, 73–82.

Ball, M. C., (2002). Interactive effects of salinity and irradiance on growth: Implications for mangrove forest structure along salinity gradients. *Trends Ecol., 16*, 126–139.

Banerjee, K., (2013). Decadal change in the surface water salinity profile of Indian Sundarbans: A potential indicator of climate change. *J. Mar. Sci. Res. Dev., 7*.

Biswas, R., & Sen, S., (2020). City greens, post amphan: A survey. *International Journal of Creative Research Thoughts (IJCRT), 8*(6), 1986–1993. ISSN: 2320-2882. Available: http:// www.ijcrt.org/papers/IJCRT2006269.pdf; http://doi.one/10.1729/Journal.23791 (accessed on 2 June 2022).

Bouillon, S. M., Frankignoulle, F., Dehairs, B., Velimirov, A., Eiler, H., Etcheber, G., & Borges, A. V., (2003). Inorganic and organic carbon biogeochemistry in the Gautami Godavari estuary (Andhra Pradesh, India) during pre-monsoon: The local impact of extensive mangrove forests. *Global Biogeochem. Cycles, 17*(4), 1114. doi: 10.1029/2002GB002026.

Bouillon, S., Borges, A. V., Eda-Moya, E. C., Diele, K., Dittmar, T., Duke, N. C., Kristensen, E., et al., (2008). Mangrove production and carbon sinks: A revision of global budget estimates. *Global Biogeochem Cycles, 22*, 1–12. doi: 10.1029/2007GB00305.

Cerón-Bretón, R. M., Cerón-Bretón, J. G., Sánchez-Junco, R. C., Damián-Hernández, D. L., Guerra-Santos, J. J., Muriel-Garcia, M., & Cordova-Quiroz, A. V., (2011). Evaluation of carbon sequestration potential in mangrove forest at three estuarine sites in Campeche Mexico. *Int. J. Energy Environ., 5*(4), 487–494.

Chakraborty, S. K., Giri, S., Chakravarty, G., & Bhattacharya, N., (2009). Impact of eco-restoration on the biodiversity of Sundarbans mangrove ecosystem, India. *Water, Air, Soil Pollution: Focu., 9*, 303–320.

Champion, H. G., & Seth, S. K., (1968). *A Revised Survey of Forest Types of India* (p. 404). Govt. of India Press, New Delhi.

Chaudhuri, A. B., & Choudhury, A., (1994). *Mangroves of the Sundarbans, India (Volume 1, 247p)*. IUCN-Bangkok. Thailand.

Chave, J., Condit, R., Aguilar, S., Hernandez, A., Lao, S., & Perez, R., (2004). Error propagation and scaling for tropical forest biomass estimates. *Philos Trans R Soc B, 359*, 409–420.

Cornforth, W. A., Fatoyinbo, T. E., Freemantle, T. P., & Pettorelli, N., (2013). Advanced land observing satellite phased array type L-band SAR (ALOS PALSAR) to inform the conservation of mangroves: Sundarbans as a case study. *Remote Sens., 5*, 224–237.

Crooks, S., Herr, D., Tamelander, J., Laffoley, D., & Vandever, J., (2011). Mitigating climate change through restoration and management of coastal wetlands and near-shore marine ecosystems: Challenges and opportunities. *Environment Department Paper, 121,* World Bank, Washington, DC.

De, T. K., Mitra, A., & Choudhury, A., (1987). Studies on phytoplankton and productivity around Sagar Island and Prentice Island, Sunderbans, West Bengal. *Records Zoological Survey of India, 81*(3, 4), 81–92.

Detwiler, R. P., & Hall, C. A. S., (1988). Tropical forests and the global carbon cycle. *Science, 239*, 42–47. doi: 10.1126/science. 239.4835.42.

Donato, D., Kauffman, J. B., Murdiyarso, D., Kurnianto, S., Stidham, M., & Kanninen, M., (2011). Mangroves among the most carbon-rich forests in the tropics. *Nat. Geosci., 4*, 293–297. doi: 10.1038/ngeo1123.

Fujimoto, K., (2004). Belowground carbon sequestration of mangrove forests in the Asia-Pacific region. In: Vannucci, M., (ed.), *Mangrove Management and Conservation* (pp. 138–146). United Nations University Press, New York.

Fujimoto, K., Imaya, A., Tabuchi, R., Kuramoto, S., Utsugi, H., & Murosushi, T., (1999). Belowground carbon storage of Micronesian mangrove forests. *Ecol. Res., 14*, 409–413. doi: 10. 1046/j.1440-1703.1999.00313.x.

Giri, C., Pengra, B., Zhu, Z., Singh, A., & Tieszen, L. L., (2007). Monitoring mangrove forest dynamics of the Sundarbans in Bangladesh and India using multi-temporal satellite data from 1973 to 2000. *Estuar. Coast. Shelf Sci., 73*, 91–100.

Gross, J., Flores, E., & Schwendenmann, L., (2014). Stand structure and aboveground biomass of a *Pelliciera rhizophorae* mangrove forest, gulf of monitjo Ramsar site, pacific coast, Panama. *Wetlands, 34*(1), 55–65. doi: 10.1007/s13157-013- 0482-1.

Gupta, A., Basu, J., & Bhatta, A., (2009). "@110/hr". In: *Down to Earth* (p. 18). Society for Environmental Communications: New Delhi, India.

Harun-or-Rashid, S., Biswas, S. R., Böcker, R., & Kruse, M., (2009). Mangrove community recovery potential after catastrophic disturbances in Bangladesh. *For. Ecol. Manag., 257*, 923–930.

Heal, O. W., & French, D. D., (1974). Decomposition of organic matter in Tundra. In: Holding, A. J., et al., (eds), *Soil Organisms and Decomposition in Tundra* (pp. 279–309). Tundra Biome Steering Committee Stockholm.

Hogarth, P. J., (2007). *The Biology of Mangroves and Seagrasses* (p. 273). Oxford University Press: Oxford, UK.

Hutchings, P. A., & Saenger, P., (1987). *Ecology of Mangroves*. University of Queensland Press, Brisbane.

Iftekhar, M. S., & Saenger, P., (2008). Vegetation dynamics in the Bangladesh Sundarbans mangroves: A review of forest inventories. *Wetl. Ecol. Manag., 16*, 291–312.

IPCC, (2007). Summary for policymakers. In: Solomon, S., Qin, D., Manning, M., Chen, Z., Marquis, M., Avery, K. B., Tignor, M., & Miller, H. L., (eds.), *Climate Change, 2007: The physical science basis*. Contribution of working group I to the fourth assessment report of the intergovernmental panel on climate change. Cambridge University Press, Cambridge.

IPCC, (2013). Summary for policymakers. In: Stocker, T. F., Qin, D., Plattner, G. K., Tignor, M., Allen, S. K., Boschung, J., Nauels, A., Xia, Y., Bex, V., & Midgley, P. M., (eds.), *Climate Change, 2013: The Physical Science Basis*. Contribution of working group I to the fifth assessment report of the intergovernmental panel on climate change. Cambridge University Press, Cambridge, United Kingdom.

IUFRO, (2009). *Adaptation of Forests and People to Climate Change* (Vol. 22, p. 224). A global assessment report. IUFRO World Series.

Karim, A., (1988). *Environmental Factors and the Distribution of Mangroves in the SUNDARBANS with Special Reference to Heritiera Fomes.* Buch.-Ham. Ph.D. Thesis. University of Calcutta, India.

Kathiresan, K., & Bingham, B. L., (2001). Biology of mangrove and mangrove ecosystems. *Adv. Mar. Biol., 40*, 81–251.

Kauffman, J. B., & Cole, T. G., (2010). Micronesian mangrove forest structure and tree responses to a severe typhoon. *Wetlands, 30*, 1077–1084. doi: 10.1007/s13157-010-0114-y.

Kennedy, H., Beggins, J., Duarte, C. M., Fourqurean, J. W., Holmer, M., Marbà, N., & Middelburg, J. J. (2010). Seagrass sediments as a global carbon sink: Isotopic constraints. *Glob. Biogeochem. Cycles, 24.*

Khan, M. N. I., Suwa, R., & Hagihara, A., (2007). Carbon and nitrogen pools in a mangrove stand of *Kandelia obovata* (S., L.) Yong: Vertical distribution in the soil–vegetation system. *Wetl. Ecol. Manag., 15*(2), 141–153. doi: 10.1007/s11273- 006-9020-8.

Klimesŏva, J., Latzel, V., De Bello, F., & Van, G. J. M., (2008). Plant functional traits in studies of vegetation changes in response to grazing and moving towards a use of more specific traits. *Preslia, 80*, 245–253.

Komiyama, A., Havanond, S., Srisawatt, W., Mochida, Y., Fujimoto, K., Ohnishi, T., Ishihara, S., & Miyagi, T., (2000). Top/root biomass ratio of a secondary mangrove (*Ceriops tagal* (Perr.) C.B. Rob.) forest. *For. Ecol. Manag., 139*, 127–134. doi: 10. 1016/S0378-1127(99)00339-4.

Komiyama, A., Ong, J. E., & Poungparn, S., (2008). Allometry, biomass, and productivity of mangrove forests: A review. *Aquat. Bot., 89*, 128–137. doi: 10.1016/j.aquabot.2007.12.006.

Krauss, K. W., McKee, K. L., Lovelock, C. E., Cahoon, D. R., Saintilan, N., Reef, R., & Chen, L., (2014). How mangrove forests adjust to rising sea level. *New Phytol, 202*, 19–34.

Kristensen, E., Bouillon, S., Dittmar, T., & Marchand, C., (2008). Organic carbon dynamics in mangrove ecosystems: A review. *Aquat. Bot., 89*, 201–219. doi: 10.1016/j.aquabot.2007.12.005.

Kumara, M. P., Jayatissa, L. P., Krauss, K. W., Phillips, D. H., & Huxham, M., (2010). High mangrove density enhances surface accretion, surface elevation change, and tree survival in coastal areas susceptible to sea-level rise. *Oecologia, 164*, 545–553.

Laffoley, D. D. A., & Grimsditch, G., (2009). *The Management of Natural Coastal Carbon Sinks*. IUCN Gland, Switzerland.

Lovelock, C. E., Feller, I. C., McKee, K. L., & Thompson, R., (2005). Variation in mangrove forest structure and sediment characteristics in Bocas del Toro, Panama. *Caribbean Journal of Science, 41*, 456–464.

Machiwa, J. F., & Hallberg, R. O., (2002). An empirical model of the fate of organic carbon in a mangrove forest partly affected by anthropogenic activity. *Ecol Model, 147*, 69–83. doi: 10. 1016/S0304-3800(01)00407-0.

Mitra, A., Gangopadhyay, A., Dube, A., Schmidt, A. C., & Banerjee, K., (2009). Observed changes in water mass properties in the Indian Sundarbans (Northwestern Bay of Bengal) during 1980–2007. *Curr. Sci. India, 97*, 1445–1452.

Mitra, A., Sengupta, K., & Banerjee, K., (2011). Standing biomass and carbon storage of aboveground structures in dominant mangrove trees in the Sundarbans. *For. Ecol. Manag., 261*, 1325–1335. doi: 10.1016/j.foreco.2011.01.012.

Nicholls, R. J., Wong, P. P., Burkett, V. R., Codignotto, J. O., Hay, J. E., McLean, R. F., Ragoonaden, S., & Woodroffe, C. D., (2007). Coastal systems and low-lying areas. In: Parry, M. L., Canziani, O. F., Palutikof, J. P., Van Der, L. P. J., & Hanson, C. E., (eds.), *Climate Change 2007: Impacts, Adaptation and Vulnerability*. Contribution of Working Group II to the Fourth Assessment Report of the Intergovernmental Panel on Climate Change, Cambridge University Press.

Odum, E. P., (1971). *Fundamentals of Ecology* (pp. 1–297). W.B. Saunders and Co., Philadelphia, U.S.A,.

Page, S. E., Siegert, F., Rieley, J. O., Boehm, H. D. V., Jaya, A., & Limin, S., (2002). The amount of carbon released from peat and forest fires in Indonesia during 1997. *Nature, 420*, 61–65. doi: 10. 1038/nature01131.

Pendleton, L., Donato, D. C., Murray, B. C., Crooks, S., Jenkins, W. A., Sifleet, S., Craft, C., et al., (2012). Estimating global "blue carbon" emissions from conversion and degradation of vegetated coastal ecosystems. *PLoS One, 7*, e43542.

Raha, A., Das, S., Banerjee, K., & Mitra, A., (2012). Climate change impacts on Indian Sunderbans: A time series analysis (1924–2008). *Biodivers. Conserv., 21*, 1289–1307.

Rahman, M. A., (1990). A comprehensive report on sundry (*Heritiera fomes*) trees with particular reference to top dying in the Sundarbans. In: Rahman, M. A., Khandakar, M. A., Ahmed, F. U., & Ali, M. O., (eds.), *Seminar on Top Dying of Sundri (Heritiera fomes) Trees* (p. 256). Bangladesh Agricultural Research Council: Dhaka, Bangladesh.

Saenger, P., (2002). *Mangrove Ecology, Silviculture and Conservation*. Kluwer Academic Publishers: Dordrecht, the Netherlands. [Google Scholar].

Sapit, D., Damrong, S., Ladawan, P., Chongrak, W., Sakhan, T., & Aor, P., (2011). An assessment of stand structure and carbon storage of a mangrove forest in Thailand. *IUFRO World Ser., 29*, 28–30.

Schelake, C. L., & Odum, E. P., (1962). Mechanisms for maintaining high productivity in Georgia estuaries. *Proceedings of Gulf Caribb Fishery Institute, 14*, 75–80.

Selvam, V., (2003). Environmental classification of mangrove wetlands of India. *Curr. Sci. India, 84*, 757–765.

Sen, S., (2020). Sunderban mangroves, post amphan: An overview. *International Journal of Creative Research Thoughts (IJCRT), 8*(6), 2751–2755. ISSN: 2320-2882.

Shams-Uddin, M., Shah, M. A. R., Khanom, S., & Nesha, M. K., (2013). Climate change impacts on the Sundarbans mangrove ecosystem services and dependent livelihoods in Bangladesh. *Asian J. Conserv. Biol., 2*, 152–156.

Smith, T. J., (1992). 'Forest structure'. In Robertson, A. I., & Alongi, D. M., (eds.), *Tropical Mangrove Ecosystems* (pp. 101–136). American Geophysical Union, Washington DC.

Twilley, R. R., & Chen, R., (1998). A water budget and hydrology model of a basin mangrove forest in Rookery Bay, Florida. *Mar. Freshw. Res., 49*, 309–323. doi: 10.1071/MF97220.

Wahid, S. M., Mukand, S. B., & Bhuiyan, A. R., (2007). Hydrologic monitoring and analysis in the Sundarbans mangrove ecosystem. *Bangladesh. J Hydrol., 332*, 381–395. doi: 10. 1016/j.jhydrol.2006.07.016.

Westoby, M., (1998). A leaf-height-seed (LHS) plant ecology strategy scheme. *Plant Soil, 199*, 213–227. doi: 10.1023/A: 1004327224729.

Westoby, M., Falster, D. S., Moles, A. T., Vesk, P. A., & Wright, I. J., (2002). Plant ecological strategies: Some leading dimensions of variation between species. *Annu. Rev. Ecol. Syst., 33*, 125–159. doi: 10.1146/annurev.ecolsys.33.010802.150452.

Woodwell, G. M., Hobbie, J. E., Houghton, R. A., Melillo, J. M., Moore, B., Peterson, B. J., & Shaver, G. R., (1983). Global deforestation: Contribution to atmospheric carbon dioxide. *Science, 222*(4628), 1081–1086. doi: 10.1126/science.222.4628. 1081.

Zaman, S., Bhattacharyya, S. B., Pramanick, P., Raha, A. K., Chakraborty, S., & Mitra, A., (2013). Rising water salinity: A threat to mangroves of Indian Sundarbans. In: Abedin, M. A., Habiba, U., & Shaw, R., (eds.), *Water Insecurity: A Social Dilemma* (pp. 167–183). Emerald Group Publishing Limited: Bingley, UK.

CHAPTER 10

Phenolic Compounds Under Stress

BABITA RANA[1] and KAVITA CHAHAL[2]

[1]*Department of Botany, G. N. Khalsa College, Matunga–400019, Mumbai, Maharashtra, India, E-mail: babitarana2009@yahoo.com*

[2]*Department of Botany, Government College, Bichhua, Chhindwara, Madhya Pradesh, India*

ABSTRACT

Plants growing under natural conditions are consistently exposed to a range of biotic and abiotic stress responsible for the heavy reduction in plant growth and productivity. Biotic stress may include a wide variety of pathogenic bacteria, viruses, fungi, insects, and herbivores, while certain unfavorable environmental conditions such as water scarcity, high salt, and heavy metal content in plant habitat serve as abiotic stress. Plants by their nature produce various primary and secondary metabolites. Plants require primary metabolites for their fundamental functions hence playing a role in the growth and development, while secondary metabolites protect plants from environmental adversities and in encountering stress constraints. Phenolic compounds, terpenes, and nitrogen-containing compounds are forming the major classes of secondary metabolites, which witness drastic seasonal, geographical, and environmental fluctuations. Phenolic compounds, one of the major class of secondary metabolites play an important role in mitigating the detrimental outcomes of biological and non-biological stresses. The present study undertakes the diverse phenolic compound involved and their mode of guiding plant defense system.

Plant Metabolites under Environmental Stress: Mechanisms, Responses, and Adaptation Strategies.
Nivas M. Desai, Manasi Patil, Umesh R. Pawar (Eds.)

10.1 INTRODUCTION

Under natural conditions, plants encounter a great deal of adverse environmental conditions, and hence they possess a multitude of defense mechanisms by which they can cope with various types of biotic and abiotic stresses (Ballhorn et al., 2009). The most manifested biotic, abiotic, and man-induced stress factors are presented in Figure 10.1.

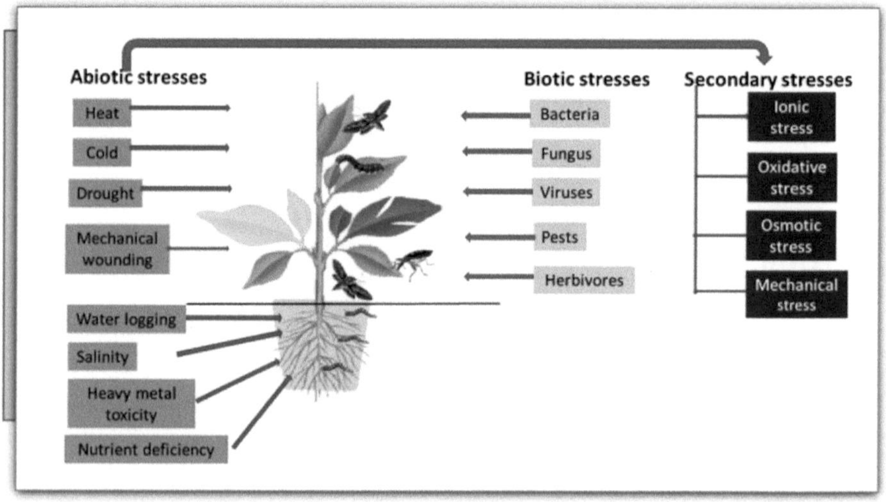

FIGURE 10.1 Category of biotic and abiotic stresses.

In the last decade, secondary metabolites, abundantly dispersed in plants, became candidates of increasing curiosity in relevance to their considerable practical applications for medicinal, nutritive, and cosmetic aspects (Akhi et al., 2021), as well as to their indubitable contribution to plant stress physiology. Plant organisms being unprovided with motility and immune system, have expansive alternative defense strategies, implying the wide range of secondary metabolites as tools to combat stress constraints. The diversity and interactions exhibited by the secondary metabolites are central to the protective role ranging from metal toxicity, salt, and drought resistance, UV protection to signal transduction (Grassmann et al., 2002; Osborn et al., 2003; Mazid et al., 2011).

There are three primary categories of secondary metabolites: phenolics, terpenes, and alkaloids. Phenolic compounds are synthesized particularly from various intermediates of the shikimic acid pathway, confer major defensive implications to the plants. Terpenes consist of 5-C isoprene as a basic structural unit and act as toxins and feeding repellents to many herbivores. Secondary metabolites of the third category, i.e., alkaloids are synthesized specifically from common amino acids (Wink, 2008), serve as efficient defensive components against pathogens and predators.

Phenolic compounds, one of the major class of secondary metabolites display a broad array of biological activities in plants, such as photo-protectant, defense against pathogen attack, coloration to plant organs to attract pollinators, and most important is the antioxidant activity that protects plants from oxygen free radicals produced as a result of impairment of metabolic processes by various adverse environmental factors Waskiewicz et al., 2013). Over the last few decades, the accessibility of molecular biology tools has facilitated the understanding of the mechanism of stress and defense response signaling the production of secondary metabolites *in-vivo* and *in-vitro* plant growth studies.

10.2 PHENOLIC COMPOUNDS AND THEIR CLASSIFICATION

Phenolics are one of the widely distributed and highly heterogeneous groups of secondary metabolites. Phenolic compounds are typically divided into two classes: simple phenols and polyphenols (Kulbat, 2016). Simple phenolics consist of a hydroxyl group-containing aromatic ring as the basic skeleton (Kant et al., 2015). The group of simple phenolic compounds includes phenolic acids and coumarins. Phenolic acids could be derivatives of hydroxybenzoic acid or hydroxycinnamic acid. The former includes gallic acid, vanillic acid, protocatechuic acid, syringic acid, and gentisic acid while hydroxycinnamic acid constitutes caffeic acid, p-coumaric acid, ferulic acid, sinapic acid, and caftaric acid (Michalak, 2006; Akhi et al., 2021). Polyphenols are complex derivatives containing several aromatic rings linked together (Figure 10.4) and are further categorized as flavonoids and non-flavonoids (Sumbul et al., 2011). Different flavonoids compounds are flavones, flavonols, isoflavones, flavanones, flavanols, anthocyanins, and chalcones. Tannins, lignans, and stilbenes constitute non-flavonoids (Tsimogiannis and Oreopoulou, 2019; Lattanzio, 2013). Detailed classification of phenolics is presented in Figure 10.2.

10.3 BIOSYNTHESIS OF PHENOLIC COMPOUNDS

Phenolic compounds are synthesized in the chloroplast, endoplasmic reticulum (ER) membrane, and cytosol of plant cells. The amino acid phenylalanine acts as the precursor for the synthesis of phenolic compounds through the shikimic acid pathway. The first step in the synthesis of all phenolic compounds is the simple uncoupling of the amino ($-NH_2$) group from phenylalanine to generate cinnamic acid. The reaction is catalyzed by the enzyme phenylalanine ammonia-lyase (PAL).

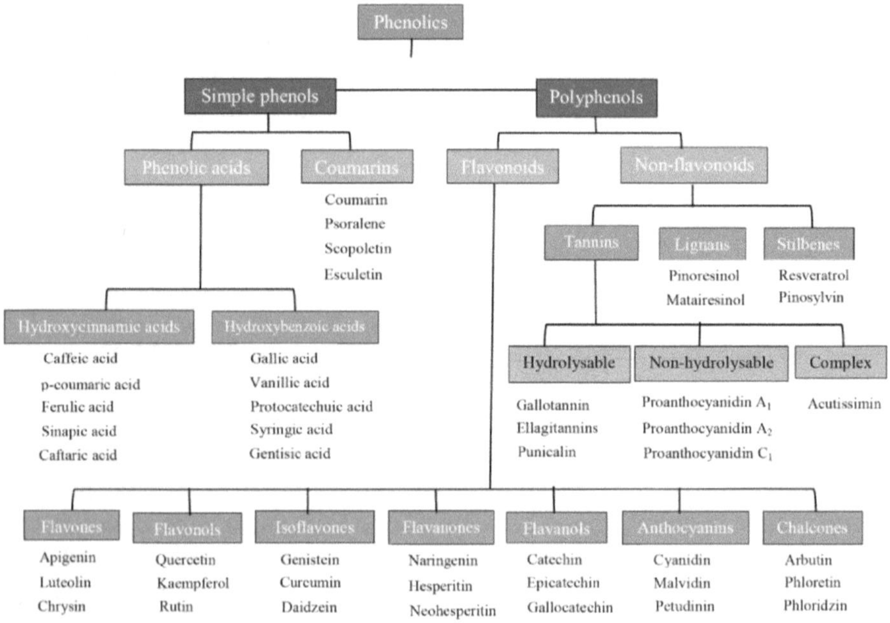

FIGURE 10.2 Classification of phenolic compounds.

Cinnamic acid is then used as the pioneer molecule for the synthesis of other phenolic compounds through the phenylpropanoid pathway (Knaggs, 2003). Flavonoids are synthesized from chalcone produced from p-coumaroyl Co-A and malonyl co-A in the presence of the enzyme chalcone synthase. Chalcone is acted upon by the enzyme chalcone isomerase to form flavanone which is the precursor for different flavonoids (Figures 10.3–10.5).

10.4 MECHANISM OF OPERATION OF DEFENSE SYSTEM BY PHENOLIC COMPOUNDS

There are various modes of operation of defense systems in plants against biotic and abiotic stresses. Under abiotic stress and pathogen attack, there is overproduction of ROS (reactive oxygen species) like superoxide anion (O_2^-), hydrogen peroxide (H_2O_2), singlet oxygen (O^-) and hydroxyl radical OH^- in the chloroplast, mitochondria, ER, peroxisomes, plasma membrane, and cell wall of plant cells. Depending on the concentration of ROS in the plants, they exhibit a binary role. At fair concentration, the ROS plays a vital role in cell multiplication, differentiation, and viability (Mittler, 2017). The excess accumulation of ROS is exceedingly deleterious to the plant cells leading to lipid peroxidation (LPO), oxidation of proteins, membrane damage, enzyme inhibition, irreversible damage to nucleic acids, and cell mortality. The detoxification of excess ROS in plants is mediated by two types of antioxidant systems: enzymatic and non-enzymatic. The enzymatic antioxidant system includes the action of enzymes like peroxidase (POX), superoxide dismutase (SOD), catalase, POX, ascorbic acid (AA) POX, and glutathione reductase (Ara et al., 2013) to nullify the deleterious effect of the ROS. The non-enzymatic antioxidants include vitamin C, tocopherols, carotenoids, glutathione, and phenol derivatives (Ahmad et al., 2010; Sharma et al., 2012). The function of ROS is to trigger plant defense genes which in turn activate the phenylpropanoid pathway for the synthesis of phenolic compounds. The "trapping" of free radicals facilitates the suppression of membrane LPO, which is the widely known phenomenon highlighting the antioxidant properties of phenolic compounds.

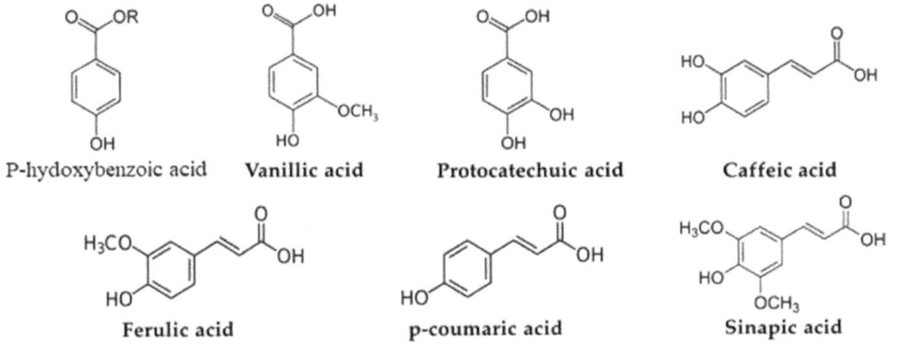

FIGURE 10.3 Chemical structure of some phenolic acids.

The molecular structure, number, and position of hydroxyl groups largely determine the phenolic compounds' action. Phenolic compounds, especially flavonoids, also play a role in the reduction of the fluidity of cell membranes, thereby limiting free radicals' diffusion, which in turn decreases the peroxidation of membrane lipids, resulting in stabilization of cell membranes (Kulbat,

2016). Another operation of the defense system involves the development of multiple barriers to prevent the entry of pathogens and enzyme-mediated damage to the pathogen cell wall.

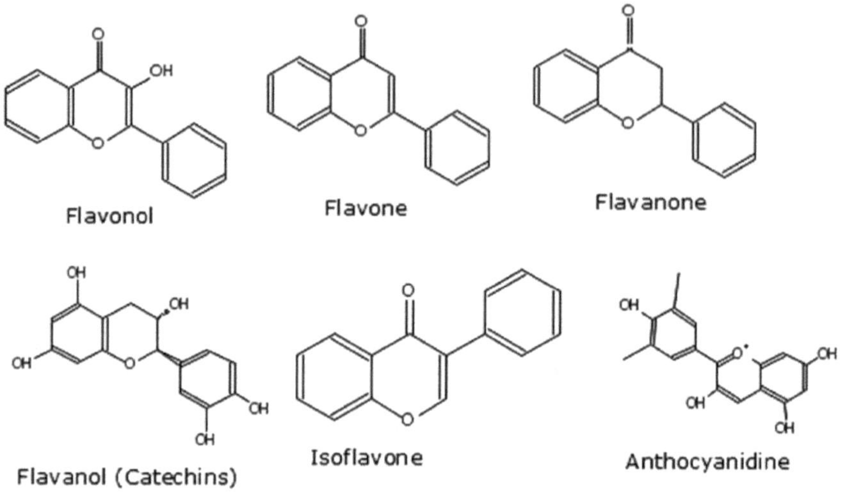

FIGURE 10.4 Chemical structure of some flavonoids.

Source: https://www.akspublication.com/paper05_jul-dec2007.htm.

Besides, plants utilize a particular recognition process permitting the quick recognition of pathogen invasion and commencement of vigorous defensive responses. Once infected, a few plants also develop immunity to succeeding microbial attacks (Schaller and Ryan, 1996; Bernays, 2017).

10.5 ROLE OF PHENOLS IN STRESS MANAGEMENT

10.5.1 PHENYLPROPANOIDS

It has been proven scientifically that phenylpropanoids and their derivatives get accumulated in plants growing under stress conditions, thus justifying the protective role in plant cells. Reyes and Cisneros-Zevallos (2003) expressed that mechanical damage induces the synthesis of phenolic compounds and surge in antioxidant and PAL activity in purple-flesh potatoes (*Solanum tuberosum* L.). Similar results were obtained by Posmyk et al. (2009) in copper-treated red cabbage seedlings. The presence of copper ions initiated oxidative stress leading to the destruction of cell membranes. The authors measured the high concentration of thiobarbituric acid reactive substances (TBARS), thus indicating the involvement of phenolic compounds in the defense system against copper ions mediated oxidative stress.

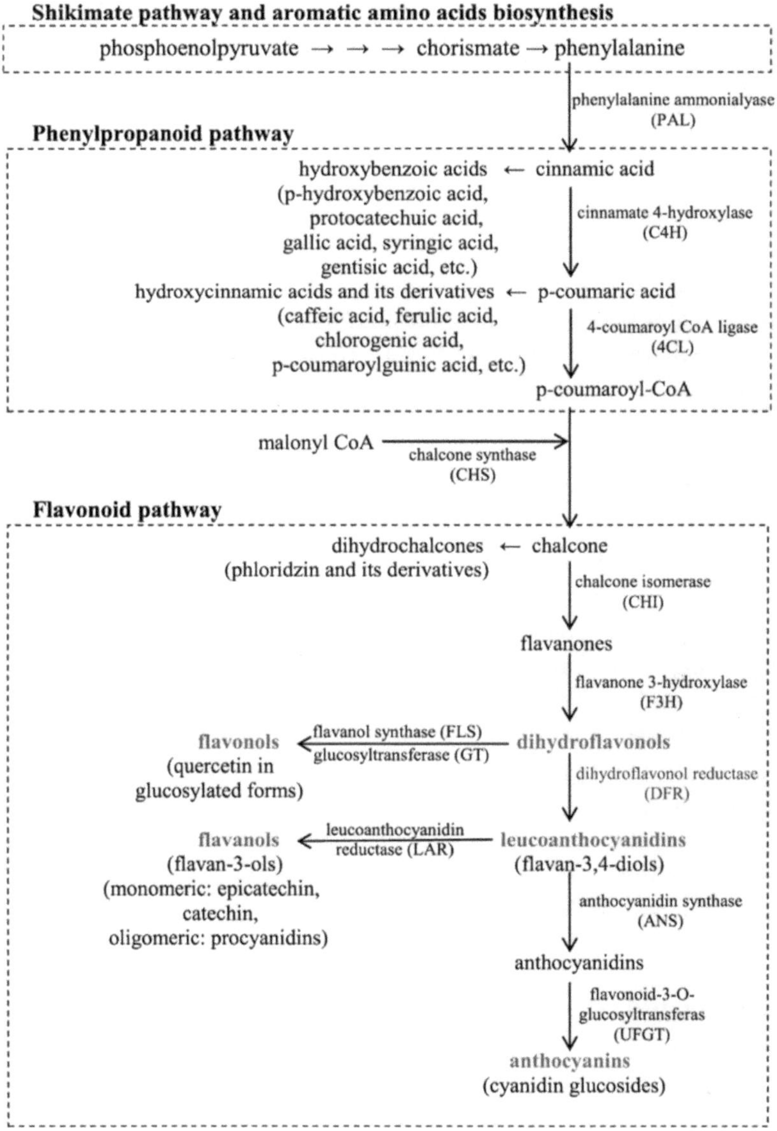

FIGURE 10.5 Biosynthesis pathway of phenolic compounds.

Source: https://www.researchgate.net/figure/Biosynthesis-pathway-of-apple-phenolic-compounds_fig2_266576498.

Salicylic acid synthesized through the phenylpropanoid pathway is a principal compound involved in the initiation of plant defense settings. Both local and systemic augmentation of salicylic acid has been reported under stress conditions. One of the volatile derivatives of salicylic acid– salicylic acid methyl ester, can defend the host plant and neighboring plants as well from pests and herbivores (Horvath and Chua, 1994).

Reports document the accumulation of some hydroxycinnamic and hydroxybenzoic acids such as gallic acid, caffeic acid, caftaric acid, cinnamyl malic acid, and vanillic acid in plants growing under saline conditions. Thus, counterbalancing the secondary stress induced by high salt concentrations (Sarker and Oba, 2018; Scagel et al., 2019).

There are many pieces of evidence of accumulation of hydroxycinnamic acid in plants undergoing wounding stress and pathogen attack. Hydroxycinnamic acids are converted to monolignols. Monolignols are transported to cell walls where they get oxidize and polymerize to form lignin and suberin, helping in wound healing (Jacobo-Velazquez and Cisneros-Zevallos, 2012).

10.5.2 COUMARINS

Coumaric acid is produced by hydroxylation of cinnamic acid or deamination of tyrosine by the tyrosine ammonia-lyase enzyme. Coumaric acid may then undergo cyclization leading to the formation of coumarins. Coumarins are exceptionally functioning molecules with a broad range of anti-microbial, antifungal, and antiherbivore activity. Some derivatives of coumarins are highly active against multifarious soil-borne plant pathogenic fungi and display more stability as compared to solo coumarin compounds (Brooker et al., 2008). A coumarin derivative: 7-hydroxylated coumarin successfully hampers with the parasitism activity of *Orobanche cernua* on *Helianthus annuus* (Serghini et al., 2001). Coumarin, dicumarol owing to its bitter taste protects plants from herbivores (Kulbat, 2016).

10.5.3 FLAVONOIDS

Flavonoids are a heterogeneous group of secondary metabolites. The chemical structure of flavonoid is based on a 15-carbon skeleton (C6-C3-C6) of diphenyl propane. Depending on the number and type of side groups the flavonoids have been classified into the following subclasses: flavones, flavonols, isoflavonols, flavanones, anthocyanins, and chalcones. Most flavonoids impart characteristic color to different plant organs. For example, quercetin (flavonol) is responsible for yellow shade and cyanidin, malvidin, delphinidin, and pelargonidin (anthocyanins) give out blue to pink tinge.

Research reports support the accumulation of flavonoids in the vacuoles of epithelial cells of UV irradiated plants playing an important role in plant cells' defense from oxidative damage. Ryan et al. (2001) demonstrated the protective role of flavonoids in UVB-exposed mutants of *Arabidopsis* plants showing less tolerance to UVB radiation than wild-type relatives. The mutants lack flavonoid 3'-hydroxylase enzyme that forms dihydroquercetin from dihydrokaempferol thus supporting the photoprotective role of quercetin. The accumulation of flavonoids in plants plays a significant role in salt tolerance. Caliskan (2017) observed high levels of rutin, quercetine, isoquercetine, quercitrine, and chlorogenic acid in response to the high concentration of salts such as magnesium sulfate and chlorides of calcium and sodium in greenhouse-grown *Hypericum pruinatum* plantlets. In response to salinity stress, the phenylpropanoid pathway gets stimulated leading to the biosynthesis of a gush of phenolic compounds displaying high antioxidant potential.

Another critical abiotic stress is drought stress that affects many physiological and biochemical processes of the plants. The increased flavonoid content under drought stress conditions has been reported in willow leaves (Akula and Ravishankar, 2011) and *Arabidopsis* (Nakabayashi et al., 2014). Nakabayashi and associates measured the anthocyanin and flavonol levels in *Arabidopsis* plants at intermittent intervals of 2 days under induced drought stress conditions and observed the high amount of anthocyanin and flavonol in response to drought. The over-expression of important enzymes of flavonoid biosynthesis such as chalcone synthase, chalcone isomerase, and flavanone-3-hydroxylase has been detected in *Arabidopsis* plants. Sanchez-Rodríguez (2011) reported the accumulation of flavonoids like kaempferol and quercetin enhancing drought tolerance in tomato plants. Flavonoid accumulation in the cytoplasm can efficiently detoxify harmful H_2O_2 radicals produced as a result of water scarcity stress.

The majority of flavonoids exhibit antimicrobial activity. The elevated levels of flavonoids in plants have been observed in response to microbial infection, mechanical injury, frost, and insufficient nutrients. Certain flavonoids impart protection against heavy metals by chelating iron, nickel, copper, and zinc. This fact is supported by an investigation on *Zea mays* L. grown in aluminum ions polluted soil, where an upsurge in levels of quercetin and catechin in the root exudates were noticed (Michalak, 2006). An increase in chlorogenic acid and rutin content is observed in *Zea mays* plant exposed to high doses of heavy metals copper, cadmium, and lead (Mira et al., 2002). Chelating heavy metals by flavonoids is likely to be an efficient mode of defense system against the deleterious effects of metal accumulation in the soil. Kostyuk et al. (2004) confirmed that the metal-flavonoid complex displays considerable high superoxide dismutation activity as compared to the native flavonoid.

Bota and Deliu (2011) reported positive parallelism between flavonoid and copper sulfate concentrations in cell cultures of *Digitalis lanata*. Similar results were depicted by Badiaa (2020) in tomato crop exposed to the high concentration of heavy metals copper and zinc. The highest accumulation of flavonoids was observed in leaves compared to roots and it was dose-dependent. These investigations revealed that micro-elements such as copper and zinc play a crucial role in the survival of the plant. However, a high concentration of these metals is stressful and elicits the defense system in crops.

10.5.4 TANNINS

Tannins have been classified into two categories: hydrolysable tannins and condensed tannins, produced by polymerization of phenolic acids and flavonoid units, respectively. Tannins occur in high concentration, particularly in the leaves and bark of the source plant. The ability of tannins to bind and denature proteins and their unpleasant bitter taste render them excellent compounds that provide plant protection against insects and herbivores (Oates et al., 1980). Tannins possibly have a special function in combating the disease-causing microbes in certain plants. Pyrocatechinic and chlorogenic acids arrest spore germination and subsequent growth of the fungus *Colletotrichum circinans*, the causal organism of smudge disease of onion (Taye and Borkataki, 2020). It is suggested that disease-resistant cultivars can oxidize chlorogenic acid into more potent antifungal compounds quinones easily as compared to susceptible ones (Mazid et al., 2011). Tannins owing to their ability to chelate heavy metal ions play a significant role in encountering heavy metal stress in plants (Lavid et al., 2001).

10.5.5 LIGNIN AND LIGNANS

Lignin is a highly branched polymer derived from the phenylpropanoid pathway, formed from three different monolignols viz., coniferyl, coumaryl, and sinapyl. These metabolites have noticeably distinct physiological roles. The heterogeneous lignin is a structural cell wall component of vascular elements providing mechanical support to terrestrial plants and enabling long-distance water transportation. Its physical and chemical toughness impedes browsing and renders it comparably resistant to digestion by herbivores and insect pathogens (Mader and Amberg-Fisher, 1982). Lignification is considered the first-line defense as it prevents the entry and subsequent growth of pathogenic organisms (Davin and Lewis, 2000). The role of lignin in the plant defense system was supported by Boudet (2006) in investigation with wheat expressing low resistance to *Puccinia graminis* when treated with a lignin synthesis inhibitor. Michalak (2006) reported the enhanced synthesis of lignin precursors in plants exposed to heavy metal stress.

Lignans are chemical compounds derived from diphenols with dibenzylbutane as a basic structural unit. Lignans are analogous to phytoestrogens, playing a significant role in plant defense. These compounds act as inhibitors of enzymes such as cellulases, glucosidases, polygalacturonases, and laccases produced as host cells and tissues degradative enzymes by microbes. Furthermore, research findings support the over-expression of the lignan synthesis gene during the lignification process in maritime pine and flax. Due to the analogy of lignans to phytoestrogens, these compounds can interfere with the normal metamorphosis process of insects by disrupting their endocrine system (Paniagua et al., 2017). The lignans occur in seed coats, flowers, sapwood, and heartwood of stem, leaves, bark, roots, etc.

10.5.6 STILBENES

A large evidentiary material is available to support the implications of stilbenes in both inducible and constitutive defense mechanisms in plants. Stilbenes, non-flavonoid phenolic compounds, are well known to produce in plants in response to pathogen attack, wounding, and oxidative stress. The widely studied stilbene, resveratrol is found in various angiosperms including grapevine (*Vitis vinifera*), cranberry (*Vaccinium macrocarpon*), bilberry (*V. myrtillus*), peanut (*Arachis hypogaea*), tomato (*Lycopersicon esculentum*), cocoa (*Theobroma cacao*), and few gymnosperms (*Pinus* and *Picea*). Stilbenes are key molecules involved in the battling of grapevine cultivars against severe fungal pathogens, *Botrytis cinerea*, *Plasmopara viticola,* and *Erysiphe necator*, the causal organisms of gray mold, downy mildew, and powdery mildew disease respectively. Resistant grapevine varieties produce elevated levels of stilbenes, d-viniferin, and pterostilbene at the infected area which are distinctly toxic to the fungal attack (Viret et al., 2018).

Conifers have developed a complex system of defense that integrates both anatomical structures and toxic stilbenes accumulated against bark beetle attacks (Kolosava and Bohlmann, 2012). Paetz et al. (2013) investigated stilbenes in Norway spruce bark with a diaryl-ethene skeleton with antifungal properties against *Ceratocystis polonica* infection. It was evident from high transcript levels of the stilbene synthase gene. However, with the progress of the infection, stilbene concentrations declined, and this was due to the metabolism of host phenolics to ring-opened, deglycosylated, and dimeric compounds which are used as a source of carbon by *C. polonica.*

10.5.7 PHYTOALEXINS

Phytoalexins are compounds belonging to flavonoids and isoflavonoids with typically protective properties. A number of them have a chemical structure resembling that of sesquiterpenoids. Isoflavonoid phytoalexins are characteristic of the family Fabaceae whereas sesquiterpenoid phytoalexins are typical for Solanaceae. Phytoalexins are synthesized in plants and accumulate around the site of infection in response to pathogen attacks by bacteria, fungi, viruses, and nematodes. This helps in mitigating the proliferation of invading pathogens (Jeandet, 2015). These compounds function like antibiotics, inhibiting the sporulation and mycelial growth of phytopathogenic bacteria and fungi. However, some pathogenic organisms can deactivate phytoalexins by hydroxylation and demethylation of aromatic rings, rendering phytoalexins more sensitive to oxidation (Kulbat, 2016).

10.6 CONCLUSIONS

Phenolic compounds have a crucial role in the plant defense system. The majority of the phenolic compounds have powerful antioxidant properties. A couple of them display the capacity to chelate substantial metal particles. Significantly, phenolic phytoalexins display anti-microbial and antifungal action. Coumarins and tannins repulse herbivores. Phenylpropanoids are threshold molecules for the production of lignin and suberin that reinforce cell walls thus impeding the invasion of pathogens to plant tissues. The resisting response of plants is apparently due to the up-regulation of enzyme levels or their activities involved in the synthesis of phenolic compounds.

The secondary metabolites synthesized under abiotic stress have potential application in the nutraceutical, pharmaceutical, cosmetics, and agrochemical markets. This methodology can be extended to horticultural crops for the production of plant metabolites of high commercial value.

To explore and capitalize on this aspect with a commercial view, the factors need to be taken into account, which includes the impact on different groups of biologically active compounds, molecular mechanisms of stress physiology, and the response to stresses of different crop plants.

KEYWORDS

- **biosynthesis**
- **cinnamic acid**
- ***Helianthus annuus***
- ***Orobanche cernua***
- **phenolics**
- **phenylalanine ammonia-lyase**

REFERENCES

Ahmad, P., Jaleel, C. A., Salem, M. A., Nabi, G., & Sharma, S., (2010). Roles of enzymatic and nonenzymatic antioxidants in plants during abiotic stress. *Critical Reviews in Biotechnology, 30*(3), 161–175.

Akhi, M. Z., Haque, M. M., & Biswas, M. S., (2021). Role of secondary metabolites to attenuate stress damages in plants. In: *Antioxidants*. IntechOpen.

Akula, R., & Ravishankar, G. A., (2011). Influence of abiotic stress signals on secondary metabolites in plants. *Plant Signaling and Behavior, 6*(11), 1720–1731.

Ara, N., Nakkanong, K., Lv, W., Yang, J., Hu, Z., & Zhang, M., (2013). Antioxidant enzymatic activities and gene expression associated with heat tolerance in the stems and roots of two cucurbit species ("*Cucurbita maxima*" and "*Cucurbita moschata*") and their interspecific inbred line "maxchata". *International Journal of Molecular Sciences, 14*(12), 24008–24028.

Badiaa, O., Yssaad, H. A. R., & Topcuoglu, B., (2020). Effect of heavy metals (copper and zinc) on proline, polyphenols and flavonoids content of tomato (*Lycopersicon esculentum* Mill.). *Plant Archives, 20*(1), 2125–2137.

Ballhorn, D. J., Kautz, S., Heil, M., & Hegeman, A. D., (2009). Cyanogenesis of wild lima bean *(Phaseolus lunatus* L.) is an efficient direct defense in nature. *Plant Signaling and Behavior, 4*(8), 335–345.

Bernays, E. A., (2017). *Insect-Plant Interactions (1993): Volume V*. CRC Press.

Bhattacharya, A., Sood, P., & Citovsky, V., (2010). The roles of plant phenolics in defense and communication during *Agrobacterium* and *Rhizobium* infection. *Molecular Plant Pathology, 11*(5), 705–719.

Bota, C., & Deliu, C., (2011). The effect of copper sulphate on the production of flavonoids in *Digitalis lanata* cell cultures. *Farmacia, 59*(1), 113–118.

Boudet, A. M., (2000). Lignins and lignification: Selected issues. *Plant Physiology and Biochemistry, 38*(1, 2), 81–96.

Brooker, N., Windorski, J., & Bluml, E., (2008). Halogenated coumarin derivatives as novel seed protectants. *Communications in Agricultural and Applied Biological Sciences, 73*(2), 81–89.

Caliskan, O., Radusiene, J., Temizel, K. E., Staunis, Z., Cirak, C., Kurt, D., & Odabas, M. S., (2017). The effects of salt and drought stress on phenolic accumulation in greenhouse-grown *Hypericum pruinatum. Italian Journal of Agronomy, 12*(918), 271–275.

Davin, L. B., & Lewis, N. G., (2000). Dirigent proteins and dirigent sites explain the mystery of specificity of radical precursor coupling in lignan and lignin biosynthesis. *Plant Physiology, 123*(2), 453–462.

Grassmann, J., Hippeli, S., & Elstner, E. F., (2002). Plant's defense and its benefits for animals and medicine: Role of phenolics and terpenoids in avoiding oxygen stress. *Plant Physiology and Biochemistry, 40*(6–8), 471–478.

Horvath, D. M., & Chua, N. H., (1994). The role of salicylic acid in systemic acquired resistance. *Current Opinion in Biotechnology, 5*(2), 131–136.

Jacobo-Velazquez, D. A., & Cisneros-Zevallos, L., (2012). An alternative use of horticultural crops: Stressed plants as biofactories of bioactive phenolic compounds. *Agriculture, 2*(3), 259–271.

Jeandet, P., (2015). Phytoalexins: current progress and future prospects. *Molecules, 20*(2), 2770–2774.

Kant, M. R., Jonckheere, W., Knegt, B., Lemos, F., Liu, J., Schimmel, B. C. J., & Alba, J. M., (2015). Mechanisms and ecological consequences of plant defense induction and suppression in herbivore communities. *Annals of Botany, 115*(7), 1015–1051.

Knaggs, A. R., (2003). The biosynthesis of shikimate metabolites. *Natural Product Reports, 20*(1), 119–136.

Kolosova, N., & Bohlmann, J., (2012). Conifer defense against insects and fungal pathogens. In book: *Growth and Defense in plants* (pp. 85–109). Springer, Berlin, Heidelberg.

Kostyuk, V. A., Potapovich, A. I., Strigunova, E. N., Kostyuk, T. V., & Afanas' Ev, I. B., (2004). Experimental evidence that flavonoid metal complexes may act as mimics of superoxide dismutase. *Archives of Biochemistry and Biophysics, 428*(2), 204–208.

Kulbat, K., (2016). The role of phenolic compounds in plant resistance. *Biotechnology and Food Science, 80*(2), 97–108.

Lattanzio, V., (2013). Phenolic compounds: Introduction 50. *Natural Products,* 1543–1580.

Lavid, N., Schwartz, A., Yarden, O., & Tel-Or, E., (2001). The involvement of polyphenols and peroxidase activities in heavy-metal accumulation by epidermal glands of the waterlily (*Nymphaeaceae*). *Planta, 212*(3), 323–331.

Mazid, M., Khan, T. A., & Mohammad, F., (2011). Role of secondary metabolites in defense mechanisms of plants. *Biology and Medicine, 3*(2), 232–249.

Michalak, A., (2006). Phenolic compounds and their antioxidant activity in plants growing under heavy metal stress. *Polish Journal of Environmental Studies, 15*(4), 523–530.

Mira, L., Fernandez, M. T., Santos, M., Rocha, R., Florencio, M. H., & Jennings, K. R., (2002). Interactions of flavonoids with iron and copper ions: A mechanism for their antioxidant activity. *Free Radical Research, 36*(11), 1199–1208.

Mittler, R., (2017). ROS are good. *Trends in Plant Science, 22*(1), 11–19.

Nakabayashi, R., Yonekura-Sakakibara, K., Urano, K., Suzuki, M., Yamada, Y., Nishizawa, T., & Saito, K., (2014). Enhancement of oxidative and drought tolerance in *Arabidopsis* by overaccumulation of antioxidant flavonoids. *The Plant Journal, 77*(3), 367–379.

Oates, J. F., Waterman, P. G., & Choo, G. M., (1980). Food selection by the south Indian leaf monkey, *Presbytis johnii*, in relation to leaf chemistry. *Oecologia, 45*(1), 45–56.

Osbourn, A. E., Qi, X., Townsend, B., & Qin, B., (2003). Dissecting plant secondary metabolism–constitutive chemical defenses in cereals. *New Phytologist, 159*(1), 101–108.

Paetz, C., Wadke, N., Brand, W. A., Wright, L. P., Gershenzon, J., Schmidt, A., & Hammerbacher, A., (2013). A common fungal associate of the spruce bark beetle metabolizes the stilbene defenses of Norway spruce. *Free Radical Research, 36*(11), 1199–1208.

Paniagua, C., Bilkova, A., Jackson, P., Dabravolski, S., Riber, W., Didi, V., & Hejatko, J., (2017). Dirigent proteins in plants: Modulating cell wall metabolism during abiotic and biotic stress exposure. *Journal of Experimental Botany, 68*(13), 3287–3301.

Posmyk, M. M., Kontek, R., & Janas, K. M., (2009). Antioxidant enzymes activity and phenolic compounds content in red cabbage seedlings exposed to copper stress. *Ecotoxicology and Environmental Safety, 72*(2), 596–602.

Reyes, L. F., & Cisneros-Zevallos, L., (2003). Wounding stress increases the phenolic content and antioxidant capacity of purple-flesh potatoes (*Solanum tuberosum* L.). *Journal of Agricultural and Food Chemistry, 51*(18), 5296–5300.

Ryan, K. G., Swinny, E. E., Winefield, C., & Markham, K. R., (2001). Flavonoids and UV photoprotection in *Arabidopsis* mutants. *Zeitschrift für Naturforschung C, 56*(9, 10), 745–754.

Sanchez-Rodríguez, E., Moreno, D. A., Ferreres, F., Del Mar Rubio-Wilhelmi, M., & Ruiz, J. M., (2011). Differential responses of five cherry tomato varieties to water stress: Changes on phenolic metabolites and related enzymes. *Phytochemistry, 72*(8), 723–729.

Sarker, U., & Oba, S., (2018). Augmentation of leaf color parameters, pigments, vitamins, phenolic acids, flavonoids and antioxidant activity in selected *Amaranthus tricolor* under salinity stress. *Scientific Reports, 8*(1), 1–9.

Scagel, C. F., Lee, J., & Mitchell, J. N., (2019). Salinity from NaCl changes the nutrient and polyphenolic composition of basil leaves. *Industrial Crops and Products, 127*, 119–128.

Schaller, A., & Ryan, C. A., (1996). Systemin–a polypeptide defense signal in plants. *BioEssays, 18*(1), 27–33.

Serghini, K., Perez De, L., A., Castejon, M. M., Garcia, T. L., & Jorrin, J. V., (2001). Sunflower (*Helianthus annuus* L.) response to broomraoe (*Orobanche cernua* loefl.) parasitism: Induced synthesis and excretion of 7-hydroxylated simple coumarins. *Journal of Experimental Botany, 52*, 227–234.

Sharma, P., Jha, A. B., Dubey, R. S., & Pessarakli, M., (2012). Reactive oxygen species, oxidative damage, and antioxidative defense mechanism in plants under stressful conditions. *Journal of Botany, 2012*. doi: 10.1155/2012/217037.

Sumbul, S., Ahmad, M. A., Mohd, A., & Mohd, A., (2011). Role of phenolic compounds in peptic ulcer: An overview. *Journal of Pharmacy and Bioallied Sciences, 3*(3), 361.

Taye, R. R., & Borkataki, S., (2020). Role of secondary metabolites in plant defense against insect herbivores. *Int. J. Chem. Stud, 8*, 200–203.

Tsimogiannis, D., & Oreopoulou, V., (2019). Classification of phenolic compounds in plants. In: *Book: Polyphenols in Plants* (pp. 263–284). Academic Press.

Viret, O., Spring, J. L., & Gindro, K., (2018). Stilbenes: Biomarkers of grapevine resistance to fungal diseases. *OENO One, 52*(3), 235–241.

Waskiewicz, A., Muzolf-Panek, M., & Golinski, P., (2013). Phenolic content changes in plants under salt stress. In: *Ecophysiology and Responses of Plants Under Salt Stress* (pp. 283–314). Springer, New York, NY.

Wink, M., (2008). Plant secondary metabolism: Diversity, function and its evolution. *Natural Product Communications, 3*(8), 1205–1216.

CHAPTER 11

Stress and Its Influence on the Generation of Reactive Oxygen Species and Oxidative Damage in Plants

J. NAVEEN,[1] G. HITHAMANI,[1] and H. G. PUSHPALATHA[2]

[1]*Department of Biochemistry, CSIR–Central Food Technological Research Institute, Mysore–570020, Karnataka, India*

[2]*Associate Professor, Maharani's Science College for Women, Mysore–570001, Karnataka, India, E-mail: pushpa_2001@rediffmail.com*

ABSTRACT

Stress is an inevitable challenge plant come across; either abiotic or biotic or both. Environmental stress causes adverse effects on plant growth and development and leads to negative effect on agriculture like lower crop yield and poor quality. Plant reacts to the stress by producing excess reactive oxygen species (ROS), which causes oxidative stress that affects the plant's own cells. Various cell organelles produce different kinds of ROS with different characteristic features. A limited quantity of ROS is required to the plant for its normal biochemical, physiological activities and is maintained by antioxidant defense systems (non-enzymatic and enzymatic antioxidants). Antioxidant components are plant secondary metabolites, play a role in defense mechanisms. Equilibrium between the ROS generation and its detoxification is much required to the plant system in order to survive under adverse environmental stresses. Hence, the field of research in this area has attracted researchers massively. A large area of research needs to be addressed/remains unexplored in understanding the ROS action as a signaling molecule. In this chapter, generation of ROS, types of ROS, site

Plant Metabolites under Environmental Stress: Mechanisms, Responses, and Adaptation Strategies.
Nivas M. Desai, Manasi Patil, Umesh R. Pawar (Eds.)

of ROS generation, response of ROS to different environmental stresses, ROS as signaling molecule and detoxification of ROS by antioxidants are highlighted and have been discussed in detail.

11.1 INTRODUCTION

Plants are known to reduce the stress of living beings (i.e., humans, and animals) by utilizing the carbon dioxide present in the air and converting it into precious oxygen. Even though the plants promote the life of the beings (by detoxifying the polluted air) they also come across various hurdles. Stress to plants is in numerous forms like extreme temperature, salinity, drought, mechanical stress, waterlogging/flood, and biotic stress like pathogen and pest attack. These stresses affect plant health which is a significant threat to sustainable agriculture and its production (Hasanuzzaman et al., 2020). Plants are very sensitive and react to the stress by means of chemical reactions, i.e., producing reactive oxygen species (ROS). ROS is a part of plant metabolism and mechanism of action to their routine need, but excess of ROS production was observed during the stress conditions (Noctor and Foyer, 1998) and are harmful to the plant health.

The primary sites of ROS generation are chloroplast, mitochondria, apoplast, peroxisomes, and plasma membranes (Singh et al., 2019). ROS have various radical molecules, hydroxyl radical ($OH^•$), superoxide anion ($O_2^{•-}$), alkoxy radical ($RO^•$), hydroperoxyl radical ($H_2O^•$) and non-radicle like singlet oxygen ($^1O_2$), hydrogen peroxide (H_2O_2) (Mehla et al., 2017; Hasanuzzaman et al., 2019). Radicals $OH^•$, $O_2^{•-}$, $^1O_2$, and H_2O_2 are the most important and destructive ROS. During the favorable condition, the ROS generated at the basal level is unable to cause damage to cell because it is constantly scavenged/managed by the antioxidants (Foyer and Noctor, 2005; Hasanuzzaman et al., 2020). But, the induced ROS acts as a toxic molecule, which helps hypersensitive reaction (HR-causes cell death) in the modulation of numerous plant metabolic and defense pathways. ROS is common in various cellular activities, and the different antioxidant enzymes quench the same. ROS also acts as a defense molecule and signaling molecule. Stress is an adverse environmental condition which we humans can judge and safeguard ourselves. Even plants can sense and react to environmental stress by enhancing ROS production as it produces the pathogen/pest attack. If ROS level exceeds the limit of the defense mechanisms, the plant cell is said to be under "oxidative stress" (Sharma et al., 2012).

Higher ROS generation due to various stresses can cause risk to the cells, i.e., causes peroxidation of lipids, oxidation of proteins, nucleic acids damage, enzyme inhibition, programmed cell death (PCD) pathway activation, and finally causes cell death (Mittler, 2002; Sharma and Dubey, 2005, 2007). ROS at higher concentrations is toxic to the foreign body (pathogen/pest), which helps to kill the pathogen at the penetration site. The role of HR-reaction, which results in cell death, is to avoid further spread of the pathogen. In addition to the destructive activity, ROS act as signaling molecules (second messengers) in various cellular processes, leading to activation of numerous defense and resistance mechanisms in the plants to environmental stresses (Desikan et al., 2001; Neill et al., 2002; Yan et al., 2007). The role of signaling and activation of defense mechanism usually depends on the equilibrium between ROS generation and scavenging activity (Figure 11.1).

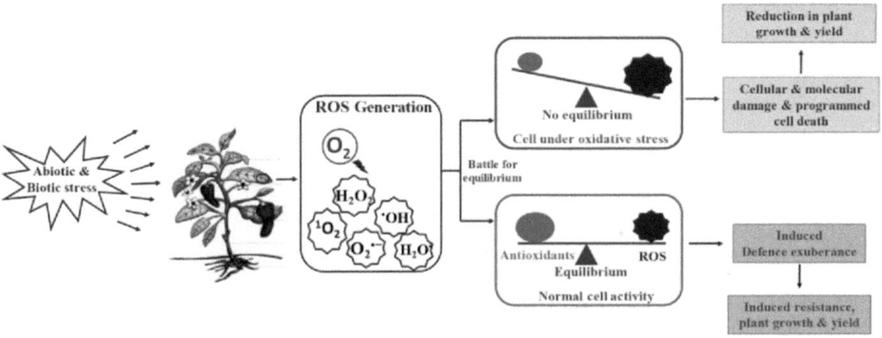

FIGURE 11.1 An overview of reactive oxygen species generation and potential antioxidant activity in the plant under normal and stress conditions.

Even though the ROS has a multifunctional role, the management of ROS levels inside the plant cells is essential to avoid oxidative damage and not eliminate the ROS completely. Hence, the role of antioxidative enzymes (catalase: CAT; superoxide dismutase: SOD; ascorbate peroxidase (POX): APX; guaiacol peroxidase: GPX; monodehydroascorbate reductase: MDHAR; glutathione reductase: GR; and dehydroascorbate reductase: DHAR) and nonenzymatic antioxidants (carotenoids, phenolics, and tocopherols) are also key to detoxify/scavenge the ROS (Noctor and Foyer, 1998). An increase in antioxidants' activity in plants combats oxidative stress induced by environmental stress and links with the increased tolerance to the environmental stresses (Zaefyzadeh et al., 2009; Chen et al., 2010).

Therefore, ROS, and antioxidants play a significant role in stress tolerance and the plant breeder can develop oxidative stress-tolerant crop by altering the transgenic lines, i.e., by improving multiple antioxidants and their level (Allen et al., 1997; Faize et al., 2011; Lee et al., 2007). Hence, the present chapter focuses types of ROS and its site production under stress, role ROS as a messenger, and the antioxidants involved to maintain ROS under environmental stress conditions (Figure 11.1).

11.2 FORMS OF REACTIVE OXYGEN SPECIES (ROS)

Generation of ROS is the primary biochemical reaction that occurs during the phototropic action of the plant. All ROS, like free radicals, ions, and reactive molecules are derived from O_2 (Figure 11.2). It is estimated that only 1–2% of O_2 is converted into ROS (Bhattacharjee, 2005). ROS acts as a secondary messenger at normal concentration, and at higher concentration, it damages the biomolecules. The different types of ROS are $^1O_2$ (singlet oxygen), $OH^•$ (hydroxyl radical), H_2O_2 (hydrogen peroxide), and $O_2^{•-}$ (superoxide radical) (Bhattacharjee, 2005).

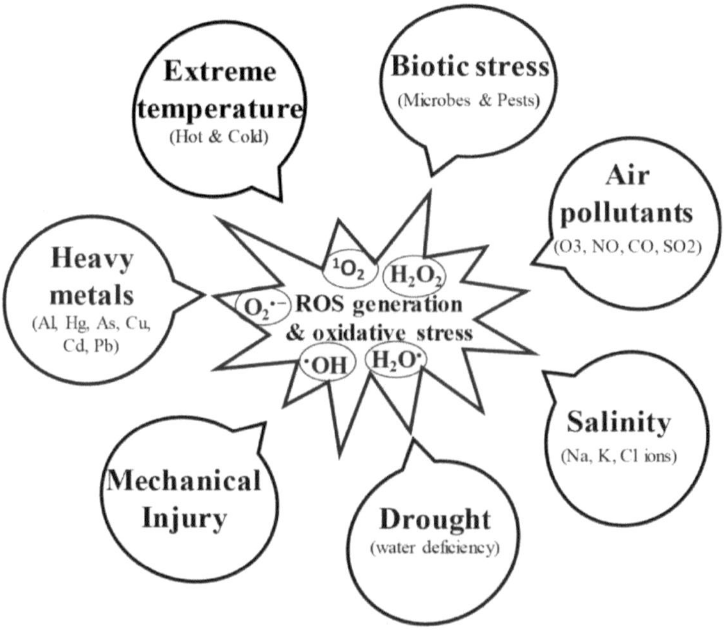

FIGURE 11.2 Environmental stress factors cause ROS generation and oxidative stress in plant.

Superoxide radical ($O_2^{\bullet-}$) is generated constantly in the chloroplasts, i.e., by the partial reduction or energy transfer of O_2. The formation of $O_2^{\bullet-}$ is mainly in the thylakoid-localized photosystem I (PSI) in the non-cyclic electron transport chain (ETC) and in some other cellular compartments. Sometimes in ETC reacts with the O_2, which gives rise to $O_2^{\bullet-}$ and is the initial ROS formation. $O_2^{\bullet-}$ also undergoes various reactions and generates other types of ROS. $O_2^{\bullet-}$ have a short half-life reactive time of 2–4 μs, which means to cause less damage by itself. Meanwhile, it undergoes a transformation into more toxic and reactive $^1O_2$ and $OH^{\bullet}$ and also causes membrane lipid peroxidation (LPO) (Halliwell, 2006).

Singlet Oxygen ($^1O_2$) is generated from the reaction of triplet state chlorophyll in the antenna system and from O_2. The stress creates insufficient intracellular CO_2 concentration, by stomatal closure, which leads to the formation of $^1O_2$. The generation of $^1O_2$ causes severe damages to the photosystems (PSI and PSII), which further affects the entire photosystem machinery. $^1O_2$ has a very little half-life of 3 μs (Hatz et al., 2007), it can diffuse through 100 nanometers from the site of generation and harms a wide range of targets like nucleic acids, pigments, lipids, and proteins (Wagner et al., 2004; Krieger-Liszkay et al., 2008). One molecule of O_2 is the key accountable factor for light-induced damage of photosystem II (PS II) action and meanwhile cause cellular death. In addition, $^1O_2$ plays a role in protecting the photo-oxidative stress by up-regulating the protective genes (Krieger-Liszkay et al., 2008). The carotenoids like tocopherol, β-carotene, and plastoquinone effectively scavenge One molecule of O_2.

H_2O_2 is formed due to the reduction and protonation reaction of one molecule of O_2. The generation of H_2O_2 is by the catalytic reaction by SOD and also non-enzymatically by being dismutation to H_2O_2. The production of H_2O_2 is detected under normal conditions, but under stress conditions due to stomatal closure and less availability of CO_2, enhanced H_2O_2 production has been noticed (Sharma et al., 2012). Noctor et al. (2002a) stated that 70% of H_2O_2 generation was due to the stress condition. The key H_2O_2 production site is mitochondria, chloroplast, cell membrane, endoplasmic reticulum (ER), produced by photorespiration and β-oxidation of fatty acid. H_2O_2 plays a twin role; at a lower concentration, it behaves as a signaling molecule for various physiological processes like photorespiration and photosynthesis (Noctor et al., 2002b), senescence (Peng et al., 2005) cell cycle, growth, and development (Tanou et al., 2009a, b) and stomatal movement (Bright et al., 2006). And at higher concentration, it damages the cells. H_2O_2 has a longer half-life of 1 ms and is the main causal factor for PCD (Dat et al., 2000). It also deactivates various enzymes like sedoheptulose 1, fructose 1,6 bisphosphotases, 7 bisphosphotases, and phospho-ribulokinase.

Hydroxyl radical ($OH^•$) is the most poisonous and reactive radical known to originate from Fenton reaction between H_2O_2 and $O_2^{•-}$ catalyzed by transition metals like Fe. $OH^•$ is generated at neutral pH and has the ability to destroy different cellular and molecular components, i.e., membrane destruction, protein damage, and LPO. $OH^•$ is considered as a most toxic radical than H_2O_2 because of the absence of an enzymatic system to quench this toxic radical, and this leads to a larger accumulation of $OH^•$ in the cell and leading to cellular death (Pinto et al., 2003).

11.3 SITE OF ROS GENERATION IN PLANT

The generation of ROS is common in normal conditions and its accumulation enhances in environmental stress conditions. The production of ROS origin was at various cell organelles like chloroplasts, mitochondria, ER, peroxisomes, plasma membranes, and cell wall. Under light conditions, ER, and peroxisomes are the major sites of ROS production, while in the dark condition; the mitochondria are the major producers of ROS (Choudhury et al., 2013).

11.3.1 CHLOROPLAST

Photosynthesis is the major plant activity that takes place in the chloroplast. The chloroplast has a light-harvesting photosystem (photosystem I and PS II), which forms the light-harvesting core in the thylakoids. Due to this reason, the increased ROS production was more in chloroplast (Pfannschmidt, 2003). In stress and highlight condition, enhanced generation of ROS limits CO_2 concentration, which leads to form $O_2^{•-}$ at the PS via the Mehler reaction (Elstner, 1991). In the Mehler reaction the NADP is reduced to NADPH, which further enters the Calvin cycle and reduces the CO_2 level (final electron acceptor). The leakage of an electron to the O_2 in the PS causes $O_2^{•-}$ radical generation. The formation of $O_2^{•-}$ from the reduction of O_2 is a rate-limiting step, and the generated $O_2^{•-}$ is considered as more aggressive ROS. The radical $O_2^{•-}$ may further protonate to $HO_2^•$ in the internal lumen membrane may undergo enzymatic dismutation (Cu/Zn SOD at the PSI) to H_2O_2 in the stromal membrane surface. Also, at the Fe-S centers, where Fe_2^+ is available through the Fenton reaction, the H_2O_2 may transform to a more dangerous $OH^•$ radical (Cleland and Grace, 1999).

11.3.2 MITOCHONDRIA

Mitochondria are an important cell organelles involved in photorespiration. Mitochondrial electron transport chain (mtETC) is the culprit, which stocks sufficiently energized electrons to reduce O_2 to ROS. Mitochondria is the one that produces the most harmful ROS like $O_2^{\bullet-}$ and H_2O_2 (Navrot et al., 2007) and is enhanced during stressful conditions (Pastore et al., 2007). Even though $O_2^{\bullet-}$ is the major ROS generated in the mitochondria. It is further converted to H_2O_2 by the enzymes ascorbate peroxidase (APX) and Mn-SOD (i.e., 4–5% of O_2 consumed is converted into H_2O_2) (Sharma et al., 2012).

Mitochondrial ETC (mtETC) consists of two major components, Complex I and Complex III, which are responsible for the ROS production (Møller et al., 2007; Noctor et al., 2007). Complex I/NADH Dehydrogenase significantly reduces O_2 to $O_2^{\bullet-}$, further Complex I enhances reverse electron flow to Complex I from Complex III (due to shortage of NAD+-linked substrates). ATP hydrolysis controls the flow of electrons in the reverse way (Turrens, 2003). In Complex III, ubiquinone, completely in its condensed form gives an electron to Cytochrome C1, leaving behind an unstable ubisemiquinone semi-radical which forms electron leakage to O_2 and thereby generates $O_2^{\bullet-}$ (Murphy, 2009). Aconitase and 1-Galactono-γ-lactone dehydrogenase (GAL) enzymes present in the mitochondrial matrix are also responsible for ROS production by directly feeding electrons to ETC (Rasmusson et al., 2008). However, mitochondrial SOD (Mn-SOD) and mitochondrial alternative oxidase (AOX) are the two enzymes that reduce the ROS stress in the mitochondria (Mittova et al., 2003).

11.3.3 PEROXISOMES

Peroxisomes are spherical shaped microbodies with a single membrane and are known as major intracellular H_2O_2 producing sites. The production of toxic H_2O_2 is by the process of oxidative metabolisms (Del Río et al., 2006; Palma et al., 2009) such as the enzymatic reaction of flavin oxidases, fatty acid β-oxidation, glycolate oxidase reaction, and the disproportionation of $O_2^{\bullet-}$ radicals (Baker and Graham, 2002). The oxidation of glycolate by glycolate oxidase accounts for major H_2O_2 production, during photorespiration (Noctor et al., 2002a). Peroxisomes also produce $O_2^{\bullet-}$ like mitochondria and chloroplast. The xanthine oxidase (XOD) (E.C.1.17.3.2) of peroxisomes produces $O_2^{\bullet-}$ as a byproduct during the metabolism of xanthine and hypoxanthine into uric acid in the peroxisomal matrix. NADPH-dependent small ETC consists of Cyt b, and NADH, reduces O_2 by donating an electron and generates $O_2^{\bullet -}$ and release it to the cytosol.

In stressful conditions (availability of low water and CO_2), stomata close and causes decreased O_2 and CO_2 concentration, leading to increased photo-respiration, which meanwhile forms glycolate. Further, oxidation of glycolate by glycolate oxidase enzyme releases H_2O_2 and is considered a major H_2O_2 producer during photorespiration (Noctor et al., 2002a). Additionally, three peroxisomal membrane polypeptides that are integral membrane proteins (m/z 18, 29, and 132 kDa), NADH as an electron donor (m/z 18 and 32 kDa), NADPH also as an electron donor (m/z 29-kDa), and metabolic processes like flavin oxidation and β-oxidation of fatty acids produce $O_2^{\bullet -}$ (Noctor et al., 2002b).

11.3.4 PLASMA MEMBRANE

The plasma membrane plays an important role in the survival of the cell by providing necessary information to the cell about the external environmental changes. In the plasma membrane, NADPH-dependent oxidases are present and under various stress conditions produce $O_2^{\bullet -}$ by transferring electrons from cytosolic NADPH to O_2 (Apel and Hirt, 2004). Further, it produce $O_2^{\bullet -}$ dismutation spontaneously to H_2O_2 by the catalyzed reaction of the enzyme SOD. Kwak et al. (2003) have documented the fact that NADPH oxidase enzyme, plays a key role in the ROS production and accumulation during abiotic stress conditions and pathogenic infection (Apel and Hirt, 2004; Torres et al., 2002; Kwak et al., 2003).

11.3.5 APOPLAST

The apoplast is the diffusible space of the plant cell membrane, through which the incoming CO_2 diffuses into a soluble form and thereafter enters the cytosol to involve in photosynthesis. During abiotic stress, cell wall located enzymes and stress signals along with abscisic acid (ABA) make apoplast a prominent site for H_2O_2 production (Hu et al., 2006; Heyno et al., 2011; Apel and Hirt, 2004). From the oxalic acid the cell wall, the associated oxalate oxidase release H_2O_2 and CO_2, and this enzyme release more apoplastic H_2O_2 during plant-pathogen interaction (Lane, 2002). In addition to this enzyme, other enzymes like cell wall-linked polyamine oxidase, peroxidase (POX), and oxalate oxidase also generate H_2O_2 (Heyno et al., 2011).

11.3.6 ENDOPLASMIC RETICULUM (ER)

Environmental stress causes the accumulation of unfolded proteins, H_2O_2 in the ER and is called ER stress. In ER, NADPH-dependent electron transport, including Cyt P450, generates $O_2^{\bullet-}$ (Mittler, 2002). Cyt P450 reacts with the organic substrate (RH), further reduced by flavoprotein to form free radical intermediate (Cyt P450 R), which readily interacts with triplet oxygen ($3O_2$) to form Cyt P450-ROO^- (oxygenate complex) that leads to generate a by-product $O_2^{\bullet-}$.

11.3.7 CELL WALL

Cell wall is also considered as an active site for ROS generation. During stress conditions, the cell wall-associated POX forms H_2O_2 in the presence of NADH. In addition, cell wall-associated lipoxygenase (LOX) affects hydroperoxidation of polyunsaturated fatty acids (PUFA) to form ROS ($OH^\bullet$, $O_2^{\bullet-}$, H_2O_2, and $^1O_2$). Cell wall localized enzyme diamine oxidase produces ROS as a by-product by utilizing the diamines/polyamines (Elstner, 1991). During biotic stress, the lignin in the cell wall undergoes repetitive cross-linking through H_2O_2 dependent pathway to form cell wall accumulated lignin (Higuchi, 2006). The cell wall POXs produced ROS during biotic stress in cotton (*Xanthomonas campestris* pv. *malvacearum*) and under abiotic stress (potassium deficiency) in Arabidopsis (Martinez et al., 1998; Kim et al., 2010).

11.4 STRESS AND ROS

Plant during its lifetime come across various environmental stresses like mechanical stress (Suzuki and Mittler, 2012), i.e., abiotic stresses like cold, heat drought, pollutants, toxic content (Wu et al., 2017) and biotic stress (microbial and pest attack) (Yang et al., 2017). To these various stresses, plants react by producing ROS as a defense mechanism that acts as a signaling molecule and triggers the signal transduction which may also cause plant cell death (Sewelam et al., 2016). The cell death helps in the morphogenesis of the plant and, importantly, induces the resistance in the plant to the stress conditions (Frederickson and Loake, 2014). The details of plant response to various stresses are given below in subsections.

11.4.1 ROS TO MECHANICAL STRESS

Mechanical stress is an unavoidable threat to the plants' survival and yield of the crop. Researchers focus on the mechanical injury and its effect above the ground level of the plant. However, the root injury is somewhat ignored even though it is the main organ for plant growth. Plant to the mechanical wounding responds through ROS production, which also acts as a signaling molecule to induce plant immunity (Mittler et al., 2011). Orozco-Cardenas and Ryan (1999) detected H_2O_2 accumulation in the tomato plant within an hour of wounding, and further, the accumulation extended to the unwounded leaves after 4–6 hours of wounding. But the tomato mutant cultivar did not show H_2O_2 accumulation even after the wounding. Transmembrane NADPH oxidases and POXs involved in the detoxification of ROS (Minibayeva et al., 2014). Calcium spikes/calcium level in the plant cell plays a major role in wound-induced oxidative stress (Monshausen et al., 2007; Kimura et al., 2012). Benelujaephajri et al. (2013) blocked the plant's calcium channel by pre-treatment with calcium oxalate and calcium chelators EGTA or with calcium channel blocker, leading to eliminating the H_2O_2 accumulation in wound-induced plant cells. Dubiella et al. (2013) stated the relationship between calcium accumulation and ROS generation in plant cells both *in vitro* and *in vivo*. The hypocotyls of the *Cakile maritima* seedlings were wounded and the response after 3 hours showed a higher accumulation of H_2O_2 in the wounded cells and meanwhile altered the level of the antioxidant enzyme (Houmani et al., 2018).

11.4.2 ROS TO ABIOTIC STRESS

Drought is the common abiotic stress to the plant, which is known to induce ROS generation and oxidative stress. In abiotic stress, soil salinity plays a major role in the plant's health and yield worldwide (Horie and Schroeder, 2004). Soil salinity poses a serious threat, and the plant's tolerances mechanisms help improve crop yield and quality. Salt stress and its tolerance is one of the major research areas (Munns and Tester, 2008). Salinity stress interferes with the carbon metabolism leading to the generation of ROS. Salinity stress causes stomatal closure, leading to a decrease in transpiration and ion redistribution, alkalization, and ABA accumulation. Stomatal closure leads to decreased CO_2 exchange resulting in lowered the intercellular CO_2 level, which transfers the electron to O_2 to generate $O^{\bullet-}$ (Geilfus et al., 2015). Further, salinity affects the plant by different complications like osmotic pressure, ion toxicity, genotoxicity, nutritional deficiency, and oxidative stress (Hasanuzzaman et al., 2020). But the ROS generation and oxidative stress vary among the cultivars. Lalarukh and Shahbaz (2020) reported that in the two genotypes of sunflower (FH-572 and FH-621), exposed to salinity stress (NaCl; 120 mM), found a varied increase of H_2O_2 in FH-572 (78%) and FH-621 (20%). Hence the authors concluded that FH-621 showed more tolerance to salinity stress.

Treatment of sodium chloride (NaCl; 100 mM and 200 mM) caused stress 2.5 and 3 fold increase in H_2O_2 production and 2 and 3 fold increase in thiobarbituric acid reactive substances (TBARS), respectively when compared to control (Rehman et al., 2019). Under salt stress, the enhanced ROS content, LPO, and electrolyte leakage were observed by Cheng et al. (2020) in rice root tissues up to two times compared to control. In the tomato plant, under 100 mM NaCl salinity stress, more accumulation of H_2O_2 (176%) and $O_2^{\bullet-}$ (157%) and increased electrolyte leakage (158%) and malondialdehyde (94%) content was observed (Ahanger et al., 2020a). An increased level of H_2O_2 and malondialdehyde by 50 and 25%, respectively, was noticed in maize plants under salinity stress (NaCl; 120 mM) compared to control (Arora et al., 2020). Ahmad et al. (2019) also found 2 fold increase in malondialdehyde, electrolyte, H_2O_2, and $O_2^{\bullet-}$ during NaCl (100 mM) exposure of Mung bean plants. In the sweet pepper cultivar, the increase in malondialdehyde and electrolyte leakage increased by 2 fold due to 0.4% of salt stress (Abdelaal et al., 2020).

Water deficit/drought causes the higher generation of ROS by stomatal closure, declined CO_2 uptake which leads to impaired photosynthetic rate, light harvest, and utilization through which the ROS overproduction occurs (Hasanuzzaman et al., 2019b; Hasanuzzaman et al., 2020). Over-production of ROS during drought was linked with the denaturation of protein from photo-respiration, TCA cycle enzyme inactivation, and reduction in carboxylation efficiency (Hasanuzzaman et al., 2014). A decrease in NADP+ regeneration leads to electro transport chain reduction and generates higher electrolyte leakage resulting in higher ROS generation and oxidative stress (Fahad et al., 2017; Hasanuzzaman et al., 2018). *Phragmites karka* grown under drought (40% water holding capacity up to 35 days) condition showed an increase in malondialdehyde by 22% compared to the control plant (Abideen et al., 2020). Campos et al. (2019) found an increase in malondialdehyde in the *Coffea arabica* plant after 20 days of drought condition. Drought stress, in finger millet plants, showed increased H_2O_2 accumulation and electrolyte leakage (Satish et al., 2018). In rice plants, drought was created without irrigating for 8 days and analyzed for the ROS. The results revealed that under drought conditions, increased generation of $O_2^{•-}$ (1.8 fold), H_2O_2 (2.1) and malondialdehyde (1.66 fold) was observed when compared to control plants (Saha et al., 2018). Tomato plant without irrigation for 6 days resulted in an increase of electrolyte leakage (39%) and malondialdehyde content (Malhotra et al., 2017). *Brassica napus* L. cv. Bina Sharisha-3 on polyethylene glycol (PEG) treatment (10 and 20%) caused hyperosmotic stress, and the resulted in increase in H_2O_2 and malondialdehyde content (Hasanuzzaman et al., 2017, 2018). Similarly, Nahar et al. (2017) in *Vigna radiata* cultivar found enhanced generation of H_2O_2 along with enhanced LOX activity and membrane peroxidation due to 48 h of hyperosmotic stress (5% PEG). Due to deficit water conditions in wheat cultivar, an increase in H_2O_2 and EL level was observed (Abbas et al., 2018). Drought exposed *Solanum lycopersicum* plant showed increased H_2O_2 (37%), $O_2^{•-}$ (75%), and malondialdehyde (83%) content (Rady et al., 2020). Filippou et al. (2011) also recorded the increase in H_2O_2 and malondialdehyde content in *M. truncatula* under drought conditions, but the parameters were reversed after watering. Drought severity and plant response as susceptible/resistant depends on the cultivars. The two different cultivars of *Phaseolus valgaris* namely Bn-16 (drought-sensitive variety) and Bn-150 were examined. The results showed two times more malondialdehyde production and ROS; H_2O_2, $O_2^{•-}$, $^{•}OH$ were more in Bn-16 than Bn-150 cultivar.

Metal toxicity is also one of the key environmental stress that causes ROS generation and oxidative stress by disrupting the mitochondrial electron transfer, chloroplast action, and oxidative metabolism of peroxisome in the plant. *Oryza sativa* seedlings under nickel toxicity (0.25 and 0.5 mM $NiSO_4$) for 72 h showed induced accumulation of ROS (H_2O_2) and LPO compared to control plants (Hasanuzzaman et al., 2019b), and a similar observation was made by El-Amier et al. (2019) with a lower concentration of Ni (100 mM; $NiCl_2$) in *Pisum sativum*. Researches have also shown that the cadmium toxicity on plants enhances the generation of ROS (H_2O_2 and O_2) and malondialdehyde content (Nahar et al., 2016; Hasanuzzaman et al., 2017; Al-Mahmud et al., 2018). Recently, Ahanger et al. (2020b) reported an increase in the production of H_2O_2, $O_2^{•-}$, electrolyte leakage, LPO, and LOX enzyme activity in *V. angularis* under cadmium stress (100 mM; $CdCl_2$) for 20 days. Similar findings were observed by Zaid et al. (2020), i.e., an increase in electrolyte leakage and H_2O_2 level in *Mentha arvensis* to cadmium stress (100 μM; $CdCl_2$) after 20 days of exposure. Toxicity of lead in the wheat plant was demonstrated by Hasanuzzaman et al. (2018); the exposure of lead (0.5 and 1.0 mM; $Pb(NO_3)_2$) caused an increase in generation $O_2^{•-}$, H_2O_2, and

MDA content when compared to control. Yadu et al. (2019) revealed that to arsenic stress causes enhanced ROS and oxidative stress. The *Cajanus cajan* seedlings exposure to arsenic (10 mM; Na_3AsO_4) for 5 days significantly showed an increase in the ROS (H_2O_2 and $O_2^{\bullet-}$) level and malondialdehyde content and also increased LOX (by 4 fold) activity compared to untreated plants. An increase in the level of TBARS and H_2O_2 was evident in two cultivars of *Chenopodium quinoa* due to the toxic effect of arsenic (150 and 300 mM; Na_3AsO_4) for 35 days (Parvez et al., 2020). Georgiadou et al. (2018) reported higher H_2O_2 and malondialdehyde production, which lead to cause cellular damage in basil plants, grown under copper stress at 1,000 ppm.

Extreme temperature (low and high temperature) causes overproduction of ROS and leads to cause oxidative stress, electrolyte leakage, alters cellular metabolism, enhancement of LPO, nicking of DNA, membrane damage, degradation of membrane fluidity, photosynthesis apparatus inhibition (Hasanuzzaman et al., 2013; Fahad et al., 2016). Under cold stress conditions (12°C for 6 d), 14 days old rice seedlings showed increased electrolyte leakage and malondialdehyde production by 180 and 49%, respectively (Han et al., 2017). Treatment of cold stress (15°C/8°C day/night) for 24 and 48 h to the *S. lycopersicum* cultivar causes a significant increase in H_2O_2 accumulation than the control (Liu et al., 2020). Xue et al. (2019) exposed *Ammopiptanthus mongolicus* (wild and transgenic variety) to various temperature and evaluated the accumulation of ROS. Both the varieties of plant were initially exposed 24 h at 4°C followed by 12 h at 0°C and finally 12 hours at -6°C. The transgenic plant showed low-temperature tolerance compared to the wild type, where more accumulation of H_2O_2 was observed. Ding et al. (2016) treated *Cucumber sativus* seedlings with hot temperature (35°C) and found an increase in generation of ROS ($O_2^{\bullet-}$) content by 79.9%. An increase in ROS ($O_2^{\bullet-}$) level and oxidative stress was observed in the tobacco plant when the plants were exposed to 50°C for 5 min. Similarly, Djanaguiraman et al. (2018) reported an increased level of $O_2^{\bullet-}$ in sorghum plants grown under high-temperature stress under field conditions. Results revealed that the level of $O_2^{\bullet-}$ increased by 2.3- and 3.5-fold in pistils and pollen, respectively. An increase in accumulation of H_2O_2 by 1.27-fold was noticed in rice seedlings compared to control when exposed to 38°C of hot stress for 5 days.

11.4.3 ROS TO BIOTIC STRESS

Biotic stress is the stress originated by the living organism, i.e., microorganisms, and pests. ROS generation is the hallmark of the plant's successful recognition of the pathogen to activate the defense mechanism, and failure of ROS generation may lead to successful infection. ROS generation in plants to the biotic stress is in two phases; first with the unspecific transitory phase within few minutes of the interaction with the organism, and second is linked with the defenses and HR reaction that takes place after hours of organism interaction with the plant (Grant and Loake, 2000). The pathogen recognition by plant, ROS generation, and oxidative burst occurs in the apoplast and other cellular organelles like chloroplast and mitochondria. Majorly, H_2O_2, OH, $O_2^{\bullet-}$ and singlet oxygen are produced during the biotic stress/pathogen infection (Foyer and Noctor, 2005).

During bean yellow mosaic virus infection to *Vicia faba* leaves, increase in H_2O_2 accumulation was reported (Radwan et al., 2010). After ROS generation, several enzymes are involved in the detoxification of generated radicals during the pathogen recognition. ROS inhibitors like NADPH oxidases (in the plasma membrane) and POXs (in the cell wall) increase as the fungal or bacteria pathogens are detected by the plant (Chittoor et al., 1997; Sasaki et al., 2004). The ROS accumulation and its correlation with the necrosis in cultivar were showed by Abdollahi and Ghahremani (2011). When the uracil was used (an inhibitor of ROS), reduced ROS accumulation and generation was observed in the host plant on interaction by *Erwinia amylovora.* Liu et al. (2007) revealed that the ROS generation and activation of the SIPK/Ntf4/WIPK cascade in chloroplast was due to the pathogen attack. The ROS generated acted as a signaling molecule and also showed HR response. Decreased antioxidant enzymes (CAT and APX) lead to HR to the pathogen in the host plant (Mittler et al., 1999). An increase in POX and CAT activity was observed in powdery mildew infected lines (Ashry and Mohamed, 2012). In *Vicia faba* leaves accumulated ROS due to bean yellow mosaic virus infection and the enhanced activity of CAT, POX, SOD, and APX managed the ROS level in the plant (Radwan et al., 2010).

Production of ROS was also observed during pest recognition, both during insect feeding with the host and the injury due to mechanical wounding (Maffei et al., 2006). During feeding, insects produce/secretes saliva, which induced plant defense and protect plants from the pests (Musser et al., 2002). In the *Nicotiana tabacum* plant, the caterpillar salivary glands induced H_2O_2 levels in the leaves and down-regulated the defense level (Musser et al., 2006). Also, herbivore insects enhanced antioxidant gene expression in the midgut was observed to protect themselves against ROS ingestion during feeding (Mittapalli et al., 2007). Hence, the oral secretions showed a prominent protective role in the pest to come across ROS generated during plant-pest interaction.

11.5 ROS AS A SIGNALING MOLECULE

ROS generated during the stress conditions are not only toxic, but acts as a secondary messenger by activating the intracellular signal cascades. ROS mediated signal helps to activate the defense mechanism of the plant both chemical (activates antioxidants level and signaling molecules to neutralize the toxic radical) and physical (stomatal closure and necrosis development; Neill et al., 2002; Yan et al., 2007; Kwak et al., 2003; Bethke and Jones, 2001; Miller et al., 2008). In other words, the plant part affected by the stress condition sends a signal to the unstressed plant part, which thereby enhances the overall stress adaptability and resistance (Miller et al., 2009). The plant can detect and respond through transduce and translate ROS signal in site-specific and cellular-specific manner (Xiong et al., 2002). ROS signals with redox-sensitive proteins, protein phosphorylation, calcium mobilization, and gene expression. During the signaling process, various cellular components like protein kinases, protein phosphatases, and transcription factors (TFs) was altered (Cheng and Song, 2006). Sensing the stress, ROS generation, and signal transduction show the plant's tolerance capacity against various environmental stress (Suzuki et al., 2012; Baxter et al., 2014). Miller et al. (2008) have documented the ROS generation and activation of various signaling pathways. Many researchers identified the key pathways that link zinc finger proteins and WRKY TFs during environmental stresses.

Pel et al. (2000); and Jannat et al. (2011) observed that ABA-induced through the cytosolic H_2O_2 resulted in stomatal closure. Jung et al. (2001) reported about the ROS generation and its correlation with gravitropism. ROS showed a key role in drought-induced ABA synthesis, confirming that the plant senses the drought stress. The metal contaminants (Cd_2^+ and Cu_2^+) generated ROS in the plant system, which activated and induced the MAP kinase pathway (Yeh et al., 2007). ROS also involved in alleviating dormancy, i.e., exogenous H_2O_2 induced gibberellic acid signaling, results in germination (Bahin et al., 2011). During wounding of the tomato plant, the ROS generated at the cell wall helped to accumulate H_2O_2 level, which acted as a secondary messenger by activating the defense related genes (Orozco-Cardenas et al., 2001).

ROS, H_2O_2 in particular played a major role in plant-pathogen interaction. Treatment of chili seedlings with cerebroside and cell wall elicitor extracted from the pathogenic fungi, induced the accumulation of H_2O_2. Further, the accumulated H_2O_2 acted as a messenger to induce the defense-related enzymes and phenolic content (Naveen et al., 2013a, b). Treatment of plant growth-promoting rhizobacteria and abiotic inducers (BABA and ASM) to chili seedlings induced pathogenesis-related protein, increased phenolic level, and defense-related enzymes when compared to untreated control seedlings (Naveen et al., 2019, 2020). Lignin, a cell wall thickening molecule, plays a key role in the plant defense response to stress. The generated H_2O_2 regulated lignin's biosynthesis in response to environmental stresses (Yuasa et al., 2001; Yeh et al., 2007).

11.6 ANTIOXIDANTS IN RESPONSE TO ROS

Antioxidants are the important self-defense molecules play an important role in the detoxification of excessive ROS generated in response to various environmental stress factors. Antioxidants include non-enzymatic and enzymatic ROS quenchers. Enzymatic antioxidants are activated based on the requirement and toxicity of ROS in the plant cell. The enzymatic antioxidant production depends on the plant stress tolerance and intolerance nature. The main enzymatic antioxidants are SOD, CAT, APX, GPX, GR, MDHAR, and DHAR.

SOD (1.15.1.1) plays a key role in ROS quenching and oxidative stress defense response in aerobic organisms (Scandalios, 1993), during the environmental stress condition mainly during drought and metal toxicity (Sharma and Dubey, 2005; Mishra et al., 2011). SOD is a group of metallo-enzymes which catalyzes the dismutation of $O_2^{\bullet-}$ to H_2O_2 and O_2 by reducing OH's formation (Vallabhaneni et al., 2008). Three different isoforms of SOD are reported in the plant, namely manganese SOD (Mn-SOD; located in mitochondria), copper/zinc SOD (Cu/Zn-SOD; located in the cytosol, chloroplast, and peroxisomes), and iron SOD (Fe-SOD; located in the chloroplast) (Fridovich, 1989; Racchi et al., 2001; Jackson et al., 1978; Kanematsu and Asada, 1989; Bueno et al., 1995; Del Rio et al., 1998). An increase in the activity of SOD exhibits stress tolerance in the plant and is used as a marker to analyze the stress tolerance (Gupta et al., 1993; Zaefyzadeh et al., 2009).

Catalase (CAT; 1.11.1.6) is a tetrameric heme-containing enzyme involved H_2O_2 detoxification, converting 26 million H_2O_2 molecules into H_2O in a minute (Del Rio et al., 2006) and was the first enzyme reported and characterized. Even though CAT has less affinity to H_2O_2 than APX, the turnover rate of CAT is much higher than any other antioxidant enzyme. H_2O_2 is produced more in peroxisomes during photorespiratory oxidation, β-oxidation, and stress condition (Del Rio et al., 2006; Corpas et al., 2008), and CAT quenches the H_2O_2 in peroxisome than in mitochondria, cytosol, and chloroplast (Mhamdi et al., 2010). CAT has three types of isoform; CAT I (expressed in photosynthetic tissues), CAT II (expressed in vascular tissues), and CAT III (expressed in seeds and young seedlings; Willekens et al., 1995). An increase in the CAT activity expression is related to the stress tolerance induced by the environmental stress (Guan et al., 2009).

APX (EC 1.1.11.1) is the vital component of the ascorbate-glutathione (ASC-GSH) cycle. APX reduces H_2O_2 into H_2O and DHA using two ASC components as a reducing agent. Than CAT, APX is the most abundant and more specific to H_2O_2 during stressful conditions (Wang et al., 1999). Antioxidant role of APX is majorly involved in cytosol and chloroplast. APX has five forms and the isoforms differ in their amino acid sequences and are localized in the cytosol, mitochondria, peroxisome, thylakoid, and stromata in higher plants (Jimenez et al., 1997; Madhusudhan et al., 2003; Sharma and Dubey, 2004). APX scavenges H_2O_2 of organelle-specific, but in the cytosol, the APX diffused from the other organelle also showed the activity (Mittler and Zilinsas, 1992). Enhanced activity of APX was observed in environmental stress conditions (Sharma and Dubey, 2005, 2007; Han et al., 2009; Maheshwari and Dubey, 2009; Hefny et al., 2009; Boo and Jung, 1999). In stress-tolerant transgenic plants, the increased expression of APX gene was reported by the researchers (Wang et al., 2005; Yabuta et al., 2002).

Guaiacol peroxidase (GPX; EC 1.11.1.7) is a heme-containing enzyme that eliminates excess H_2O_2 liberated during metabolic and stress condition. GPX is located in the cytosol, vacuoles, and cell wall and contains four disulfide bridges and two structural Ca_2^+ ions (Schuller et al., 1996; Asada, 1992). Cell wall lignification, ethylene biosynthesis, IAA degradation, wound healing, and defense during environmental stress has been related to the GPX enzyme activity (Kobayashi et al., 1996). GPX induced activity was observed under various stressful conditions by several researchers (Sharma and Dubey, 2005; Han et al., 2009; Verma and Dubey, 2003; Mishra et al., 2011, Moussa and Abdel-Aziz, 2008; Radotic et al., 2000; Tayefi-Nasrabadi et al., 2011).

Glutathione reductase (GR; EC 1.6.4.2) is an oxidoreductase flavoprotein, reduces GSSG to reduced glutathione (GSH) using NADPH. GR maintains GSH/GSSG ratio by catalyzing the disulfide bone in glutathione in the ASC-GSH cycle. GR is more prominent in chloroplast and cytosol and mitochondria in lesser quantity (Ghisla and Massey, 1989; Edwards et al., 1990). About 80% of GR activity is found in chloroplast and detoxifies H_2O_2 produced during Mehler reaction. Reduced GR expression in tomato chloroplast was correlated with the susceptibility to chilling stress (Shu et al., 2011) and improved GR expression in *Nicotiana tabacum* showed tolerance to the oxidative stress (Foyer et al., 1995; Aono et al., 1993).

Dehydroascorbate reductase (DHAR, EC 1.8.5.1) catalyzes the DHA reduction to ascorbic acid (AA) using GSH as a reducing substrate and plays a key role in maintaining AA. DHA is a short-lived chemical that irreversibly converts DHAR to AA. DHAR have a role in the symplast and apoplast to maintain the redox state in the plant cell during the stressful condition (Sharma and Dubey, 2005, 2007; Maheshwari and Dubey, 2009; Boo and Jung, 1999; Yoshida et al., 2006). The transgenic stress (drought, salinity, and toxicity) tolerant potato plant showed more DHAR activity (Eltayeb et al., 2011), and the salinity stress tolerance was observed in the *L. japonicas*, which also showed more cytosolic DHAR (Rubio et al., 2009). DHAR is found richly in roots, seeds, in green and etiolated shoots.

Monodehydroascorbate reductase (MDHAR, E.C.1.6.5.4) is located mainly in cytosol, chloroplast, peroxisomes, and mitochondria (Jimenez et al., 1997; Hossain et al., 1984). MDHAR enzyme catalyzes the regeneration of AA from MDHA radical with electron donor NAD(P)H. MDHA radical generated in APX catalyzed reaction has a short lifetime, and if not quickly reduced, it unequally reduces to DHA and AA (Ushtmaru et al., 1997). During the environmental stresses, MDHAR activity increases in plants (Sharma and Dubey, 2005, 2007; Maheshwari and Dubey, 2009). The upregulation of the MDHAR gene was enhanced by the PEG and salt stress in the tobacco plant. Li et al. (2010) found higher chloroplastic MDHAR expression in temperature tolerant transgenic tomato plants.

11.7 CONCLUSION

The plant is an important and foremost need of human beings, and protecting the plants from various obstacles is the major concern of today. In reply to the plant protection, human beings can get good environment (fresh air and favorable environmental condition) and food. ROS is an inevitable plant process during its lifetime, which is a normal cellular activity in various cell organelles (mitochondria, chloroplast, apoplast, and plasma membrane) to protect plants from adverse environmental conditions. In normal conditions, the ROS generation is low or moderate, and at this level, the ROS acts as a messenger that signals various activities of the plant. But, under the environmental stress condition, production of ROS increases drastically, which causes exacerbating damage to the cell and cellular components includes, damages DNA, inhibits protein synthesis, and enhances lipid oxidation which causes adverse effect on enzyme activity. To overcome this tough situation of ROS generation, the plant produces antioxidant enzymes and components to combat oxidative stress due to unfavorable environmental conditions. Antioxidants help to quench the toxic radicles generated at a higher level and save plant cells from necrosis. Hence, understanding the plant genomics, proteomics, metabolomics, and the plant's mechanism to various environmental stresses of biotic and abiotic stress is very much needed. There is no doubt that breeding plants or transgenic varieties for over-expression of antioxidant components can enhance stress tolerance. The improved antioxidant levels in the plant (as in-built) using biotechnological aspects will enhance the ROS and oxidative stress tolerance level of the plant to multiple biotic and abiotic stresses.

ACKNOWLEDGMENTS

This study was funded by the University Grant Commission (UGC), Government of India. Dr. Naveen, J. (Award No. F./PDFSS-2015-17-KAR-11600) greatly acknowledges the University Grant Commission (UGC), Government of India, and for financial assistance.

KEYWORDS

- **abiotic stress**
- **antioxidants**
- **biotic stress**
- **oxidative damage**
- **plant defense**
- **reactive oxygen species**

REFERENCES

Abbas, T., Rizwan, M., Ali, S., Adrees, M., Mahmood, A., Zia-ur-Rehman, M., Ibrahim, M., et al., (2018). Biochar application increased the growth and yield and reduced cadmium in drought stressed wheat grown in an aged contaminated soil. *Ecotoxicol. Environ. Saf., 148*, 825–833.

Abdelaal, K. A., EL-Maghraby, L. M., Elansary, H., Hafez, Y. M., Ibrahim, E. I., El-Banna, M., El-Esawi, M., & Elkelish, A., (2020). Treatment of sweet pepper with stress tolerance-inducing compounds alleviates salinity stress oxidative damage by mediating the physio-biochemical activities and antioxidant systems. *Agronomy, 10*, 26.

Abdollahi, H., & and, Ghahreman, Z., (2011). Ghahremani. The role of chloroplasts in the interaction between *Erwinia amylovora* and host plants. *Acta Horticulturae, 896*, 215–221.

Abideen, Z., Koyro, H. W., Huchzermeyer, B., Ansari, R., Zulfiqar, F., & Gul, B., (2020). Ameliorating effects of biochar on photosynthetic efficiency and antioxidant defense of *Phragmites karka* under drought stress. *Plant Biol., 22*, 259–266.

Ahanger, M. A., Aziz, U., Sahli, A. A., Alyemeni, M. N., & Ahmad, P., (2020). Combined kinetin and spermidine treatments ameliorate growth and photosynthetic inhibition in *Vigna angularis* by up-regulating antioxidant and nitrogen metabolism under cadmium stress. *Biomolecules, 10,* 147.

Ahanger, M. A., Mir, R. A., Alyemeni, M. N., & Ahmad, P., (2020). Combined effects of brassinosteroid and kinetin mitigates salinity stress in tomato through the modulation of antioxidant and osmolyte metabolism. *Plant Physiol. Biochem., 147*, 31–42.

Ahmad, P., Ahanger, M. A., Alam, P., Alyemeni, M. N., Wijaya, L., Ali, S., & Ashraf, M., (2019). Silicon (Si) supplementation alleviates NaCl toxicity in mung bean [*Vigna radiata*(L.) Wilczek] through the modifications of physio-biochemical attributes and key antioxidant enzymes. *J. Plant Growth Regul., 38,* 70–82.

Al Mahmud, J., Hasanuzzaman, M., Nahar, K., Bhuyan, M. B., & Fujita, M., (2018). Insights into citric acid-induced cadmium tolerance and phytoremediation in *Brassica juncea* L. Coordinated functions of metal chelation, antioxidant defense and glyoxalase systems. *Ecotoxicol. Environ Safe., 147*, 990–1001.

Allen, R. D., Webb, R. P., & Schake, S. A., (1997). Use of transgenic plants to study antioxidant defenses. *Free Radical Biology and Medicine, 23*, 473–479.

Aono, M., Kubo, A., Saji, H., Tanaka, K., &. Kondo, N.,., (1993). Enhanced tolerance to photo-oxidative stress of transgenic *Nicotiana tabacum* with high chloroplastic glutathione reductase activity. *Plant and Cell Physiology, 34,* 129–135.

Apel, K., & Hirt, H., (2004). Reactive oxygen species: Metabolism, oxidative stress, and signal transduction. *Annual Review of Plant Biology, 55*, 373–399.

Arora, M., Saxena, P., Abdin, M., & Varma, A., (2020). Interaction between *Piriformospora indica* and *Azotobacterchroo coccumdiminish* the effect of salt stress in *Artemisia annua* L. by enhancing enzymatic and non-enzymatic antioxidants. *Symbiosis, 80*, 61–73.

Asada, K., (1992). Ascorbate peroxidase: A hydrogen peroxide scavenging enzyme in plants. *Physiologia Plantarum, 85*, 235–241.

Ashry, N. A., & Mohamed, H. I., (2012). Impact of secondary metabolites and related enzymes in flax resistance and/or susceptibility to powdery mildew. *African Journal of Biotechnology, 5*, 1073–1077.

Bahin, E., Bailly, C., Sotta, B., Kranner, I., Corbineau, F., & Leymarie, J., (2011). Crosstalk between reactive oxygen species and hormonal signaling pathways regulates grain dormancy in barley plant. *Cell and Environment, 34*, 980–993.

Baker, A., & Graham, A. I., (2002). *Plant Peroxisomes: Biochemistry, Cell Biology and Biotechnological Applications*. Kluwer Academic Publishers, Dordrecht, The Netherlands, 2002.

Baxter, A., Mittler, R., & Suzuki, N., (2014). ROS as key players in plant stress signaling. *J. Exp. Bot., 65*, 1229–1240.

Beneloujaephajri, E., Costa, A., L'Haridon, F., Metraux, J. P., & Binda, M., (2013). Production of reactive oxygen species and wound-induced resistance in *Arabidopsis thaliana* against *Botrytis cinerea* are preceded and depend on a burst of calcium. *BMC Plant Biol., 13*, 160.

Bethke, P. C., & Jones, R. L., (2001). Cell death of barley aleurone protoplasts is mediated by reactive oxygen species. *Plant Journal, 1*, 19–29.

Bhattacharjee , S. 2005. Reactive oxygen species and oxidative burst: Roles in stress, senescence and signal transduction in plants Soumen. Current science, vol. 89, no. 7, 10.

Boo, Y. C., & Jung, J., (1999). Water deficit-induced oxidative stress and antioxidative defenses in rice plants. *Journal of Plant Physiology, 155*, 255–261.

Bright, J., Desikan, R., Hancock, J.T., Weir,I. S. and Neill, S. J. 2006. ABA-induced NO generation and stomatal closure in Arabidopsis are dependent on H2O2 synthesis. Plant J 45: 113–122.

Bueno, P., Varela, J., Gimenez-Gallego, G., & Del Rio, L. A., (1995). Peroxisomal copper, zinc superoxide dismutase. Characterization of the isoenzyme from watermelon cotyledons. *Plant Physiology, 108*, 1151–1160.

Campos, C. N., Ávila, R. G., De Souza, K. R. D., Azevedo, L. M., & Alves, J. D., (2019). Melatonin reduces oxidative stress and promotes drought tolerance in young *Coffea arabica* L. plants. *Agric. Water Manag., 211*, 37–47.

Chen, Q., Zhang, M., & Shen, S., (2010). Effect of salt on malondialdehyde and antioxidant enzymes in seedling roots of Jerusalem artichoke (*Helianthus tuberosus* L.). *Acta Physiologiae Plantarum, 33*, 273–278.

Cheng, Y. W., Kong, X. W., Wang, N., Wang, T. T., Chen, J., & Shi, Z. Q., (2020). Thymol confers tolerance to salt stress by activating anti-oxidative defense and modulating Na^+ homeostasis in rice root. *Ecotoxicol. Environ. Safe, 188*, 109894.

Cheng, Y., & Song, C., (2006). Hydrogen peroxide homeostasis and signaling in plant cells. Science in China. *Series C, Life Sciences, 49*, 1–11.

Chittoor, J. M., Leach, J. E., & White, F. F., (1997). Differential induction of a peroxidase gene family during infection of rice by *Xanthomonas oryzae* pv. *oryzae*. *Molecular Plant-Microbe Interactions, 10*, 861–871.

Choudhury, S., Panda, P., Sahoo, L., & Panda, S. K. (2013). Reactive oxygen species signaling in plants under abiotic stress. *Plant Signal. Behav*. 23681.

Cleland, R. E., & Grace, S. C., (1999). Voltammetric detection of superoxide production by photosystem II. *FEBS Letters, 457*, 348–352.

Corpas, F. J., Palma, J. M., Sandalio, L. M., Valderrama, R., Barroso, J. B., & Del Río, L. A., (2008). Peroxisomal xanthine oxidoreductase: Characterization of the enzyme from pea (*Pisum sativum* L.) leaves. *Journal of Plant Physiology, 165*(13), 1319–1330.

Dat, J., Vandenabeele, S., Vranová, E., Van, M. M., Inzé, D., & Van, B. F., (2000). Dual-action of the active oxygen species during plant stress responses. *Cell. Mol. Life Sci., 57*, 779–795.

Del Río, L. A., Pastori, G. M., & Palma, J. M., (1998). The activated oxygen role of peroxisomes in senescence. *Plant Physiology, 116*(4), 1195–1200.

Del Río, L. A., Sandalio, L. M., Corpas, F. J., Palma, J. M., & Barroso, J. B., (2006). Reactive oxygen species and reactive nitrogen species in peroxisomes. Production, scavenging, and role in cell signaling. *Plant Physiology, 141*(2), 330–335.

Desikan, R., A-H-Mackerness, S., Hancock, J. T., & Neill, S. J., (2001). Regulation of the *Arabidopsis* transcriptome by oxidative stress. *Plant Physiology, 127*, 159–172.

Ding, X., Jiang, Y., He, L., Zhou, Q., Yu, J., Hui, D., & Huang, D., (2016). Exogenous glutathione improves high root-zone temperature tolerance by modulating photosynthesis, antioxidant and osmolytes systems in cucumber seedlings. *Sci. Rep., 6*, 35424.

Djanaguiraman, M., Perumal, R., Jagadish, S., Ciampitti, I., Welti, R., & Prasad, P., (2018). Sensitivity of sorghum pollen and pistil to high-temperature stress. *Plant Cell Environ., 41*, 1065–1082.

Dubiella, U., Seybold, H., Durian, G., Komander, E., Lassig, R., & Witte, C. P., (2013). Calcium-dependent protein kinase/NADPH oxidase activation circuit is required for rapid defense signal propagation. *Proc. Natl. Acad. Sci. U.S.A., 110*, 8744–8749.

Edwards, E. A., Rawsthorne, S., & Mullineaux, P. M., (1990). Subcellular distribution of multiple forms of glutathione reductase in leaves of pea (*Pisum sativum* L.). *Planta, 180*, 278–284.

El-Amier, Y., Elhindi, K., El-Hendawy, S., Al-Rashed, S., & Abd-ElGawad, A., (2019). Antioxidant system and biomolecules alteration in *Pisum sativum* under heavy metal stress and possible alleviation by 5-aminolevulinic acid. *Molecules, 24*, 4194.

Elstner, E. F., Pell, E. J., & Steffen, K. L., (1991). Mechanisms of oxygen activation in different compartments of plant cells. In: *Active Oxygen/Oxidative Stress and Plant Metabolism* (pp. 13–25). American Society of Plant Physiologists, Rockville, USA.

Eltayeb, A. E., Yamamoto, S., Habora, M. E. E., Yin, L., Tsujimoto, H., & Tanaka, K., (2011). Transgenic potato overexpressing *Arabidopsis* cytosolic AtDHAR1 showed higher tolerance to herbicide, drought and salt stresses. *Breeding Science, 61*, 3–10.

Fahad, S., Bajwa, A. A., Nazir, U., Anjum, S. A., Farooq, A., Zohaib, A., Sadia, S., et al., (2017). Crop production under drought and heat stress: Plant responses and management options. *Front. Plant Sci., 8*, 1147.

Fahad, S., Hussain, S., Saud, S., Khan, F., Hassan, S., Nasim, W., Arif, M., et al., (2016). Exogenously applied plant growth regulators affect heat-stressed rice pollens. *J. Agron. Crop Sci., 202*, 139–150.

Faize, M., Burgos, L., & Faize, L., (2011). Involvement of cytosolic ascorbate peroxidase and Cu/Zn-superoxide dismutase for improved tolerance against drought stress. *Journal of Experimental Botany, 62*, 2599–2613.

Filippou, P., Antoniou, C. and Fotopoulos, V. 2011. Effect of drought and rewatering on the cellular status and antioxidant response of *Medicago truncatula* plants. Plant Signaling & Behavior 6:2, 270–277.

Foyer, C. H., & Noctor, G., (2005). Redox homeostasis and antioxidant signaling: A metabolic interface between stress perception and physiological responses. *Plant Cell, 17*, 1866–1875.

Foyer, C. H., Souriau, N., & Perret, S., (1995). Overexpression of glutathione reductase but not glutathione synthetase leads to increases in antioxidant capacity and resistance to photoinhibition in poplar trees. *Plant Physiology, 109*, 1047–1057.

Frederickson Matika, D. E., and Loake, G. J. (2014). Redox regulation in plant immune function. Antioxid. Redox Signal. 21, 1373–1388. doi: 10.1089/ars.2013. 5679

Fridovich, I., (1989). Superoxide dismutases. An adaptation to a paramagnetic gas. *Journal of Biological Chemistry, 264*, 7761–7764.

Geilfus, C. M., Mithöfer, A., Ludwig-Müller, J., Zörb, C., & Muehling, K. H., (2015). Chloride-inducible transient apoplastic alkalinizations induce stomata closure by controlling abscisic acid distribution between leaf apoplast and guard cells in salt-stressed *Vicia faba*. *New Phytol., 208*, 803–816.

Georgiadou, E. C., Kowalska, E., Patla, K., Kulbat, K., Smolinska, B., Leszczynska, J., & Fotopoulos, V., (2018). Influence of heavy metals (Ni, Cu, and Zn) on nitro-oxidative stress responses, proteome regulation and allergen production in basil (*Ocimum basilicum* L.) plants. *Front. Plant Sci., 9*, 862.

Ghisla, S., & Massey, V., (1989). Mechanisms of flavoprotein-catalyzed reactions. *European Journal of Biochemistry, 181*, 1–17.

Grant, J. J., & Loake, G. J., (2000). Role of reactive oxygen intermediates and cognate redox signaling in disease. *Plant Physiol., 124*, 21–29.

Guan, Z., Chai, T., Zhang, Y., Xu, J., & Wei, W., (2009). Enhancement of Cd tolerance in transgenic tobacco plants over-expressing a Cd-induced catalase cDNA. *Chemosphere, 76*, 623–630.

Gupta, A. S., Heinen, J. L., Holaday, A. S., Burke, J. J., & Allen, R. D., (1993). Increased resistance to oxidative stress in transgenic plants that overexpress chloroplastic Cu/Zn superoxide dismutase. *Proceedings of the National Academy of Sciences of the United States of America, 90*, 1629–1633.

Halliwell, B., (2006). Reactive species and antioxidants. Redox biology is a fundamental theme of aerobic life. *Plant Physiol., 141*, 312–322.

Han, C., Liu, Q., & Yang, Y., (2009). Short-term effects of experimental warming and enhanced ultraviolet-B radiation on photosynthesis and antioxidant defense of *Picea asperata* seedlings. *Plant Growth Regulation, 58*, 153–162.

Han, Q. H., Huang, B., Ding, C. B., Zhang, Z. W., Chen, Y. E., Hu, C., Zhou, L. J., Huang, Y., Liao, J. Q., & Yuan, S., (2017). Effects of melatonin on antioxidative systems and photosystem II in cold-stressed rice seedlings. *Front. Plant Sci., 8*, 785.

Hasanuzzaman, M., Alam, M. M., Nahar, K., Mohsin, S. M., Bhuyan, M. B., Parvin, K., Hawrylak-Nowak, B., & Fujita, M., (2019). Silicon-induced antioxidant defense and methyl glyoxal detoxification works coordinately in alleviating nickel toxicity in *Oryza sativa* L. *Ecotoxicology, 28*, 261–276.

Hasanuzzaman, M., Bhuyan, M., Anee, T. I., Parvin, K., Nahar, K., Mahmud, J. A., & Fujita, M., (2019). Regulation of ascorbate-glutathione pathway in mitigating oxidative damage in plants under abiotic stress. *Antioxidants, 8*, 384.

Hasanuzzaman, M., Borhannuddin, B. M. H. M., Zulfiqar, F., Raza, A., Mohsin, S. M., Al Mahmud, J., Fujita, M., & Fotopoulos, V., (2020). Reactive oxygen species and antioxidant defense in plants under abiotic stress: Revisiting the crucial role of a universal defense regulator. *Antioxidants, 9*, 1–52.

Hasanuzzaman, M., Nahar, K., Alam, M., Roychowdhury, R., & Fujita, M., (2013). Physiological, biochemical, and molecular mechanisms of heat stress tolerance in plants. *Int. J. Mol. Sci., 14*, 9643–9684.

Hasanuzzaman, M., Nahar, K., Anee, T. I., & Fujita, M., (2017). Exogenous silicon attenuates cadmium-induced oxidative stress in *Brassica napus* L. by modulating As A-GSH pathway and glyoxalase system. *Front. Plant Sci., 8*, 1061.

Hasanuzzaman, M., Nahar, K., Anee, T., Khan, M., & Fujita, M., (2018). Silicon-mediated regulation of antioxidant defense and glyoxalase systems confers drought stress tolerance in *Brassica napus* L. *African. J. Bot., 115*, 50–57.

Hasanuzzaman, M., Nahar, K., Gill, S. S., & Fujita, M., (2014). Drought stress responses in plants, oxidative stress, and antioxidant defense. In: Tuteja, N., & Gill, S. S., (eds.), *Climate Change and Plant Abiotic Stress Tolerance* (pp. 209–250). Wiley: Weinheim, Germany.

Hasanuzzaman, M., Nahar, K., Khan, M. I. R., Al Mahmud, J., Alam, M. M., & Fujita, M., (2020). Regulation of reactive oxygen species metabolism and glyoxalase systems by exogenous osmolytes confers thermo tolerance in *Brassica napus*. *Gesunde Pflanz., 72*, 3–16.

Hasanuzzaman, M., Nahar, K., Rahman, A., Inafuku, M., Oku, H., & Fujita, M., (2018). Exogenous nitric oxide donor and arginine provide protection against short-term drought stress in wheat seedlings. *Physiol. Mol. Biol. Plants, 24*, 993–1004.

Hasanuzzaman, M., Nahar, K., Rahman, A., Mahmud, J. A., Alharby, H. F., & Fujita, M., (2018). Exogenous glutathioneat tenuates lead-induced oxidative stress in wheat by improving antioxidant defense and physiological mechanisms. *J. Plant Interact., 13*, 203–212.

Hatz, S., Lambert, J. D., & Ogilby, P. R., (2007). Measuring the lifetime of singlet oxygen in a single cell: Addressing the issue of cell viability. *Photochem. Photobiol. Sci., 6*, 1106–1116.

Hefny, M., & Abdel-Kader, D. Z., Ashraf, M., Ozturk, M., & Athar, H. R., (2009). Antioxidant-enzyme system as selection criteria for salt tolerance in forage sorghum genotypes (*Sorghum bicolour* L. Moench). In: *Salinity and Water Stress* (pp. 25–36). The Netherlands.

Heyno, E., Mary, V., Schopfer, P., & Krieger-Liszkay, A., (2011). Oxygen activation at the plasma membrane: Relation between superoxide and hydroxyl radical production by isolated membranes. *Planta, 234*, 35–45.

Higuchi, E. T., (2006). Look back over the studies of lignin biochemistry. *J. Wood Sci., 52*, 2–8.

Horie, T., & Schroeder, J. I., (2004). Sodium transporters in plants. Diverse genes and physiological functions. *Plant Physiol., 136*, 2457–2462.

Hossain, M. A., Nakano, Y., & Asada, K., (1984). Monodehy-droascorbate reductase in spinach chloroplasts and its participation in regeneration of ascorbate for scavenging hydrogen peroxide. *Plant and Cell Physiology, 25*, 385–395.

Houmani, H., Rodríguez-Ruiz, M., Palma, J. M., & Corpas, F. J., (2018). Mechanical wounding promotes local and long distance response in the halophyte Cakile maritime through the involvement of the ROS and RNS metabolism. *Nitric Oxide*.

Hu, X., Zhang, A., Zhang, J., & Jiang, M., (2006). Abscisic acid is a key inducer of hydrogen peroxide production in leaves of maize plants exposed to water stress. *Plant Cell Physiol., 47*, 1484–1495.

Jackson, C., Dench, J., Moore, A. L., Halliwell, B., Foyer, C. H., & Hall, D. O., (1978). Subcellular localization and identification of superoxide dismutase in the leaves of higher plants. *European Journal of Biochemistry, 91*, 339–344.

Jannat, R., Uraji, M., & Morofuji, M., (2011). Roles of intracellular hydrogen peroxide accumulation in abscisic acid signaling in *Arabidopsis* guard cells. *Journal of Plant Physiology, 168*, 1919–1926.

Jimenez, A., Hern´andez, J. A., Del, L. A. R., & Sevilla, F., (1997). Evidence for the presence of the ascorbate-glutathione cycle in mitochondria and peroxisomes of pea leaves. *Plant Physiology, 114*, 275–284.

Jung, H. J., Yun, S. B., & June, S. L., (2001). Role of auxin-induced reactive oxygen species in root gravitropism. *Plant Physiology, 126*, 1055–1060.

Kanematsu, S., & Asada, K., (1989). Cuzn-superoxide dismutases in rice: Occurrence of an active, monomeric enzyme and two types of isozyme in leaf and non-photosynthetic tissues. *Plant and Cell Physiology, 30*, 381–391.

Kim, M. J., Ciani, S., & Schachtman, D. P., (2010). A peroxidase contributes to ROS production during *Arabidopsis* root response to potassium deficiency. *Molecular Plant, 3*, 420–427.

Kimura, S., Kaya, H., Kawarazaki, T., Hiraoka, G., Senzaki, E., & Michikawa, M., (2012). Protein phosphorylation is a prerequisite for the Ca^{2+}-dependent activation of *Arabidopsis* NADPH oxidases and may function as a trigger for the positive feedback regulation of Ca^{2+} and reactive oxygen species. *Biochim. Biophys. Acta, 1823*, 398–405.

Kobayashi, K., Kumazawa, Y., Miwa, K., & Yamanaka, S., (1996). γ-glutamyl lysine cross-links of spore coat proteins and transglutaminase activity in *Bacillus subtilis*. *FEMS Microbiology Letters, 144*, 157–160.

Krieger-Liszkay, A., Fufezan, C., & Trebst, A., (2008). Singlet oxygen production in photosystem II and related protection mechanism. *Photosyn. Res., 98*, 551–564.

Kwak, J. M., Mori, I. C., Pei, Z. M., Leonhardt, N., Torres, M. A., & Dangl, J. L., (2003). NADPH oxidase *AtrbohD* and *AtrbohF* genes function in ROS–dependent ABA signaling in *Arabidopsis*. *EMBO J., 22*, 2623–2633.

Lalarukh, I., & Shahbaz, M., (2020). Response of antioxidants and lipid peroxidation to exogenous application of alpha-tocopherol in sunflower (*Helianthus annuus* L.) under salt stress. *Pak. J. Bot., 52*, 75–83.

Lane, B. G., (2002). Oxalate germins and higher-plant pathogens. *IUBMB Life, 53*, 67–75.

Lee, Y. P., Kim, S. H., Bang, J. W., Lee, H. S., Kwak, S. S., & Kwon, S. Y., (2007). Enhanced tolerance to oxidative stress in transgenic tobacco plants expressing three antioxidant enzymes in chloroplasts. *Plant Cell Reports., 26*, 591–598.

Li, F., Wu, Q. Y., Sun, Y. L., Wang, L. Y., Yang, X. H., & Meng, Q. W., (2010). Over expression of chloroplastic monodehy-droascorbate reductase enhanced tolerance to temperature and methyl viologen-mediated oxidative stresses. *Physiologia Plantarum, 139*, 421–434.

Liu, T., Ye, X., Li, M., Li, J., Qi, H., & Hu, X., (2020). H_2O_2 and NO are involved in trehalose-regulated oxidative stress tolerance in cold-stressed tomato plants. *Environ. Exp. Bot.*, 171.

Liu, Y., Ren, D., Pike, S., Pallardy, S., Gassmann, W., & Zhang, S., (2007). Chloroplast-generated reactive oxygen species are involved in hypersensitive response-like cell death mediated by a mitogen-activated protein kinase cascade. *Plant Journal, 51*, 941–954.

Madhusudhan, R., Ishikawa, T., Sawa, Y., Shigeoka, S., & Shibata, H., (2003). Characterization of an ascorbate peroxidase in plastids of tobacco by-2 cells. *Physiologia Plantarum, 117*, 550–557.

Maffei, M. E., Mithofer, A., Arimura, G. I., Uchtenhagen, H., Bossi, S., Bertea, C. M., Cucuzza, L. S., et al., (2006). Effects of feeding *Spodoptera littoralis* on lima bean leaves. III. Membrane depolarization and involvement of hydrogen peroxide. *Plant Physiology*, (140), 1022–1035.

Maheshwari, R., & Dubey, R. S., (2009). Nickel-induced oxidative stress and the role of antioxidant defense in rice seedlings. *Plant Growth Regulation, 59*, 37–49.

Malhotra, C., Kapoor, R. T., Ganjewala, D., & Singh, N., (2017). Sodium silicate mediated response of antioxidative defense system in *Lycopersicon esculentum* Mill. under water stress. *Int. J. Phytomed., 9*, 364–378.

Martinez, C., Montillet, J. L., & Bresson, E., (1998). Apoplastic peroxidase generates superoxide anions in cells of cotton cotyledons undergoing the hypersensitive reaction to *Xanthomonas campestris* pv. *malvacearum* 18. *Molecular Plant-Microbe Interactions, 11*, 1038–1047.

Mehla, N., Sindhi, V., Josula, D., Bisht, P., & Wani, S. H., (2017). An introduction to antioxidants and their roles in plant stress tolerance. In: Khan, M. I. R., & Khan, N., (eds.), *Reactive Oxygen Species and Antioxidant Systems in Plants: Role and Regulation under Abiotic Stress* (pp. 1–23). Springer: Singapore.

Mhamdi, A., Queval, G., Chaouch, S., Vanderauwera, S., Van, B. F., & Noctor, G., (2010). Catalase function in plants: A focus on Arabidopsis mutants as stress-mimic models. *Journal of Experimental Botany, 61*, 4197–4220.

Miller, G., Schlauch, K., Tam, R., Cortes, D., Torres, M. A., Shulaev, V., Dangl, J. L., & Mittler, R., (2009). The plant NADPH oxidase RbohD mediates rapid systemic signaling in response to diverse stimuli. *Sci Signal, 2*, 45.

Miller, G., Shulaev, V., & Mittler, R., (2008). Reactive oxygen signaling and abiotic stress. *Physiologia Plantarum, 133*, 481–489.

Minibayeva, F., Beckett, R. P., & Kranner, I., (2014). Roles of apoplastic peroxidases in plant response to wounding. *Phytochemistry*. doi: 10.1016/j.phytochem.2014.06.008.

Mishra, S., Jha, A. B., & Dubey, R. S., (2011). Arsenite treatment induces oxidative stress, upregulates antioxidant system and causes phytochelatin synthesis in rice seedlings. *Protoplasma, 248*, 565–577.

Mittapalli, O., Neal, J. J., & Shukle, R. H., (2007). Antioxidant defense response in a galling insect. *Proc Natl Acad Sci USA, 104*, 1889–1894.

Mittler, R., & Zilinskas, B. A., (1992). Molecular cloning and characterization of a gene encoding pea cytosolic ascorbate peroxidase. *Journal of Biological Chemistry, 267*, 21802–21807.

Mittler, R., (2002). Oxidative stress, antioxidants and stress tolerance. *Trends in Plant Science, 7*, 405–410.

Mittler, R., (2002). Oxidative stress, antioxidants and stress tolerance. *Trends Plant Sci, 7*, 405–410.

Mittler, R., Herr, E. H., & Orvar, B. L., (1999). Transgenic tobacco plants with reduced capability to detoxify reactive oxygen intermediates are hyper responsive to pathogen infection. *Proceedings of the National Academy of Sciences of the United States of America, 96*, 14165–14170.

Mittler, R., Vanderauwera, S., Suzuki, N., Miller, G., Tognetti, V. B., Vandepoele, K., et al., (2011). ROS signaling: The new wave? *Trends Plant Sci., 16*, 300–309.

Mittova, V., Tal, M., Volokita, M., & Guy, M., (2003). Up−regulation of the leaf mitochondrial and peroxisomal antioxidative systems in response to salt−induced oxidative stress in the wild salt−tolerant tomato species *Lycopersicon pennellii. Plant Cell Environ., 26*, 845–856.

Møller, I. M., Jensen, P. E., & Hansson, A., (2007). Oxidative modifications to cellular components in plants. *Annu. Rev. Plant Biol., 58*, 459–481.

Monshausen, G. B., Bibikova, T. N., Messerli, M. A., Shi, C., & Gilroy, S., (2007). Oscillations in extracellular pH and reactive oxygen species modulate tip growth of Arabidopsis root hairs. *Proc. Natl. Acad. Sci. U.S.A., 104*, 20996–21001.

Moussa, R., & Abdel-Aziz, S. M., (2008). Comparative response of drought tolerant and drought-sensitive maize genotypes to water stress. *Australian Journal of Crop Sciences, 1*, 31–36.

Munns, R., & Tester, M., (2008). Mechanisms of salinity tolerance. *Annu. Rev. Plant Physiol., 59*, 651–681.

Murphy, M. (2009). How mitochondria produce reactive oxygen species. Biochem. J. 417, 1–13. doi: 10.1042/BJ20081386

Musser, R. O., Hum-Musser, S. M., Eichenseer, H., Peiffer, M., Ervin, G., Murphy, J. B., & Felton, G. W., (2002). Herbivory: Caterpillar saliva beats plant defenses. *Nature, 416*, 599–600.

Musser, R., Farmer, E., Peiffer, M., Williams, S., & Felton, G., (2006). Ablation of caterpillar labial salivary glands: Technique for determining the role of saliva in insect-plant interactions. *J. Chem Ecol., 32*, 981–992.

Nahar, K., Hasanuzzaman, M., Alam, M. M., Rahman, A., Mahmud, J. A., Suzuki, T., & Fujita, M., (2017). Insights into spermine-induced combined high temperature and drought tolerance in mung bean: Osmo-regulation and roles of antioxidant and glyoxalase system. *Protoplasma, 254*, 445–460.

Nahar, K., Rahman, M., Hasanuzzaman, M., Alam, M. M., Rahman, A., Suzuki, T., & Fujita, M., (2016). Physiological and biochemical mechanisms of spermine-induced cadmium stress tolerance in mung bean (*Vigna radiata* L.) seedlings. *Environ. Sci. Pollut. Res., 23*, 21206–21218.

Naveen, J., Hariprasad, P., Chandra, N. S., & Niranjana, S. R., (2013a). Cerebroside mediated elicitation of defense response in chilli *(Capsicum annuum* L.) against *Colletotrichum capsici* infection. *Journal of Plant Interactions, 8*(1), 65–73.

Naveen, J., Hariprasad, P., Navya, H. M., & Niranjana, S. R., (2013b). Induction of resistance in chilli seedlings against anthracnose disease by crude cell wall elicitors from *Colletotrichum capsici* isolate uom-18. *Int. J. Pharm. Bio. Sci., 4*(4), 304–314.

Naveen, J., Navya, H. M., Hariprasad, P., Hithamani, G., & Niranjana, S. R., (2019). Rhizobacteria *Bacillus* spp. induce resistance against anthracnose disease in chili (*Capsicum annuum* L.) through activating host defense response. *Egyptian Journal of Biological Pest Control, 29*, 451–459.

Naveen, J., Navya, H. M., Hariprasad, P., Hithamani, G., & Niranjana, S. R., (2020). Acibenzolar-S-methyl and β-amino butyric acid-induced upregulation of biochemical defense against *Colletotrichum capsici* infection in chilli (*Capsicum annuum*). *Archives of Phytopathology and Plant Protection, 53*, 141–161.

Navrot, N., Rouhier, N., Gelhaye, E., & Jacquot, J. P., (2007). Reactive oxygen species generation and antioxidant systems in plant mitochondria. *Physiol. Plant, 129*, 185–195.

Neill, S., Desikan, R., & Hancock, J., (2002). Hydrogen peroxide signaling. *Current Opinion in Plant Biology, 5*, 388–395.

Noctor, G., & Foyer, C. H., (1998). Ascorbate and glutathione: Keeping active oxygen under control. *Annual Review of Plant Biology, 49*, 249–279.

Noctor, G., De Paepe, R., & Foyer, C. H., (2007). Mitochondrial redox biology and homeostasis in plants. *Trends Plant Sci., 12*, 125–134.

Noctor, G., Veljovic-Jovanovic, S., Driscoll, S., Novitskaya, L., & Foyer, C. H., (2002a). Drought and oxidative load in the leaves of C_3 plants: A predominant role for photo-respiration. *Annals of Botany, 89*, 841–850.

Noctor, G., Veljovic-Jovanovic, S., Driscoll, S., Novitskaya, L., & Foyer, C. H., (2002b). Drought and oxidative load in the leaves of C_3 plants: A predominant role for photorespiration. *Ann. Bot., 89*, 841–850.

Orozco-Cárdenas, M. L., Narváez-Vásquez, J., & Ryan, C. A., (2001). Hydrogen peroxide acts as a second messenger for the induction of defense genes in tomato plants in response to wounding, systemin, and methyl jasmonate. *Plant Cell, 13*, 179–191.

Orozco-Cardenas, M., & Ryan, C. A., (1999). Hydrogen peroxide is generated systemically in plant leaves by wounding and system in via the octadecanoid pathway. *Proc. Natl. Acad. Sci. U.S.A., 96*, 6553–6655.

Palma, J. M., Corpas, F. J., & Del Río, L. A., (2009). Proteome of plant peroxisomes: New perspectives on the role of these organelles in cell biology. *Proteomics, 9*, 2301–2312.

Parvez, S., Abbas, G., Shahid, M., Amjad, M., Hussain, M., Asad, S. A., Imran, M., & Naeem, M. A., (2020). Effect of salinity on physiological, biochemical and photostabilizing attributes of two genotypes of quinoa (*Chenopodium quinoa* Wild.) exposed to arsenic stress. *Ecotoxicol. Environ. Saf., 187*, 109814.

Pastore, D., Trono, D., Laus, M. N., Di Fonzo, N., & Flagella, Z., (2007). Possible plant mitochondria involvement in cell adaptation to drought stress a case study: Durum wheat mitochondria. *J. Exp. Bot., 58*, 195–210.

Pel, Z. M., Murata, Y., & Benning, G., (2000). Calcium channels activated by hydrogen peroxide mediate abscisic acid signaling in guard cells. *Nature, 406*, 731–734.
Peng, C. L., Ou, Z. Y., Liu, N., & Lin, G. Z., (2005). Response to high temperature in flag leaves of super high-yielding rice Pei'ai 64S/E32 and Liangyoupeijiu. *Rice Sci., 12*, 179–186.
Pfannschmidt, T., (2003). Chloroplast redox signals: How photosynthesis controls its own genes. *Trends Plant Sci., 8*, 33–41.
Pinto, E., Sigaud-kutner, T., Leitao, M. A., Okamoto, O. K., Morse, D., & Colepicolo, P., (2003). Heavy metal-induced oxidative stress in algae. *J. Phycol., 39*, 1008–1018.
Racchi, M. L., Bagnoli, F., Balla, I., & Danti, S., (2001). Differential activity of catalase and superoxide dismutase in seedlings and in vitro micro-propagated oak (*Quercus robur* L.). *Plant Cell Reports, 20*, 169–174.
Radotic, K., Dučić, T., & Mutavdžić, D., (2000). Changes in peroxidase activity and iso-enzymes in spruce needles after exposure to different concentrations of cadmium. *Environmental and Experimental Botany, 44*, 105–113.
Radwan, D. E. M., Fayez, K. A., Mahmoud, S. Y., & Lu, G., (2010). Modifications of antioxidant activity and protein composition of bean leaf due to Bean yellow mosaic virus infection and salicylic acid treatments. *Acta Physiologiae Plantarum, 32*, 891–904.
Rady, M. M., Belal, H. E., Gadallah, F. M., & Semida, W. M., (2020). Selenium application in two methods promotes drought tolerance in *Solanum lycopersicum* plant by inducing the antioxidant defense system. *Sci. Hortic., 266*, 109290.
Rasmusson, A. G., Geisler, D. A., & Møller, I. M., (2008). The multiplicity of dehydrogenases in the electron transport chain of plant mitochondria. *Mitochondrion, 8*, 47–60.
Rehman, S., Abbas, G., Shahid, M., Saqib, M., Farooq, A. B. U., Hussain, M., Murtaza, B., Amjad, M., Naeem, M. A., & Farooq, A., (2019). Effect of salinity on cadmium tolerance, ionic homeostasis and oxidative stress responses in Conocarpus exposed to cadmium stress: Implications for phytoremediation. *Ecotoxicol. Environ. Safe, 171*, 146–153.
Rubio, M. C., Bustos-Sanmamed, P., Clemente, M. R., & Becana, M., (2009). Effects of salt stress on the expression of antioxidant genes and proteins in the model legume *Lotus japonicas*. *New Phytologist, 181*, 851–859.
Saha, I., De, A. K., Sarkar, B., Ghosh, A., Dey, N., & Adak, M. K., (2018). Cellular response of oxidative stress whensub1A QTL of rice receives water deficit stress. *Plant Sci. Today, 5*, 84–94.
Sasaki, K., Iwai, T., Hiraga, S., et al., (2004). Ten rice peroxidases redundantly respond to multiple stresses including infection with rice blast fungus. *Plant and Cell Physiology, 45*(10), 1442–1452.
Satish, L., Rency, A. S., & Ramesh, M., (2018). Spermidine sprays alleviate the water deficit-induced oxidative stress in finger millet (*Eleusine coracana* L.) plants. *3 Biotech, 8*, 63.
Scandalios, J. G., (1993). "Oxygen stress and superoxide dismutases. *Plant Physiology, 101*(1), 7–12.
Schuller, D. J., Ban, N., Van, H. R. B., McPherson, A., & Poulos, T. L., (1996). The crystal structure of peanut peroxidase. *Structure, 4*, 311–321.
Sewelam, N., Kazan, K., & Schenk, P. M., (2016). Global plant stress signaling: Reactive oxygen species at the cross-road. *Front Plant Sci., 7*, 187.
Sharma, P., & Dubey, R. S., (2005). Induces oxidative stress and enhances the activities of antioxidant enzymes in growing rice seedlings. *Plant Growth Regulation, 46*, 209–221.

Sharma, P., & Dubey, R. S., (2004). Ascorbate peroxidase from rice seedlings: Properties of enzyme isoforms, effects of stresses and protective roles of osmolytes. *Plant Science, 167*, 541–550.

Sharma, P., & Dubey, R. S., (2007). Involvement of oxidative stress and role of antioxidative defense system in growing rice seedlings exposed to toxic concentrations of aluminum. *Plant Cell Reports, 26*, 2027–2038.

Sharma, P., Jha, A. B., Dubey, R. S., & Pessarakli, M., (2012). Reactive oxygen species, oxidative damage, and antioxidative defense mechanism in plants under stressful conditions. *Journal of Botany*, 1–26.

Shu, D. F., Wang, L. Y., Duan, M., Deng, Y. S., & Meng, Q. W., (2011). Antisense-mediated depletion of tomato chloroplast glutathione reductase enhances susceptibility to chilling stress. *Plant Physiology and Biochemistry, 49*, 1228–1237.

Singh, A., Kumar, A., Yadav, S., & Singh, I. K., (2019). Reactive oxygen species-mediated signaling during abiotic stress. *Plant Gene., 18*, 100173.

Suzuki, N., & Mittler, R., (2012). Reactive oxygen species-dependent wound responses in animals and plants. *Free Radical Biology and Medicine, 53*, 2269–2276.

Suzuki, N., Koussevitzky, S., Mittler, R., & Miller, G., (2012). ROS and redox signaling in the response of plants to abiotic stress. *Plant Cell Environ., 35*, 259–270.

Tanou, G., Molassiotis, A., & Diamantidis, G., (2009a). Hydrogen peroxide-and nitric oxide-induced systemic antioxidant prime-like activity under NaCl-stress and stress-free conditions in citrus plants. *J. Plant Physiol., 166*, 1904–1913.

Tanou, G., Molassiotis, A., & Diamantidis, G., (2009b). Induction of reactive oxygen species and necrotic death-like destruction in strawberry leaves by salinity. *Environ. Exp. Bot., 65*, 270–281.

Tayefi-Nasrabadi, H., Dehghan, G., Daeihassani, B., Movafegi, A., & Samadi, A., (2011). "Some biochemical properties of guaiacol peroxidases as modified by salt stress in leaves of salt-tolerant and salt-sensitive safflower *(Carthamus tinctorius* L. cv.) cultivars. *African Journal of Biotechnology, 10*, 751–763.

Torres, M. A., Dangl, J. L., & Jones, J. D. G., (2002). *Arabidopsis* $gp91^{phox}$ homologues Atrbohd and Atrbohf are required for accumulation of reactive oxygen intermediates in the plant defense response. *Proceedings of the National Academy of Sciences of the United States of America, 99*, 517–522.

Turrens, J. F., (2003). Mitochondrial formation of reactive oxygen species. *J. Physiol., 552*, 335–344.

Ushimaru, T., Maki, Y., Sano, S., Koshiba, K., Asada, K., & Tsuji, H., (1997). Induction of enzymes involved in the ascorbate-dependent antioxidative system, namely, ascorbate peroxidase, monodehydro ascorbate reductase and dehydroascorbate reductase, after exposure to air of rice (*Oryza sativa*) seedlings germinated under water. *Plant and Cell Physiology, 38*, 541–549.

Vallabhaneni, R., Li, F., Yu, J., Rocheford, T., & Wurtzel, E. T., (2008). The maize phytoene synthase gene family: Overlapping roles for carotene genesis in endosperm, photomorphogenesis and thermal stress tolerance. *Plant Physiology, 147*, 1334–1346.

Verma, S., & Dubey, R. S., (2003). Lead toxicity induces lipid peroxidation and alters the activities of antioxidant enzymes in growing rice plants. *Plant Science, 164*, 645–655.

Wagner, D., Przybyla, D., Op Den Camp, R., Kim, C., Landgraf, F., & Lee, K. P., (2004). The genetic basis of singlet oxygen-induced stress responses of *Arabidopsis thaliana. Science, 306*, 1183–1185.

Wang, J., Zhang, H., & Allen, R. D., (1999). Over-expression of an *Arabidopsis* peroxisomal ascorbate peroxidase gene in tobacco increases protection against oxidative stress. *Plant and Cell Physiology, 40*, 725–732.

Wang, Y., Wisniewski, M., Meilan, R., Cui, M., Webb, R., & Fuchigami, L., (2005). Over-expression of cytosolic ascorbate peroxidase in tomato confers tolerance to chilling and salt stress. *Journal of the American Society for Horticultural Science, 130*, 167–173.

Willekens H., Inzé D., van Montagu M., & van Camp W., (1995). Catalases in plants. *Molecular Breeding, 1*, 207–228.

Xiong, L., Schumaker, K. S., & Zhu, J. K., (2002). Cell signaling during cold, drought, and salt stress. *Plant Cell, 14*, 165–183.

Xue, M., Guo, T., Ren, M., Wang, Z., Tang, K., Zhang, W., & Wang, M., (2019). Constitutive expression of chloroplast glycerol-3-phosphate acyl-transferase from *Ammopiptanthus mongolicus* enhances unsaturation of chloroplast lipids and tolerance to chilling, freezing and oxidative stress in transgenic *Arabidopsis*. *Plant Physiol. Biochem., 143*, 375–387.

Yabuta, Y., Motoki, T., Yoshimura, K., Takeda, T., Ishikawa, T., & Shigeoka, S., (2002). Thylakoid membrane-bound ascorbate peroxidase is a limiting factor of antioxidative systems under photo-oxidative stress. *Plant Journal, 32*, 915–925.

Yadu, B., Chandrakar, V., Tamboli, R., & Keshavkant, S., (2019). Dimethyl thiourea antagonizes oxidative responses by up-regulating expressions of pyrroline-5-carboxylate synthetase and antioxidant genes under arsenic stress. *Int. J. Environ. Sci. Technol., 16*, 8401–8410.

Yan, J., Tsuichihara, N., Etoh, T., & Iwai, S., (2007). Reactive oxygen species and nitric oxide are involved in ABA inhibition of stomatal opening. *Plant Cell and Environment, 30*, 1320–1325.

Yang, D.-L., Shi, Z., Bao, Y., Yan, J., Yang, Z., Yu, H., et al. (2017). Calcium pumps and interacting BON1 protein modulate calcium signature, stomatal closure, and plant immunity. Plant Physiol. 175, 424–437. doi: 10.1104/pp.17.00495.

Yeh, C. M., Chien, P. S., & Huang, H. J., (2007). Distinct signaling pathways for induction of MAP kinase activities by cadmium and copper in rice roots. *Journal of Experimental Botany, 58*, 659–671.

Yoshida, S., Tamaoki, M., Shikano, T., et al., (2006). Cytosolic dehydroascorbate reductase is important for ozone tolerance in *Arabidopsis thaliana*. *Plant and Cell Physiology, 47*(2), 304–308.

Yuasa, T., Ichimura, K., Mizoguchi, T., & Shinozaki, K., (2001). Oxidative stress activates ATMPK6, an *Arabidopsis* homologue of map kinase. *Plant and Cell Physiology, 42*, 1012–1016.

Zaefyzadeh, M., Quliyev, R. A., Babayeva, S. M., & Abbasov, M. A., (2009). The effect of the interaction between genotypes and drought stress on the superoxide dismutase and chlorophyll content in durum wheat landraces. *Turkish Journal of Biology, 33*, 1–7.

Zaid, A., Mohammad, F., & Fariduddin, Q., (2020). Plant growth regulators improve growth, photosynthesis, mineral nutrient and antioxidant system under cadmium stress in menthol mint (*Mentha arvensis* L.). *Physiol. Mol. Biol. Plants, 26*, 25–39.

CHAPTER 12

Phenolic Compounds: Stress Markers of Plants and Their Protective Role

G. HITHAMANI,[1] J. NAVEEN,[1] and H. G. PUSHPALATHA[2]

[1]*Department of Biochemistry, CSIR–Central Food Technological Research Institute, Mysore–570020, Karnataka, India*

[2]*Associate Professor, Maharani's Science College for Women, Mysore–570001, Karnataka, India, E-mail: pushpa_2001@rediffmail.com*

ABSTRACT

Phenolic compounds are plant secondary metabolites play a key role in plants' physiology like growth and development, as major antioxidant molecules, including hormonal regulation, nutrient mineralization, and photosynthetic activities. Phenolic compounds synthesized from Phenylpropanoid biosynthetic pathway, consist of aromatic benzene ring with one/more hydroxyl groups. Phenolics accumulate during both normal and stress conditions in plants. Excess phenolics are produced by plants mainly during the environmental stress condition to protect the plant cell from ROS and oxidative stress. Phenolic compounds quench the excess toxic radicals produced during the environmental stress. In addition to the antioxidant potential, phenolics, and its derivatives play role in signal transduction to activate various defense mechanisms of plant during the environmental stress condition. Phenolics and its derivatives act as toxic material to the pathogen and pests and restrict their growth, development, dispersion, and habitat. Accumulation of lignin which is a complex polyphenol, thickens the plant cell wall and make it more rigid and helps to avoid entry of pests and pathogen into the plant system. Scientific community has great research interest in the field of polyphenols and its

Plant Metabolites under Environmental Stress: Mechanisms, Responses, and Adaptation Strategies.
Nivas M. Desai, Manasi Patil, Umesh R. Pawar (Eds.)

response to the environmental stresses. This chapter provides information regarding the phenolic compounds and its derivatives and their response to biotic and abiotic stresses in plants.

12.1 INTRODUCTION

Environmental stress is the factor that affects plants health and yield. Environmental stress is divided into abiotic and biotic stress. Abiotic stresses are caused by the non-living source, namely, salinity, drought, UV irradiation, toxicity, and flooding. The biotic stresses are caused by living organisms like pathogens (viruses, bacteria, and fungi) and pests. Environmental stress is a significant risk to food security and is the predominant and the leading cause of crop and yield loss worldwide. It has been estimated that about 50% of crop and yield loss is due to environmental stress (Francini and Sebastiani, 2019; Arora et al., 2019). Plants recognize the risk by their receptors, which triggers the protective mechanism against environmental stresses.

Plants react to various stresses by producing reactive oxygen species (ROS) as an initial response leading to cause oxidative stress and hypersensitive reaction (HR). The excess generation of ROS is toxic to plants, which kills the plant cell and such plants are called susceptible cultivar. The plants in replay activate the ROS quenching antioxidants components and reduces the toxicity; in this condition, the plants are said to be resistant/tolerant cultivar. ROS acts as a messenger, which activates various biochemical defense mechanisms in the plant through activating numerous defense genes. Antioxidants components produced in response to ROS are of two types, namely enzymatic antioxidants, and non-enzymatic antioxidants. The non-enzymatic antioxidants are the secondary metabolites and play an essential role in the ROS detoxification generated during stress condition and also help in regular plant activity.

Among the non-enzymatic antioxidants, phenolics are the essential that play a specialized role in physiological function throughout the plant lifetime. Phenolic compounds are widespread in plants and they vary both quantitatively and qualitatively. Phenolic compounds are known to exhibit vast biological processes. Phenolics are produced through the activation of the phenylpropanoid biosynthetic pathway, which typically upregulated during the stress conditions like extreme temperatures, drought, heavy metal toxicity, salinity, ultraviolet radiations, and microbial pathogen attack (Lamaoui et al., 2018; Mittler, 2006; Naveen et al., 2019, 2020). Phenolics are the most studied group of plant metabolites which have more than 8,000 molecules (González-Sarrías et al., 2020). Phenolics are aromatic benzene ring compounds with one or more hydroxyl groups produced by plants mainly for protection against stress (Bhattacharya et al., 2020). These compounds are monomeric and polymeric; their structure varies extensively as simple phenolics/polyphenolics. Based on their structure, they are divided into phenolic acids, flavonoids, stilbenoids, and lignins (Samec et al., 2021). Increased synthesis and accumulation of phenolics, and their related components guarantees the surveillance, strengthens, and increases the plant's competitiveness against environmental stress (Kumar et al., 2020).

An increase in plant phenolic accumulation is a good sign of the plant in response to environmental stress tolerance. Also, phenolic compounds further help in lignification, which is the critical factor of stress tolerance by thickening the plant cell wall. Thus, transgenic cultivar with altered phenolics level/upregulated phenolic level in response to various environmental stress is required and mandate (Faize et al., 2011; Lee et al., 2007). Hence, the present chapter focuses on the role of phenolic compounds in response to various environmental stresses.

12.2 ABIOTIC STRESS AND PHENOLIC ACCUMULATION

Abiotic stress is a part of environmental stress includes, flooding/waterlogging, drought, salinity, toxic metals/metalloids, extreme temperature (Mittler and Blumwald, 2010; Bhattacharjee, 2010). These diverse abiotic stresses cause a major threat to sustainable agriculture. Abiotic stresses induce ROS generation that leads to oxidative stress (damages plant cell, organelles, and its components). To overcome these challenges plant produces antioxidants (Figure 12.1). Phenolics are considered as vital antioxidants that quench the toxic radical generated and protects the plant cell from various stresses, including pathogen and pests (Isah, 2019). Phenolics are also considered as an antimicrobial compounds, which convert into various other forms, i.e., phenolics derived (lignin) compound and as major antioxidant enzymes (phenylpropanoid pathway); (Mishra and Sangwan, 2019). Major abiotic stresses and response of phenolics and their derivatives including enzymes has been reviewed in subsections.

12.2.1 SALINITY

Salinity/salt stress is the major abiotic stress causes severe crop and yield loss by altering the plant metabolism/secondary metabolites. To overcome the salinity stress, the plant should be in dormant condition/it should adjust to the environmental conditions. The response of the plant to salinity depends on the following aspects plant age, species type, genotype, morphology, ion movement, characteristics of membrane, tolerance of Na/K concentration in the cytosol and salt glands (Yokoi et al., 2002; Flowers and Flowers, 2005; Ahmad and Sharma, 2008). In various plant cultivars, researchers found the accumulation of polyphenols, phenolic acids, flavonoids, lignin, anthocyanin, and other secondary metabolites in response to salinity stress. During the stress condition, plant produces antioxidant phenolics in response to ROS and oxidative stress, the quality and ability of plant survival and plant health increases with the presence of phenolics. Additionally, the increased phenolic accumulation leads to increase the yield of the plant under salinity stress (Khan et al., 2011).

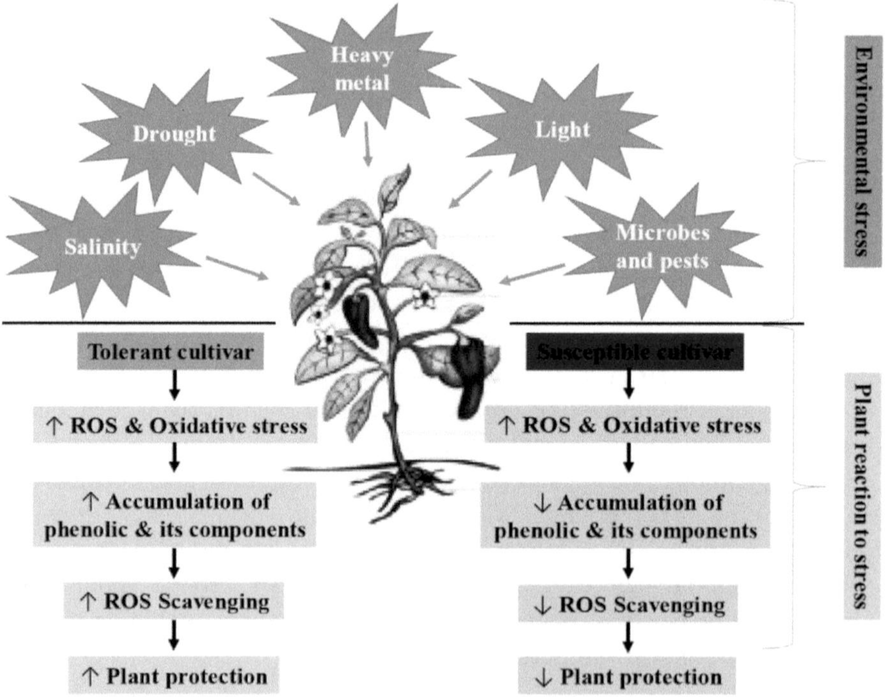

FIGURE 12.1 Environmental stress and response of stress susceptible and tolerant plant.

The antioxidant phenolic accumulation in the plant depends on the time of salinity stress, dose, and source of salt treatment. Telesiński et al. (2008)

observed increased level of peroxidase (POX) and catalase with increase in salinity stress in bean cultivar. Increase in both the enzyme concentration was observed till 21 days and further POX concentration increased till 28 days. Kim et al. (2008) reported that 100 mM NaCl treatment, enhanced total phenolic (1.3 fold) and flavonoid (14.5 fold) content in Romaine lettuce after 12 days. The treatment of 77 mM bisodium sulfate increased the flavonoid level by 4.5 folds in Romaine lettuce but not the total phenolic concentration (Mahmoudi et al., 2010). Treatment of 40 and 80 mM NaCl in soil increase the concentration of total phenolic content in strawberries cv. Korona, by 10 and 16% respectively. In case of cv. Elsanta (salinity sensitive variety) the increase in phenolic content was observed more, i.e., 14 and 23% in 40 and 80 mM NaCl treatment, respectively (Keutgen and Paweltzik, 2008). Even the anthocyanin content was increased as the concentration of salt (NaCl) increases in both cv. Korona and Elsanta. Anthocyanin biosynthesis enhanced during the saline stress condition promotes the plant growth (Aloisi et al., 2016). Keutgen and Paweltzik (2008) found an increase in superoxide dismutase (SOD) level in salt (NaCl) treated and the SOD concentration increased with increased level of salinity, i.e., dose-dependently.

The decreased and increased phenolic level in salt sensitive and tolerant plant was observed respectively. Zhou et al. (2018) investigated the alterations in plant growth, antioxidant capacity and volatile exudate of Chinese herd (Schizonepeta tenuifolia Briq.) on treatment with different concentrations of NaCl (0, 25, 50, 75, 100 mM). Results revealed that the decrease in biomass of plants was observed in salt treated plants. The total phenolic and flavonoids content increased at lower (25 mM) and moderate (50 mM) concentrations of NaCl and declined at higher concentrations (75 and 100 mM) of NaCl. Study concludes that salinity has great influence on the growth and production of secondary plant metabolites.

The accumulation of phenolics in response to the salinity varies with the plant parts, maturity stage (Zrig et al., 2011; Petridis et al., 2012; Navarro et al., 2006; Lopez-Berenguer et al., 2009). In relation to the above statement, Zrig et al. (2011) found varied levels of total phenolic concentration in root and leaves in bitter almonds after treatment of low concentration of NaCl. Similar findings were revealed by Petridis et al. (2012) that the increase in phenolic content in olive tree leaves by 129% compared to control was observed after salt treatment. Salt treatment for 3 days to the different stage fruits of pepper revealed that the concentration of phenolic level was more in red pepper fruit followed by green and turning fruit (Navarro et al., 2006). Valifarda et al. (2014) found that the treatment of *Salvia mirzayanii* an aromatic and medicinal plant, showed increased phenolic content (3.8 mg gallic acid equivalent/gm dry weight) after treatment of moderate levels of NaCl compared to control with IC_{50} value of 2.13 mg/ml compared to control in DPPH assay. Linic et al. (2019) studied the short time salinity stress on phenolic acids accumulation in three different Brassica crops (Chinese cabbage, white cabbage, and kale). Reduction in root growth and biomass with increased ROS and glutathione level was observed in Chinese cabbage. In kale and white cabbage, the accumulation of hydroxycinnamic acids and ferulic acid level was more. Reduced levels of caffeic, 4-coumaric acid and salicylic acid was observed in Chinese cabbage and kale cultivar upon treatment. Authors also mentioned that the activation/upregulation of particular phenolic acids is species-specific and only some phenolic acids participated in salt tolerance. The effect of salinity (NaCl, $MgSO_4$ and $CaCl_2$) on phenolic (rutin, chlorogenic acid, quercetin, iso-quercetin, hyperoside) content of *Hypericum pruinatum* in green house condition was revealed by Caliskan et al. (2017). The increase in the phenolics depended on the salinity concentration in a dose-dependent manner. Authors further conclude that the salinity has a marked influence on the content of phenolics.

12.2.2 HEAVY METAL/MINERAL TOXIFICATION

Toxification by heavy metals is one of the major abiotic stresses plants encounter. Volcanoes and continental dusts are the natural sources of heavy metals. Agricultural activities which include soil contamination by the use of pesticides, fertilizers, and herbicides results in heavy metal accumulation (Kisa et al., 2016). In addition, refineries, textiles, plastics, and preservative industries, etc., are other factors that cause heavy metal deposition in soil. Though few heavy metals (Fe, Cu, Zn, Co, Mo, Mn, Ni) are of biological importance, most of them are toxic elements, to name a few; Cd, Pb, Hg, As, Be, and Al. Heavy metal accumulated places are of great concern in view of their harmful effects on the human health and also on the environment.

Heavy metals are known to affect the normal growth, morphology, and physiology of plants. Accumulation of heavy metals in the ecosystem influences the metabolism of plant including photosynthesis, defense response and mineral distribution (Nazar et al., 2012). Heavy metals are known to induce oxidative stress by means of generating ROS resulting in cell death (Shahid et al., 2015). In order to cope up with the heavy metal toxicity and maintain their productivity, plants reduce uptake of heavy metals, sequester them into their vacuoles. Also, there are several metal chelators and antioxidant phytochemicals in the plants that are activated in response to heavy metals in the environment.

Influence of heavy metals on the antioxidant enzyme activities including lipid peroxidation (LPO) in two mangrove plant seedlings; *Kandelia candel* and *Bruguiera gymnorrhiza*, was studied by Zhang et al. (2007). Seedlings of *K. candel* and *B. gymnorrhiza* were exposed to different concentrations of heavy metals (Pb^{2+}, Cd^{2+}, and Hg^{2+}). After two months of exposure, leaves, and roots of treated seedlings were harvested and antioxidant enzymes and LPO were examined and compared with control seedlings. Catalase (CAT) activity increased in the leaves of heavy metal stressed *K. candel.* Superoxide dismutase (SOD) and peroxidase activities (POD) increased in leaves and roots of both the plants on heavy metal treatment, however the increase was significant in roots of *K. candel* than in *B. gymnorrhiza.* LPO measured as a means of malondialdehyde content increased only in heavy metal stressed leaves of *B. gymnorrhiza.*

Influence of excess copper (3, 60, and 120 μM) on the physiological attributes of four-week-old *Matricaria chamomilla* (Chamomile) plants was studied for 10 days (Kováčik et al., 2008). A concentration of 120 μM Cu in the nutrient solution used for the cultivation of chamomile, decreased the production of dry mass, and also chlorophyll, water, and nitrogen content. There was altered phenolic profile of chamomile in Cu excess, with increased concentration of detected phenolic compounds at 60 μM. Consequence of cadmium and copper uptake (60 and 120 μM) on the accumulation and metabolism of selected phenolic compounds and stress related parameters in chamomile plants was recorded for 7 days (Kováčik et al., 2009a). Among the phenolic compounds detected, cinnamic acids were significantly accumulated on exposure to Cd. However, salicylic acid content increased notably in all variants. Activities of shikimate dehydrogenase, polyphenol oxidase, cinnamyl alcohol dehydrogenase, and ascorbate peroxidase (APX), which are enzymes of phenolic metabolism, increased profoundly, especially in the roots in response to Cu excess. An increase in phenolic accumulation along with an increase in the phenolic metabolism-related enzyme activity corresponds to the probable antioxidant defense mechanism. The results confirmed marked Cu toxicity in comparison to Cd for chamomile plants, since Cd is unable to produce ROS by virtue of its non-redox nature. In comparison to Cd and Cu, nickel (Ni) caused significantly lower toxicity in chamomile plants (Kováčik et al., 2009b). However, enzymes related to phenolic metabolism, phenylalanine ammonia-lyase (PAL) in particular increased at the highest Ni concentration along with phenolic acids and flavonoids.

Molecular mechanism of the effects of Cd toxicity on auxin homeostasis in Arabidopsis seedlings was studied by Hu et al. (2013). On Cd treatment, indole acetic acid (IAA) content in the seedlings decreased significantly with increased activity of IAA oxidase. There was significant influence on the expression of auxin biosynthetic genes and also on the root architecture. Phenolic compounds chelate metals, which is attributable to their aromatic rings that are nucleophilic in nature. The carboxyl and hydroxyl groups of these compounds are known to bind heavy metals. Salicylic acid is a phenolic acid, endogenously produced in plants as a defense response and also known as a plant growth regulator. Role of salicylic acid during arsenic toxicity in wheat seedlings was examined by Zengin (2015). It was observed that the toxic effects of As treatment such as decreased germination parameters, increased proline, and MDA content, was partially alleviated on pre-treatment of wheat seedlings with salicylic acid. In addition, the chlorophyll and protein contents were also increased as the beneficial effects of salicylic acid treatment.

12.2.3 DROUGHT

Drought is considered as an important and major environmental stress. Drought negatively effects the plant growth, physiological changes and very importantly causes yield loss. It has been stated that under drought conditions, the accumulation of phenolics increased in plant to survive or to overcome the adverse condition. The upregulation of phenolic compounds accumulation during the drought condition is due to the modulation of phenylpropanoid biosynthesis pathway (Hernandez et al., 2009). Drought stress induces various crucial genes encoding main enzymes of the phenylpropanoid pathway, which further increases the biosynthesis of phenolic compounds *Hypericum brasiliense* plant grown under drought condition showed a 10% higher phenolic concentration than control (de Abreu and Mazzafera, 2005). Drought causes ROS generation and oxidative stress in plants, which meanwhile activates the phenolics accumulation. Researchers revealed, that the accumulation of phenolic, flavonoids, and anthocyanins are usual in drought stress, and those plants can be considered as a drought-tolerant varieties. Nogs et al. (1998) found enhanced phenolic concentration in pea plants (*Pisum sativum*) due to drought stress. Whereas, in *Tridax procumbens* plant a significant increase in total phenolic concentration was detected in both flowers and leaves grown under the drought stress (Gnanasekaran and Kalavathy, 2017). The improved level of quercetin, kaempferol, and flavonols was found in stress tolerance white clover plants grown under the drought stress conditions (Nichols et al., 2015).

Larson (1988) found enhanced phenolic acid and flavonoids content in willow leaves due to drought stress. The defense mechanism of plant against the drought condition was by the upregulation of phenolic molecules in the leaves (Akula and Ravishankar, 2011). Further, Nichols et al. (2015) found, higher phenolic acid and flavonoids content accumulation in plants due to drought stress and these antioxidant molecules accumulation acted as sunshield against drought condition. In white clover plant the flavonoid (quercetin) increased significantly under the drought stress and was more in the drought resistant genotypes in comparison with the control (Ballizany et al., 2012). Similarly, in *Crataegus laevigata* and *C. mongyna* species also the flavonoid content increased due to the drought stress (Kirakosyan et al., 2003). Hernandez et al. (2004) reported increased flavanol content in the drought-treated *Cistus clusii* plant, field plant of summer season and the plants of high temperature with lack of rain in the Mediterranean climate.

Chalker-Scott (1999) observed an increased accumulation of anthocyanins during the drought condition, and the authors concluded that the anthocyanin accumulation in plant gives ability to fight over the adverse (drought) condition. Nakabayashi et al. (2014) found increased flavonoid accumulation in wild-type and mutant *Arabidopsis thaliana*, which showed improved plant tolerance to drought conditions. Drought stress tolerance increased in plant with increased flavonoids accumulation (Del Moral, 1972). The upregulation of chlorogenic acid content by 6 and 10-fold was found in *Crataegus* spp. and *Helianthus annuus*, respectively during drought condition (Kirakosyan et al., 2004). Studies of Sanchez-Rodriguezet et al. (2011) showed the enhanced accumulation of quercetin and kaempferol in the drought resistant tomato cultivar under stress. Ma et al. (2014) found a higher amount of phenolic acids and flavonoids as antioxidant in wheat leaves that detoxify the generated radicals under drought conditions. In rice varieties (black-hulled and red-hulled) the quantity of anthocyanins and proanthocyanidins showed the ROS scavenging, which produced during drought condition (Oki et al., 2002) and also the anthocyanins presence in plant tissue showed tolerance to drought stress (Chalker-Scott, 1999).

12.2.4 UV LIGHT

Light plays a key role in the biosynthesis of primary and secondary metabolites in plants; promotes plant growth, and acts as an energy source to the plants. Higher plants need visible solar radiation to provide energy for photosynthetic mechanism (Xu et al., 2017; Koyama et al., 2012). During light harvesting, the photosynthetic system is inevitably exposed to relatively high doses of harmful solar radiation/UV radiation. Sunlight consists of UV radiation (i.e., 280–320 nm), which causes adverse effect on the plants. UV light exposure causes photo-induced chlorophyll and DNA damage, which inhibits plant growth by disturbing the photosynthesis (affects photosystem II reaction center, electron transport chain (ETC), light-harvesting complex, photosystem II donor and acceptor site).

UV-B radiation is considered as a stress factor to the plants because it may mutate DNA, it causes adverse effect on DNA, protein, and membranes, it alters metabolism process and also it generates excess toxic ROS and causes oxidative stress (Xu et al., 2017). Due to harmful light intensity, various enzymatic/non-enzymatic secondary metabolite biosynthesis pathway take place. Plant tolerate the UV stress by activating the defense mechanism by accumulating the phenolic compounds (Koyama et al., 2012). Phenolic compounds produced in response to the UV light, quenches the generated toxic radicals and protect the plant cell components by making shield inside the epidermal cell layer. Phenolic compounds accumulated helps in the development of plants and protects from high intensity light radiation like gamma, UV, blue light, far-red light, and red light (Chung et al., 2020).

Phenolics and flavonoids have high absorption capacity at 250–270 and 335–360 nm, respectively, and neutralize the UV radiation, hence called as a good UV shield (Lattanzio, 2013; Winkel-Shirley, 2002; Carletti et al., 2003). Burchard et al. (2000); and Rozema et al. (2002) reported polyphenols as a defense mechanism that acts as a shield and also possess important biological function against UV radiation. Li et al. (1993) reported that the phenolic and flavonoids has significant role in the protection of plant cells from UV radiation. Daayf and Lattanzio (2009) found the plant phenolic accumulation on UV radiation stress. Phenolic accumulation enhances in the cell due to UV stress, which helps in protecting the cell and cell organelles, i.e., by interrupting the mutagenesis and cell death by dimerization of thymine units in the DNA. Polyphenols protect the coenzymes NADP or NAD from photo destruction. It is observed that the percentage of flavonoids was more in the high-altitude plants than the temperate plants. Variation in the proportion of flavonoids composition in plants was due to long time exposure of UV radiation which activated the biosynthetic genes of flavonoids (Kolb et al., 2001). Researcher found that the excess accumulation of flavonoid composition in plant leaves was due to the excess light or UV radiation exposure (Olsson et al., 1998; Kolb et al., 2001). Anthocyanin accumulated with in the epidermis acts against the radiation as a darkening filter and also protect the mesophyll (Larcher, 1995).

Accumulation of phenolic derivatives like phenolic acid esters, flavonoids (kaempferol derivatives) or isoflavonoids and psoralens protects the plant from the harmful radiation and prevent UV-B stress (Stapleton, 1992). UV-B radiation induced flavonoid in barley (Liu et al., 1995), polyamines in cucumber (Kramer et al., 1991) and flavonols in Norway spruce (Fischbach et al., 1999) was reported. In *Catharanthus roseaus* induced phenolic compounds (vinblastine and vincristine) were observed after exposure of UV-B radiations (Bernard et al., 2009). On exposure of UV (300 to 400 nm) on Pea plants roots showed the increased amount of flavonoids (Shiozaki et al., 1999) and increase in stimulation of flavonols was observed in silver birch and grape leaves after the treatment of UV-B radiation (Tegelberg et al., 2004). Markham et al. (1998) study revealed that the more C-glycosylflavone concentration in UV-tolerant rice cultivar than the susceptible cultivar after treatment with various UV-B light range. Koes et al. (1989) proved the exposure of UV light activates the CHS transcriptional factor, which upregulates the first enzyme of flavonoid biosynthesis pathway (Koes et al., 1989; Schulze-Lefert et al., 1989). Authors also found higher flavonoid content in the higher altitude plant species (Bachereau et al., 1998; Zidorn et al., 2005; Rieger et al., 2008; Spitaler et al., 2008; Murai et al., 2009).

12.3 BIOTIC STRESS AND PHENOLICS

Plant suffers from biotic stresses and is another kind of hurdle that causes severe yield loss Due to the biotic stresses, the decrease in quality and quantity of the crop and its products was witnessed (Šamec et al., 2019). Biotic stresses are of two types: (1) microbial pathogens (virus, bacteria, fungi), and (2) pests (nematodes and insects). Pathogens and pests causes different types of diseases/infection and damages the plant at various stage of its life time. Microbial pathogens cause disease to different parts of the plant like leaves, roots, branches, stem, flowers, vegetables, and fruits. Similarly, pests also cause damage to various plant parts. Diversity of microbial pathogens and pests and their attack on various plant parts make it difficult for the plants to overcome these situations by defense action. But plant sense and reacts to the pathogen and pests by its basic mechanism like ROS generation, H_2O_2 accumulation, defense related enzymes production, PR-protein accumulation, and antioxidant molecule production (Naveen et al., 2013a, b, 2019, 2020).

The genetic mechanism of pathogen, pests, and plant have been studied in the recent past. The genetic information of biotic agents and the plants may help plant breeders to breed resistant crop varieties. Recently Naveen et al. (2021) found the biochemical and molecular diversity of the *Colletotrichum tracutatum* and results conclude that the information of pathogen and plants is more essential to breed resistant plant varieties. Besides, plant in response to the biotic agents responds with the biochemical mechanism of action by producing ROS, H_2O_2, defense related enzymes, PR-protein, and antioxidant molecule (phenolic). Among antioxidant molecules, phenols and their derived compounds are considered as antimicrobial and also toxic to pests. Hence, we reviewed: (i) microbial pathogen and phenolic response of the plant; and (ii) pests and plant phenolic in subsections.

12.3.1 MICROBIAL PATHOGENS

Plant come across various microbial pathogens (virus, fungi, and bacteria). During the infection of the microbial pathogens, plants respond biochemically by a broad range of secondary metabolites. Phenolic secondary metabolites are considered as toxic compounds to the microbial pathogens/foreign bodies and hence are called antimicrobial (antiviral, antifungal, and antibacterial). Researcher found that phenol is the primary defense against the microbial pathogens at the site of infection (Matern and Kneusal, 1988). Phenolic during the plant-pathogen interaction acts as an antioxidant molecule, as a messenger molecule and as cell wall thickening molecule. Normally, plant produces phenolics as its normal physiological function, but the enhanced accumulation of phenolics was observed during the stress conditions. These phenolic component helps to restrict the pathogen from infection and invasion. Cell wall acts as a barrier to the cell with its rigid nature, which is composed of cellulose. But during the microbial infection the phenolic derived compounds lignin and tannin (a heterogenous phenolic) gives more strength to the cell wall. Naveen et al. (2013a, b) found that elicitor treatment to the chili seedlings induced the phenolic accumulation, but in fungal pathogen (*C. capsici*) challenge inoculated seedlings showed more accumulation of phenolics than other treatments. Also, the defense related enzymes of phenylpropanoid pathway (peroxidase (POX), polyphenol oxidase, and phenylalanine lyase) were activated more in the elicitor treated and pathogen challenge inoculated seedlings than other treatments.

Cirak et al. (2014) studied the effect of fungal pathogen and plant growth-promoting bacteria on phenolic accumulation in two plant varieties

(*Hypericum perforatum* L. and *Hypericum triquetrifolium* Turra). Authors found increased accumulation of phenolics in both the cultivar after the treatment of fungus and bacteria. Also, the accumulated phenolic showed defense activity against pathogen and herbivore. The phenolic profile of infected and healthy raspberry canes was analyzed (Mikulic-Petkovsek et al., 2014). The *Leptosphaera coniothyrium* and *Didymella applanata* infection increased the phenolic metabolism in raspberry canes. The author observed more phenolic content during infection and reduced phenolic concentration during the diseased condition, and they further concluded that based on the phenolic accumulation the cultivar can be differentiated into susceptible and resistant (Mikulic-Petkovsek et al., 2014). Mikulic-Petkovsek et al. (2009) found enhanced phenolics metabolism during the infection of *Venturia inaequalis* fungus in apple leaves and fruit. In particular, hydroxycinnamic acid (7.6 folds), flavan-3-ols (2.6 folds), flavanols (2.9 folds), and total phenolics (2.4 folds) increase were observed in infected tissue than healthy tissue. Ganthaler et al. (2017) found variation of foliar phenolic content in Norway spruce tree after *Chrysomyxa rhododendri* infection. During infection, the flavonoid resveratrol increased up to 1.8 fold. The flavonoid inhibited the hyphal growth and also prevented the development of disease symptoms in Norway Spruce.

Naveen et al. (2019) showed the plant growth-promoting rhizobacteria (PGPR) treatment induces the phenolic content in chili seedlings. However, the combination of PGPR treatment and pathogen Colletotrichum capsici challenge inoculated seedlings showed higher phenolic accumulation than other treatments. The defense-related enzymes (PAL, POX, and PPO) were also found more in the PGPR treated and pathogen challenge inoculated seedlings. Subsequently, the author showed the role of abiotic inducer (beta-amino butyric acid and acibenzolar S methyl) effect on chilli seedlings after spray treatment (Naveen et al., 2020). The spray treated seedlings and *C. capsici* challenge inoculation induces the phenolic compound and enzyme (PAL, POX, and PPO) activity. Akanksha et al. (2014) evaluate the effect of three rhizosphere microbes (*Trichoderma harzianum* (THU0816), *Pseudomonas aeruginosa* (PHU094), and *Mesorhizobium* sp. (RL091)) and challenge inoculation of *Sclerotium rolfisii* pathogen on chickpea phenolic acid production. Results revealed the maximum accumulation of phenolics (syringic acid, quercetin, shikimic acid, ferulic acid, *t*-chlorogenic acid and myricetin) in the biocontrol treatment and pathogen challenge inoculated plant when compared to control and pathogen alone inoculated plants. Doehlemann et al. (2008) found an increase in phenolic (hydroxycinnamic acid) in maize leaves during *U. maydis* infection. Ponts et al. (2011) revealed that the phenolic concentration in the cereal grains helps to inhibit *F. graminearum* and its mycotoxin production. Wojakowska et al. (2013) claimed that plant secondary metabolites (phenolics and flavonoids) produced due to fungal pathogen, show a key role in managing the fungal infection.

Researchers found the accumulation of phenolics and its related components in resistant variety of the plant that enhanced during the PGPR treatment and in plants challenge inoculated with the bacterial pathogen (Vanitha and Umesha, 2011; Vanitha et al., 2009; Kavitha and Umesha, 2008). PGPR treatment is known to induce the systemic resistance by activating the phenylpropanoid pathway. The treatment of pea (*Pisum sativum*) with the PGPR induced the phenolic accumulation (gallic, cinnamic, and ferulic acids) by many fold than the untreated control (Singh et al., 2002). Around 20 different tomato cultivars were divided into four groups (highly resistant, resistant, susceptible, and highly susceptible) based on the pathogenicity. Authors found that highly resistant and resistant variety showed maximum accumulation of phenolics, lignin, and defense related enzymes (POX, PPO, and PAL) when compared to susceptible and highly susceptible cultivars. The same trend was observed in the PGPR treatment alone and PGPR treatment + challenge inoculated tomato cultivar (Kavitha and Umesha, 2008). Lee et al. (2019) reported the resistant plant produces lignin in response to bacterial infection and helps in the prevention of pathogen entry into the cell. Abdel-Monaim (2017) found the biological agents and abiotic inducers alone or in combination induced the disease protection against Fusarium Wilt Disease in Faba Bean caused by *Fusarium oxysporum*. Bacterial strains and abiotic inducers enhanced the PR-protein, total phenolic and flavonoids, which well correlated with the reduced disease incidence and severity.

PGPR treatment of tomato cultivar and challenge inoculation with *R. solanacearum* induced phenylpropanoid pathway enzymes (PPO, PAL, and POX) and the same was confirmed by using reverse transcription-polymerase chain reaction (RT-PCR) (Vanitha and Umesha, 2011). The same research team also confirmed more expression or accumulation of defense-related enzymes (PPO, POX, and PAL) and antimicrobial phenolic compound in resistant and highly resistant cultivar than in the susceptible cultivar (Vanitha et al., 2009). Lorena del et al. (2017) reported that the treatment of PGPR (*B. subtilis*, *P. fluorescens*, and *P. putida*) to the *Mentha piperita* increased the phenolic and PAL enzyme content in the leaves than in the controls. Bhattacharyya et al. (2020) authenticated 120 PGPR isolates by partial 16S rRNA gene sequencing and the treatment of it to rice and maize crops. The authors found enhanced accumulation of PAL enzyme and polyphenolic content than the control. Induced phenolic content reduced the ROS stress and proved as an antioxidant. Kanika et al. (2019) observed the reduced antioxidants in nematode infected seedlings while it was maximum in the PGPR (*P. aeruginosa* and *B. gladioli*) treated seedlings. Bacterial infection induces both passive and active defense genes. Bacterial blight causal organism and walnut interaction revealed, the upregulation of *JrPPO-1* and *P14a* genes and the increased activity of PPO enzyme (Fatemeh et al., 2016).

Christopher et al. (2020) interpreted the phenolic accumulation with the resistant variety of grapevine. During the interaction between *Xylella fastidiosa* and PdR1-resistant gene of grapevine a significant downregulation of PdR1 gene and phenolic content was observed and the expression of the gene was lesser than the susceptible cultivar. Sajjad et al. (2020) showed that the treatment of PGPR induced the defense enzyme (PAL, PPO, and POX), phenolic content and also growth parameters. Inoculation of bacteria *X. axonopodis* and *R. solanacearum* was made to potato (*Solanum tuberosum* L. cv. Agata) and metabolites alteration was analyzed (Vera et al., 2009). Multiple leaves inoculation (apical, intermediate, and basal leaves) showed more accumulation of phenolic, flavonoid, and enzymes (POX and PPO) activity increased in basal leaves compared to the apical leaves. Authors concluded that the susceptibility and resistance depends on the plant/leaf age and also on the interaction between the plant and pathogen. The fungi and bacteria include both beneficial (PGPR and PGPF) and pathogenic strains. PGPR and PGPF induces systemic resistance, growth of the plant. While in the case of pathogenic fungi and bacteria, the stress created within the plant induces the mechanism of the host (resistant/tolerant variety). In some cases, the virulent pathogen come across those hurdles by biochemical reaction or the plant does not come with enough defense mechanism (susceptible variety).

Viral infection to the plants and its disease severity become threat to the cultivation and its yield. Virus are obligate parasites have a wide range of host and management of virus is economically important because it infects wide range of crops. Viruses are extremely small and are intracellular (inside cells) pathogenic particles can only be seen with an electron microscope. The important viral pathogens are cucumber mosaic virus, tobacco mosaic virus and potato virus Y. Viruses consist of RNA/DNA as nucleic acid material covered by protein material externally called capsid. If plants cell wall is more rigid, the virus cannot come over it, hence it affects from the part of mechanical injury/by branches of lateral roots, through other means like nematode, parasites, insects, or fungi. Plant viruses once enter the cell it releases its protein coat and nucleic acid and multiply to form new viral particles. The movement of virus from cell to cell was through cytoplasmic bridges between cells, i.e., plasmodesmata, and enters infected plants systematically and further complete infection of plant is through phloem. After infection, viruses cause disease by disrupting the photosynthetic system and normal cellular processes.

Telosma mosaic virus (TeMV; 700 nm linear particle) infection to the *Passiflora edulis* changes the phytochemical components of the plant. Using RT-PCR, the small and abundant small interference RNAs (siRNAs) were generated and was treated to the fruits. Infected fruits showed maximum accumulation of phenolics, SOD, and CAT than the healthy fruits. Authors conclude that the increased accumulation of secondary metabolites (phenolics) can increase the stress tolerance level of *P. edulis* plant against TeMV (Chen et al., 2018). Zhao and Li (2021) found the interaction of virus with plant during infection induced accumulation of signaling molecule and phytohormone, which leads to promote disease protection. Abo-Zaid et al. (2020) evaluated the *Streptomyces cellulosae* isolate (Actino 48) on Tobacco Mosaic Virus in tomato. Inoculation of Actino 48 (2 × 107 cfu/mL) and challenge inoculation of TMV after 48 hours induced the enzymes (POX and chitinase) and total phenolic content. In addition, significant increase in transcripts (PAL, CHS, PR-1 and PR-3) and the fresh weight and dry weight of the Actino 48 and TMV challenge inoculated plant was observed when compared to TMV alone treated and control plant. The increase in enzyme activity, phenolic content and transcripts decreases the TMV infection and severity in tomato cultivar.

Alfalfa mosaic virus (AMV) causes severe crop losses in various plants. Antiviral activity of PGPR *Bacillus licheniformis* strain POT1 was evaluated in potato (*Solanum tuberosum* L.). Abdelkhalek et al. (2020) treated *B. licheniformis* strain POT1 and challenge inoculated the AMV after 24 hours, result revealed maximum phenolic accumulation in treated seedlings (7,218.86) than the untreated control (1606.49 mg/kg). HPLC analysis showed 20 different phenolics. The bacterial treatment followed by AMV treatment upregulated the phenylpropanoid, flavonoid (588.133-fold increase), anthocyanin transcription factor (97.005-fold) and chlorogenic acid biosynthetic pathways, which correlates with the reduction (86.79%) of viral accumulation in potato cultivar. Arpita and Subrata (2008) found a decrease in phenolic content with increase in virus disease (Yellow vein mosaic) in *Hibiscus cannabinus* L.; and *H. sabdariffa* L. plant. Higher phenols accumulation was observed in resistant cultivar than susceptible cultivar during virus-plant interactions (Singh et al., 2010). Phenolics accumulation in the plant cell wall plays a vital role in virus resistance by increasing the mechanical strength of the plant cell wall (Boudet, et al., 1995) and also increases cell wall lignification as lignin precursors (Lyon et al., 1992).

Cotton leaf curl Burewala virus (CLCuBuV) is a serious disease of cotton in India. Zeeshan et al. (2014) studied the changes in defense mechanism (biochemically) in susceptible and resistant cotton cultivars. The increase in activity of POX, PAL, CAT, SOD, and PPO enzymes and the phenolic level was observed in the resistant cultivar than the susceptible cultivar. Results correlated with the disease symptoms less symptoms in resistant variety and more symptoms in susceptible variety. The author concluded the biochemical enzymes and phenol content were used as markers to judge the resistant and susceptible cultivar. Barley yellow dwarf virus-PAV (BYDV-PAV) caused severe disease in cereal cultivation. The resistant cultivar showed more phenolic accumulation and activated the defense genes (*Bdv2* and *RYd2*) and meanwhile increases the root growth and dry mass of the plant, which was absent in the susceptible cultivar (Shormin et al., 2018). Virus infection, may activate or suppress the plant defense mechanism (Lewsey et al., 2010; Whitham, et al., 2006). Chili plants treated with pepper leaf curl virus (PepLCV) showed enhanced the total phenolic content in resistant cultivars and was reduced in susceptible cultivars (Rai et al., 2010). A positive correlation was observed with increased total phenolic content and disease resistance against cocoa swollen shoot virus disease (CSSVD) for 3 months after inoculation (Ofori et al., 2015).

12.3.2 PESTS AND PHENOLICS

Pests attack on plant in various stages like seedling, flowering, and fruiting stage, which causes severe loss of crop and its yield. The pests and plants interaction occurs in two forms: (i) physical interaction and (ii) chemical interaction. Plant produces sufficient chemical constituents that make host unpalatable to pests. By such physiological changes, plant enters the new adaptive zone was made by sensing the pests. The approaches activated by plants to resist or to avoid pests vary in diverse ways. Some plant species produce various levels of biochemical defense mechanisms like toxic compounds, which minimize the pest damages, its growth, development, dispersion, or habitat. Biochemical defense also promotes the formation of physical barrier by means of phenolic components are converted into lignin through lignification or resin. Phenolics damages the pests by causing oxidative damage to the midgut, when pests feed on the plant (Bi and Felton,

1995). Also, the plant phenolics worsens the pest's digestive enzymes, leading to cause delay in the development, finally molting occurs reducing the insect's fitness.

Even though the plant has a diverse defense mechanisms, the activation of the specific biochemical activity depends on the type of pests. The pests attack induces the quantity of the biochemicals, i.e., toxins which negatively affect the pest's infection (Tayal et al., 2020; Singh and Kariyat, 2020). The variety of mechanisms of toxicity revealed by diverse plant defensive components includes inhibition of transport, membrane disruption, or inhibition of metabolism, signal transduction, and even disruption of hormonal control of developmental processes (Gatehouse, 2002). Specific and possible plant responses through primary and secondary metabolism depends on the pests feeding behavior (Katarzyna et al., 2017).

PPO, considered as anti-nutritive enzymes, reduces the nutritive value of the plant by cross-linking proteins/catalyzing the oxidation of phenolic to quinones, which further avoids pest growth, development, and dispersion. As an effect of 50 flea beetles effect attack on cabbage plant, a higher accumulation of PAL, phenolics, and flavonoids was noticed. Authors concluded that the phenolic acids and flavonoids may play a vital role in the defense against pests (Kovalikova et al., 2019).

The two-spotted spider mite (*Tetranychus urticae* Koch) and grape mealybug (*Pseudococcus maritimus* Ehrh.) infection in strawberry and orchid respectively caused higher phenolic (during first week of insect feeding) compounds and PAL (during first 5 h of insect feeding). The activation of phenolics and PAL correlated with the reduced pest infection (Katarzyna et al., 2017). The infestation of grain aphid (*Sitobiaon avenae* F.) to winter triticale seedlings induced the phenolic acid content and helped to protect from further infestation (Chrzanowski and Leszczynski, 2008). The increased concentration of 3-deoxyanthocynadin (flavonoids) present in sorghum caused significant mortality and reduced corn leaf aphid population (Kariyat et al., 2019). Tayal et al. (2020b) showed polyphenol rich pericarp extract of purple corn (*Zea mays*) inversely affected the growth, development, and adult fitness traits in tobacco hornworm. Increase in phenolic content protects the plant with wide range of pests; Leiss et al. (2009) showed that higher chlorogenic acids accumulated inhibited thrips in chrusanthemum (*Dendranthema grandiflora* (Ramat.)). Flavonoid (pisatin) in pea inversely affected pea aphid (*Acyrthosiphon pisum* Harris) (Morkunas et al., 2015) and in rice, increase in ferulic acid concentration reduced the brown plant hopper (*Nilaparvata lugens* Stål) (Yang et al., 2017).

12.4 CONCLUSION

Phenolics by nature have a diverse biological activity in both plant and animals. Phenolics are considered as antioxidants, signaling molecules, cell wall thickening molecules and very importantly act as toxic substances to inhibit the stress factors (abiotic and biotic). As an antioxidant molecule, polyphenol reduces the oxidative stress of the plant cell by quenching the ROS generated due to abiotic and biotic stress. This chapter completely covers the overall role of phenolics against environmental (abiotic and biotic) stress. Phenolics in the plant cell help to save the cell from the adverse conditions created in nature. Induced phenolic content acts as self-medication to the plant to overcome the tough situation. Hence, the phenolic-rich cultivar is much needed whose accumulation helps plant to tolerate the tough conditions. The current chapter details the role of phenolic compounds in the maintenance of sustainable cultivation by stress management; in future, breeding phenolic enriched cultivars would help in this direction.

KEYWORDS

- **abiotic stress**
- **biotic stress**
- **indole acetic acid**
- **peroxidase activities**
- **phenolic compounds**
- **plant defense**

REFERENCES

Abdelkhalek, A., Al-Askar, A. A., & Behiry, S. I., (2020). Bacillus licheniformis strain POT1 mediated polyphenol biosynthetic pathways genes activation and systemic resistance in potato plants against Alfalfa mosaic virus. *Sci Rep., 10*, 16120.

Abdel-Monaim, M. F., (2017). Evaluation of the accumulation of pathogenesis related (PR) proteins and phenolic compounds in response to biotic and abiotic elicitors as mechanism for immune response to fusarium wilt disease in faba bean. *Plant Pathol. Microbiol., 8*, 2.

Abo-Zaid, G. A., Matar, S. M., and Abdelkhalek, A. 2020. Induction of Plant Resistance against Tobacco Mosaic Virus Using the Biocontrol Agent Streptomyces cellulosae Isolate Actino 48. Agronomy 2020, 10, 1620.

Abreu, I. N., & Mazzafera, P., (2005). Effect of water and temperature stress on the content of active constituents of *Hypericum brasiliense* choisy. *Plant Physiol. Biochem., 43*, 241–248.

Ahmad, P., & Sharma, S., (2008). Salt stress and phyto-biochemical responses of plants: A review. *Plant Soil Environ., 54,* 89–99.

Akanksha, S., Akansha, J., Birinchi, K. S., Ram, S. U., & Harikesh, B. S., (2014). Rhizosphere competent microbial consortium mediates rapid changes in phenolic profiles in chickpea during *Sclerotium rolfsii* infection. *Microbiological Research, 169*(5, 6), 353–360.

Akula, R., & Ravishankar, G. A., (2011). Influence of abiotic stress Signalson secondary metabolites in plants. *Plant Signal. Behav., 6*, 17201731.

Alessandra, F., & Luca, S., (2019). Abiotic stress effects on performance of horticultural. *Crops Horticulturae, 5*, 67.

Aloisi, I., Parrotta, L., Ruiz, K. B., Landi, C., Bini, L., Cai, G., Biondi, S., & Del Duca, S., (2016). New insight into quinoa seed quality under salinity: Changes in proteomic and amino acid profiles, phenolic content, and antioxidant activity of protein extracts. *Front. Plant Sci., 7*, 656.

Amita, B., Priyanka, S., & Vitaly, C., (2010). The roles of plant phenolics in defense and communication during *Agrobacterium* and *Rhizobium* infection. *Mol. Plant Pathol., 11*(5), 705–719.

Amita, B., Priyanka, S., & Vitaly, C., (2010). The roles of plant phenolics in defense and communication during Agrobacterium and Rhizobium infection. *Mol. Plant Pathol., 11*(5), 705–719.

Arora, N. K., (2019). Impact of climate change on agriculture production and its sustainable solutions. *Environmental Sustainability, 2*, 95–96.

Arpita, C., & Subrata, K. G., (2008). Alterations in biochemical components in Mesta plants infected with yellow vein mosaic disease. *Braz. J. Plant Physiol., 20*(4). Londrina Oct./Dec. 2008.

Bachereau, F., Marigo, G., & Asta, J., (1998). Effect of solar radiation (UV and visible) at high altitude on CAM-cycling and phenolic compounds biosynthesis in sedum album. *Physiol. Plant, 104*, 203210.

Ballizany, W. L., Hofmann, R. W., Jahufer, M. Z. Z., & Barrett, B. A., (2012). Multivariate associations of flavonoid and biomass accumulation in white clover (*Trifolium repens*) under drought(Article). *Functional Plant Biology, 39*(2), 167–177.

Bernard, Y. K., Binder, C. A. M., Peebles, J. V., & Shanks, K. Y. S., (2009). The effects of UV-B stress on the production of terpenoid indole alkaloids in *Catharanthus roseus* hairy roots. *Biotechnol. Prog., 25*, 8615.

Bhattacharjee, S., (2010). Sites of generation and physicochemical basis of formation of reactive oxygen species in plant cell. In: Dutta, G. S., (ed.), *Reactive Oxygen Species and Antioxidants in Higher Plants* (pp. 1–30). CRC Press, New York, NY, USA.

Bhattacharyya, C., Banerjee, S., Acharya, U., et al., (2020). Evaluation of plant growth promotion properties and induction of antioxidative defense mechanism by tea rhizobacteria of Darjeeling, India. *Sci. Rep., 10*, 15536 (2020).

Bi, J. L., & Felton, G. W., (1995). Foliar oxidative stress and insect herbivory: Primary compounds, secondary metabolites, and reactive oxygen species as components of induced resistance. *J. Chem. Ecol., 21*, 1511–1530.

Boudet, A., Lapierre, C., & Grima-Pettenati, J., (1995). Biochemistry and molecular biology of lignification. *New Phytologist, 129*, 203–236.

Burchard, P., Bilger, W., & Weissenbock, G., (2000). Contribution of hydroxycinnamates and flavonoids to epidermal shielding of UV-A and UV-B radiation in developing rye primary leaves as assessed by ultraviolet-induced chlorophyll fluorescence measurements. *Plant Cell Environ., 23*, 13731380.

Caliskan,O., Radusiene, J., Temizel, K. E., Staunis, Z., Cirak, C., Kurt, D. and Odabas, M. S. 2017. The effects of salt and drought stress on phenolic accumulation in greenhouse-grown Hypericum pruinatum. Italian Journal of Agronomy; volume 12:918.

Carletti, P., Masi, A., Wonisch, A., Grill, D., Tausz, M., and Ferretti, M. 2003, Environ. Experim. Bot. 50, 149.

Chalker-Scott, L., (1999). Environmental significance of anthocyanins in plant stress responses. *Photochemistry and Photobiology, 70*(1), 1–9.

Chen, S., Yu, N., Yang, S., et al., (2018). Identification of *Telosma* mosaic virus infection in *Passiflora edulis* and its impact on phytochemical contents. *Virol J., 15*, 168.

Christian, Z., Birthe, S., & Hermann, S., (2005). Altitudinal differences in the contents of phenolics in flowering heads of three members of the tribe lactuceae (*Asteraceae*) occurring as introduced species in New Zealand. *Biochemical Systematics and Ecology, 33*(9), 855–872.

Christopher, M. W., Adam, R. Z., Anne, S., Dylan, J. B., Andrew, W. M., & Rodrigo, P. P. A. (2020). Impact of phenolic compounds on progression of *Xylella fastidiosa* infections in susceptible and PdR1-locus containing resistant grapevines. *PLoS One, 15*(8):e0237545.

Chrzanowski, G., & Leszczynski, B., (2008). Induced accumulation of phenolic acids in winter triticale (*Triticosecale wittm.*) under insects feeding. *Herba Polonica, 54*, 33–40.

Chung, I. M., Paudel, N., Kim, S. H., et al., (2020). The influence of light wavelength on growth and antioxidant capacity in *Pachyrhizus erosus* (L.) Urban. *J. Plant Growth Regul., 39*, 296–312.

Cirak, C., Radusiene, J., Aksoy, H. M., Mackinaite, R., Stanius, Z., Camas, N., & Odabas, M. S., (2014). Differential phenolic accumulation in two hypericum species in response to inoculation with *Diploceras hypericinum* and *Pseudomonas putida. Plant Protect. Sci., 50*, 119–128.

Colsson, L., Veit, M., Weissenböck, G., & Bornman, J. F., (1998). Differential flavonoid response to enhanced UV-B radiation in brassica napus. *Phytochemistry, 49*(4), 1021–1028.

Daayf, F., & Lattanzio, V., (2009). *Recent Advances in Polyphenol Research.* John Wiley and Sons, West Sussex, UK.

Del, M. R., (1972). On the variability of chlorogenic acid concentration. *Oecologia, 9*, 289–300.

Doehlemann G, Wahl R, Horst RJ, Voll L, Usadel B, Poree F, Stitt M, PonsKuehnemann J, Sonnewald U, Kahmann R,. (2008) Reprogramming a maize plant: transcriptional and metabolic changes induced by the fungal biotroph Ustilago maydis. Plant J 56: 181–195.

Dunja, Š., Erna, K., Ivana, Š., Valerija, V. B., & Salopek-Sondi, B., (2021). The role of polyphenols in abiotic stress response: The influence of molecular structure. *Plants, 10*, 118.

Faize, M., Burgos, L., Faize, L., Piqueras, A., Nicolas, E., Barba-Espin, G., Clemente-Moreno, M. J., et al., (2011). Involvement of cytosolic ascorbate peroxidase and Cu/Zn-superoxide dismutase for improved tolerance against drought stress. *J. Exp. Bot.* doi: 10.1093/jxb/erq432.

Fatemeh, K., Masoud, T., Mehdi, M., Abhaya, M. D., Charles, A. L., Daniel, A. K., Timothy, B., & Kourosh, V., (2016). Induction of polyphenol oxidase in walnut and its relationship to the pathogenic response to bacterial blight. *Journal of the American Society for Horticultural Science J. Amer. Soc. Hort. Sci., 141*(2), 119–124. 2016.

Fischbach, R. J., Kossmann, B., Panten, H., Steinbrecher, R., Heller, W., & Seidlitz, H. K., (1999). Seasonal accumulation of ultraviolet-B screening pigments in needles of Norway spruce (*Picea abies* (L.)Karst). *Plant Cell Environ., 22*, 2737.

Flowers, T. J., & Flowers, S. A., (2005). Why does salinity pose such a difficult problem for plant breeders? *Agric. Water Manag., 78*, 15–24.

Gaber, A. A., Saleh, M. M., & Ahmed, A., (2020). Induction of plant resistance against tobacco mosaic virus using the biocontrol agent streptomyces cellulosae isolate actino 48. *Agronomy, 10*, 1620.

Ganthaler, A., Stoggl, W., Kranner, I., & Mayr, S., (2017). Foliar phenolic compounds in Norway spruce with varying susceptibility to *Chrysomyxa rhododendri*: Analyses of seasonal and infection-induced accumulation patterns. *Frontiers in Plant Science, 8*, 1173.

Gatehouse, J. A., (2002). Plant resistance towards insect herbivores: A dynamic interaction. *New Phytologist, 156*, 145–169.

Gnanasekaran, N., & Kalavathy, S., (2017). Drought stress signal promote the synthesis of more reduced phenolic compounds (chloroform insoluble fraction) in *Tridax procumbens*. *Free Radicals and Antioxidants, 7*(1).

González-Sarrías, A., Tomás-Barberán, F. A., & García-Villalba, R., (2020). Structural diversity of polyphenols and distribution in foods. In: Tomás-Barberán, F. A., González-Sarrías, A., & Rocío Gar-cía-Villalba, R.,(eds.), *Dietary Polyphenols: Their Metabolism and Health Effects*. Wiley: Chichester, UK.

Gunther, D., Ramon, W., Robin, J. H., Lars, M. V., Björn, U., Fabien, P., Mark, S., et al. (2008). *Reprogramming a Maize Plant: Transcriptional and Metabolic Changes Induced by the Fungal Biotroph Ustilago Maydis, 56*(2), 181–195.

Hernandez, I., Alegre, L., Van, B. F., & Munne-Bosch, S., (2009). How relevant are flavonoids as antioxidants in plants? *Trends Plant Sci., 14*, 125–132.

Hernandez, F., et al. (2004) Influence of Two Plant Extracts on Broilers Performance, Digestibility, and Digestive Organ Size. Poultry Science, 83, 169–174.

Hu, Y. F., Hou, G., Na, X. F., Yang, L., Nan, W. B., Liu, X., Zhang, Y. Q., Li, J. L., Yu Rong Bi, Y. R. (2013). Cadmium interferes with maintenance of auxin homeostasis in Arabidopsis seedlings. *Journal of Plant Physiology*, 170, 965–975.

Ida, L., Dunja, Š., Jiří, G., Valerija, V. B., Miroslav, S., & Salopek-Sondi, B., (2019). Involvement of phenolic acids in short-term adaptation to salinity stress is species-specific among brassicaceae. *Plants, 8*(6). 10.3390/plants8060155.

Iker, H., Leonor, A., & Munné-Bosch, S., (2004). Drought-induced changes in flavonoids and other low molecular weight antioxidants in *Cistus clusii* grown under Mediterranean field conditions *Tree Physiology, 24*(11), 1303–1311.

Isah, T., (2019). Stress and defense responses in plant secondary metabolites production. *Biol. Res., 52*, 39.

Kanika, K., Vijay, L. J., Anket, S., Sumit, G. G., Puja, O., Renu, B., Al-Huqail, A. A., et al., (2019). Evaluation of the role of rhizobacteria in controlling root-knot nematode infection in *Lycopersicon esculentum* plants by modulation in the secondary metabolite profiles. *AoB Plants, 11*(6).

Kariyat, R. R., Gaffoor, I., Sattar, S., Dixon, C. W., Frock, N., Moen, J., Moraes, C. M. D., et al., (2019). Sorghum 3-deoxyanthocyanidin flavonoids confer resistance against corn leaf aphid. *J. Chem. Ecol., 45*, 502–514.

Katarzyna, G., Cezary, S., Górska-Drabik, E., Paweł, C., Bożena, Ł., Izabela, K., Katarzyna, K., et al., (2017). Accumulation of amino acids and phenolic compounds in biochemical plant responses to feeding of two different herbivorous arthropod pests. *Arthropod-Plant Interactions 11*, 675–682.

Kavitha, R., & Umesha, S., (2008). Regulation of defense-related enzymes associated with bacterial spot resistance in tomato. *Phytoparasitica, 36*(2), 144–159.

Keutgen, A. J., & Paweltzik, E., (2008). Quality and nutritional value of strawberry fruit under long term salt stress. *Food Chem., 107*, 1413–1420.

Khan, T. A., Mazid, M., & Mohammad, F., (2011). Status of secondary plant products under abiotic stress: An overview. *J. Stress Physiol. Biochem., 7*, 75–98.

Kim, H. J., Fonseca, J. M., Choi, J. H., Kubota, C., & Kwon, D. Y., (2008). Salt in irrigation water affects the nutritional and visual properties of romaine lettuce (*Lactuca sativa* L.). *J Agric. Food Chem., 56*, 3772–3776.

Kirakosyan, A., Kaufman, P., Warber, S., Zick, S., Aaronson, K., Bolling, S., & Chul, C. S., (2004). Applied environmental stresses to enhance the levels of polyphenolics in leaves of hawthorn plants. *Physiol. Plant, 121*(2), 182–186.

Kirakosyan, A., Seymour, E., Kaufman, P. B., Warber, S., Bolling, S.,Chang, S. C., (2003). Antioxidant capacity of polyphenolic extracts from leaves of *Crataegus laevigata* and *Crataegus monogyna* (Hawthorn) subjected to drought and cold stress. *J. Agric. Food Chem., 51*, 39733976.

Kisa, D., Elmastas, M., Öztürk, L., & Kayir, Ö., (2016). Responses of the phenolic compounds of Zea mays under heavy metal stress. *Appl. Biol. Chem., 59*(6), 813–820.

Koes, R. E., Spelt, C. E., & Mol, J. N. M., (1989). The chalcone synthase multigene family of *Petunia hybrida* (V30): Differential, light regulated expression during flower development and UV light induction. *Plant Mol. Biol., 12*, 213–225.

Kolb, C. A., Kaäser, M. A., Kopecký, J., Zotz, G., Riederer, M., & Pfündel, E. E., (2001). Effects of natural intensities of visible and ultraviolet radiation on epidermal ultraviolet screening and photosynthesis in grape leaves. *Plant Physiol., 127*, 863875.

Kováčik, J., Grúz, J., Bačkor, M., Tomko, J., Strnad, M., & Repčák, M., (2008). Phenolic compounds composition and physiological attributes of *Matricaria chamomilla* grown in copper excess. *Environmental and Experimental Botany, 62*(2), 145–152.

Kováčik, J., Klejdus, B., & Bačkor, M., (2009b). Phenolic metabolism of *Matricaria chamomilla* plants exposed to nickel. *Journal of Plant Physiology, 166*(13), 1460–1464.

Kováčik, J., Klejdus, B., Hedbavny, J., Štork, F., & Bačkor, M., (2009a). Comparison of cadmium and copper effect on phenolic metabolism, mineral nutrients and stress-related parameters in *Matricaria chamomilla* plants. *Plant Soil, 320*, 23.

Kovalikova, Z., Kubes, J., Skalicky, M., Kuchtickova, N., Maskova, L., Tuma, J., Vachova, P., & Hejnak, V., (2019). Changes in content of polyphenols and ascorbic acid in leaves of white cabbage after pest infestation. *Molecules, 24*(14), 2622.

Koyama, K., Ikeda, H., Poudel, P. R., & Goto-Yamamoto, N., (2012). Light quality affects-flavonoidbiosynthesis in young berries of cabernet sauvignon grape. *Phytochemistry, 78*, 54–64.

Kumar, M., Tak, Y., Potkule, J., Choyal, P., Tomar, M., Meena, N. L., & Kaur, C., (2020). Phenolics as plant protective companion against abiotic stress 12 springer nature Singapore Pte Ltd. 2020R. In: Lone, et al., (eds.), *Plant Phenolics in Sustainable Agriculture* (pp. 277–308).
Lamaoui, M., Jemo, M., Datla, R., & Bekkaoui, F., (2018). Heat and drought stresses in crops and approaches for their mitigation. *Front. Plant Sci., 6*, 26.
Larcher, W., (1995). *Physiological Plant Ecology*. Springer, Berlin, Heidelberg.
Larson, R. A., (1988). The antioxidants of higher plants. *Phytochemistry, 27*, 969–978.
Lattanzio, V., (2013). Phenolic compounds: Introduction. In: Ramawat, K. G., & Mérillon, J. M., (eds.), *Natural Products: Phytochemistry, Botany and Metabolism of Alkaloids, Phenolics and Terpenes* (pp. 1543–1580). Springer: Berlin/Heidelberg, Germany.
Lee, Y. P., Kim, S. H., Bang, J. W., Lee, H. S., Kwak, S. S., & Kwon, S. Y., (2007). Enhanced tolerance to oxidative stress in transgenic tobacco plants expressing three antioxidant enzymes in chloroplasts. *Plant Cell Rep., 26*, 591–598.
Leiss, K. A., Maltese, F., Choi, Y. H., Verpoorte, R., & Klinkhamer, P. G., (2009). Identification of chlorogenic acid as a resistance factor for thrips in chrysanthemum. *Plant Physiol., 150*, 1567–1575.
Lewsey, M. G., Murphy, A. M., Maclean, D., Dalchau, N., Westwood, J. H., Macaulay, K., Bennett, M. H., et al., (2010). Disruption of two defensive signaling pathways by a viral RNA silencing suppressor. *Mol. Plant-Microbe. Interact., 23*(7), 835–845.
Li, J., Ou-Lee, T. M., Raba, R., Amundson, R. G., & Last, R. L., (1993). *Arabidopsis* flavonoid mutants are hypersensitive to UV-B irradiation. *Plant Cell, 5*, 171–179.
Lini´c, I., Šamec, D., Grúz, J., Vujˇci´c Bok, V., Strnad, M., Salopek Sondi, B. 2019. Involvement of Phenolic Acids in Short-Term Adaptation to Salinity Stress is Species-Specific among Brassicaceae. Plants, 8, 155.
Liu, L., Gitz, III. D. C., & McClure, J. W., (1995). Effects of UV-B on flavonoids, ferulic acid, growth and photosynthesis in barley primary leaves. *Physiol. Plant, 93*, 734–738.
Lopez-Berenguer, C., Del Carmen, M., Martinez-Ballesta, C., Moreno, D. A., Carvajal, M., & Garcia-Viguera, C., (2009). Growing hardier crops for better health: Salinity tolerance and the nutritional value of broccoli. *J. Agric. Food Chem., 57*, 572–578.
Lorena Del, R. C., Julieta, C., Maricel, V. S., Walter, G., & Erika, B., (2017). Inducing phenolic production and volatile organic compounds emission by inoculating *Mentha piperita* with plant growth-promoting rhizobacteria. *Scientia Horticulturae, 220*, 193–198.
Lyon, G., Heilbronn, J., Forrest, R., & Johnston, D., (1992). The biochemical basis of resistance of potato to soft rot bacteria. *Netherlands Journal of Plant Pathology, 98*, 127–133.
Ma, J. F., Ryan, P. R., & Delhaize, E., (2001). Aluminum tolerance in plants and the complexing role of organic acids. *Trends Plant Sci., 6*(6), 273–278.
Mahmoudi, H., Huang, J., Gruber, M. Y., Kaddour, R., Lachaal, M., Ouerghi, Z., & Hannoufa, A., (2010). The impact of genotype and salinity on physiological function, secondary metabolite accumulation, and antioxidative responses in lettuce. *J. Agric. Food Chem., 58*, 5122–5130.
Markham, K. R., Tanner, G. J., Caasi-Lit, M., Whitecross, M. I., Nayudu, M., & Mitchell, K. A., (1998). Possible protective role for 31,41-dihydroxyflavones induced by enhanced UV-B in a UV-tolerant rice cultivar. *Phytochemistry, 49*, 1913–1919.
Matern, U., & Kneusal, R. E., (1988). Phenolic compounds in plant disease resistance. *Phytopathogy, 16*, 153–170.
Mikulic-Petkovšek M., Štampar, F., & Veberič. R., (2009). Accumulation of phenolic compounds in apple in response to infection by the scab pathogen, *Venturia inaequalis. Physiological and Molecular Plant Pathology, 74*(1), 60–67.

Mikulic-Petkovsek, M., Schmitzer, V., Stampar, F., Veberic, R., & Koron, D., (2014). Changes in phenolic content induced by infection with *Didymella applanata* and *Leptosphaeria coniothyrium*, the causal agents of raspberry spur and cane blight. *Plant Pathology, 63*, 185–192.

Mishra, B., & Sangwan, N. S., (2019). Amelioration of cadmium stress in *Withania somnifera* by ROS management: Active participation of primary and secondary metabolism. *Plant Growth Regul., 87*, 403–412.

Mittler, R., & Blumwald, E., (2010). "Genetic engineering for modern agriculture: Challenges and perspectives. *Annual Review of Plant Biology, 61*, 443–462.

Mittler, R., (2006). Abiotic stress, the field environment and stress combination. *Trends Plant Sci., 11,* 15–19.

Morkunas, I., Wózniak, A., Formela, M., Mai, V. C., Marczak, Ł., Narożna, D., Borowiak-Sobkowiak, B., et al., (2015). Peaaphid infestation induces changes in flavonoids, antioxidative defense, soluble sugars and sugar transporter expression in leaves of pea seedlings. *Protoplasma, 253*, 1063–1079.

Murai, Y., Takemura, S., Takeda, K., Kitajima, K., & Iwashina, T., (2009). Altitudinal variation of UV-absorbing compounds in *Plantago asiatica. Biochem. Syst. Ecol., 37*, 378384.

Myoung-Hoon, L., Hwi, S. J., Seu Ha, K., Joo, H. C., Daniele, R., Hye-Jung, L., Hong, J. C., Yuki, T., John, R., & Ohkmae, K. P., (2019). Lignin-based barrier restricts pathogens to the infection site and confers resistance in plants. *EMBO J., 38*, 101948.

Nakabayashi, R., Yonekura-Sakakibara, K., Urano, K., Suzuki, M., Yamada, Y., Nishizawa, T., et al., (2014). Enhancement of oxidative and drought tolerance in *Arabidopsis* by overaccumulation of antioxidant flavonoids. *Plant J., 77*, 367–379.

Navarro, J. M., Folres, P., Garrido, C., & Martinez, V., (2006). Changes in the contents of antioxidant compounds in pepper fruits at different ripening stages, as affected by salinity. *Food Chem., 96*, 66–73.

Naveen, J., Hariprasad, P., Chandra, N. S., & Niranjana, S. R., (2013). Cerebroside mediated elicitation of defense response in chilli (*Capsicum annuum* L.) against *Colletotrichum capsici* infection. *Journal of Plant Interaction, 8*(1), 65–73.

Naveen, J., Hariprasad, P., Navya, H. M., & Niranjana, S. R., (2013). Induction of resistance in chilli seedlings against anthracnose disease by crude cell wall elicitors from *C. capsici* isolate UOM-18. *International Journal of Pharma and Biosciences, 4*(4), (B) 304–314.

Naveen, J., Navya, H. M., Hariprasad, P., Hithamani, G., & Niranjana, S. R., (2019). *Rhizobacteria bacillus* spp. induce resistance against anthracnose disease in chili (*Capsicum annuum* L.) through activating host defense response. *Egyptian Journal of Biological Pest Control, 29*(45), 1–9.

Naveen, J., Navya, H. M., Hariprasad, P., Hithamani, G., & Niranjana, S. R., (2020). Acibenzolar-S-methyl and β-amino butyric acid-induced upregulation of biochemical defense against *Colletotrichum capsici* infection in chilli (*Capsicum annuum*). *Archives of Phytopathology and Plant Protection, 53*(3, 4), 141–161.

Naveen, J., Navya, H. M., Hithamani, G., Hariprasad, P., & Niranjana, S. R., (2021). Pathological, biochemical and molecular variability of *Colletotrichum truncatum* incitant of anthracnose disease in chilli (*Capsicum annuum* L.). *Microbial Pathogenesis*, 149.

Nazar, R., Iqbal, N., Masood, A., Khan, M. I. R., Syeed, S., & Khan, N. A., (2012). Cadmium toxicity in plants and role of mineral nutrients in its alleviation. *Am. J. Plant Sci., 3*, 1476–1489.

Nichols, S. N., Hofmann, R. W., & Williams, W. M., (2015). Physiological drought resistance and accumulation of leaf phenolics in white clover interspecific hybrids. *Environ. Exp. Bot., 119*, 40–47.

Nogues, S., Allen, D. J., Morison, J. I. L., & Baker, N. R., (1998). Ultraviolet-B radiation effects on water relations, leaf development and photosynthesis in droughted pea plants. *Plant Physiol., 117*, 173–181.

Ofori, A., Padi, F. K., Ameyaw, G. A., Dadzie, A. M., & Lowor, S., (2015). Genetic variation among cocoa (*Theobroma cacao* L.) families for resistance to cocoa swollen shoot virus disease in relation to total phenolic content. *Plant Breeding., 134*, 477–484.

Oki, T., Masuda, M., Kobayashi, M., Nishiba, Y., Furuta, S., Suda, I., & Sato, T., (2002). Polymeric procyanidins as radical-scavenging components in red-hulled rice(Article). *Journal of Agricultural and Food Chemistry, 50*(26), 7524–7529.

Olsson, L. C., Veit, M., Weissenböck, G., and Bornman, J. F. 1998. Differential flavonoid response to enhanced UV-B radiation in *Brassica napus.* Phytochemistry 49, 1021–1028.

Petridis, A., Therios, I., Samouris, G., & Tananaki, C., (2012). Salinity-induced changes in phenolic compounds in leaves and roots of four olive cultivars (*Olea europaea* L.) and their relationship to antioxidant activity. *Environ. Exp. Bot., 79*, 37–43.

Ponts, N., Pinson-Gadais, L., Boutigny, A. L., Barreau, C., & Richard-Forget, F., (2011). Cinnamic-derived acids significantly affect *Fusarium graminearum* growth and *in vitro* synthesis of type B trichothecenes. *Phytopathology, 101,* 929–934.

Rai, V. P., Jaiswal, N., Kumar, S., Singh, S., Kumar, R., & Rai, A., (2010). Response of total phenols and peroxidase activity in chilli exposed to pepper leaf curl virus disease. *Vegetable Science, 37*, 78–80.

Rieger, G., Muller, M., Guttenberger, H., & Bucar, F., (2008). Influence of altitudinal variation on the content of phenolic compounds in wild populations of *Calluna vulgaris, Sambucus nigra,* and *Vaccinium myrtillus. J. Agric. Food Chem., 58*, 90809086.

Rozema, J., Björn, L. O., Bornman, J. F., Gaberscik, A., Häder, D. P., Trost, T., Germ, M., Klisch, M., Gröniger, A., Sinha, R. P., Lebert, M., He, Y. Y., Buffoni-Hall, R., de Bakker, N. V., van de Staaij, J., & Meijkamp, B. B. (2002). The role of UV-B radiation in aquatic and terrestrial ecosystemsâ€"an experimental and functional analysis of the evolution of UV-absorbing compounds. *Journal of Photochemistry and Photobiology B: Biology, 66*(1), 2–12.

Sajjad, H., Amjad, S. G., Zarrin, F. R., Muhammad, I. S. H., & Inam-ul-Haq, M. (2021). *Biological Control of Chili Damping-off Disease, Caused by Pythium Myriotylum,* 1–49.

Šamec, D., Karalija, E., Šola, I., Vujčić Bok, V., & Salopek-Sondi, B., (2021). The role of polyphenols in abiotics tress response: The influence of molecular structure. *Plants, 10,* 118.

Sánchez-Rodríguez, E., Moreno, D. A., Ferreres, F., Del Mar Rubio-Wilhelmi, M., & Ruiz, J. M., (2011). Differential responses of five cherry tomato varieties to water stress: Changes on phenolic metabolites and related enzymes. *Phytochemistry, 72*(8), 723–729.

Schulze-Lefert, P., Dangl, J. L., Becker-Andre, M., Hahlbrock, K., & Schulz, W., (1989). Inducible in vivo DNA footprints define sequences necessary for UV light activation of the parsley chalcone synthase gene. *EMBO J., 8*, 651656.

Shahid, M., Khalid, S., Abbas, G., Shahid, N., Nadeem, M., Sabir, M., Aslam, M., & Dumat, C., (2015). Heavy metal stress and crop productivity. In: *Crop Production and Global Environmental Issues*. ISBN: 978-3-319-23161-7.

Shiozaki, N., Hattori, I., Gojo, R., & Tezuka, T., (1999). Activation of growth and nodulation in symbiotic system between pea plants and leguminous bacteria by near UV radiation. *J. Photochem. Photob. B. Biol., 50*, 3337.

Shormin, C., Hongliang, H., Philip, L., Holger, M., Sergey, S., Ibrahim, A., & Meixue, Z., (2018). Agronomical, biochemical and histological response of resistant and susceptible wheat and barley under BYDV stress. *Peer J.,* 6, e4833.

Shyanadrahalli, C. V., Siddapura, R. N., & Sharanaiah, U., (2009). *Role of Phenylalanine Ammonia Lyase and Polyphenol Oxidase in Host Resistance to Bacterial Wilt of Tomato, 157*, 552–557.

Singh, A., Singh, K., Rajwar, G., & Singh, U., (2010). Phenolic acid content—a criterion for selection of resistant apple cultivars against *Podosphaera leucotricha* (Ell. and Ev.) salmon. *Archives of Phytopathology and Plant Protection, 43*, 1138–1143.

Singh, S., & Kariyat, R. R., (2020). Exposure to polyphenol-rich purple corn pericarp extract restricts fall armyworm (*Spodoptera frugiperda*) growth. *Plant Signal. Behav., 15*, 1784545.

Singh, U. P., Sarma, B. K., Singh, D. P., & Amar, B., (2002). Plant growth-promoting rhizobacteria-mediated induction of phenolics in pea (*Pisum sativum*) after infection with *Erysiphe pisi. Current Microbiology, 44*, 396–400.

Spitaler, R., Winkler, A., Lins, I., Yanar, S., Stuppner, H., & Zidorn, C., (2008). Altitudinal variation of phenolic contents in flowering heads of *Arnica montana cv.* ARBO: A 3-year comparison. *J. Chem. Ecol., 34*, 369375.

Stapleton, A. E., (1992). Ultraviolet radiation and plants: Burning questions. *Plant Cell, 4*, 1353–1358.

Tayal, M., Somavat, P., Rodriguez, I., Martinez, L., & Kariyat, R., (2020a). Cascading effects of polyphenol-rich purple corn pericarp extract on pupal, adult, and offspring of tobacco hornworm (*Manduca sexta L.*). *Commun. Integr. Biol., 13*, 43–53.

Tayal, M., Somavat, P., Rodriguez, I., Thomas, T., Christoffersen, B., & Kariyat, R., (2020b). Polyphenol-rich purple corn pericarp extract adversely impacts herbivore growth and development. *Insects, 11*, 98.

Tegelberg, R., Julkunen-Tiitto, R., & Aphalo, P. J., (2004). Red: Far-red light ratio and UV-B radiation: Their effects on leaf phenolics and growth of silver birch seedlings. *Plant Cell Environ., 27*, 10051013.

Telesiński, A., Nowak, J., Smolik, B., Dubowska, A., & Skrzypiec, N., (2008). Effect of soil salinity on activity of antioxidant enzymes and content of ascorbic acid and phenols in bean (*Pheseolus vulgaris* L.) plants. *J. Elementol., 13*, 401–409.

Valifarda, M., Mohsenzadeha, S., Kholdebarina, B., & Rowshan, V., (2014). Effects of salt stress on volatile compounds, total phenolic content and antioxidant activities of *Salvia mirzayanii. South African Journal of Botany, 93*, 92–97.

Vanitha, S. C., & Umesha, S., (2011). Pseudomonas fluorescens mediated systemic resistance in tomato is driven through an elevated synthesis of defense enzymes *Biologia Plantarum, 55*, 317–322.

Vanitha, S.C., Niranjana, S.R., Mortensen, C.N., Umesha, S. 2009. Bacterial wilt of tomato in Karnataka and its management by *Pseudomonas fluorescens*. Biocontrol 54: 685–695.

Vera, A. D. P., Fernando, R. D., & Leandro, V., (2009). Astarita defense mechanisms of *Solanum tuberosum* L. in response to attack by plant-pathogenic bacteria. *Biol. Res., 42*, 205–215.

Whitham, S. A., Yang, C., & Goodin, M. M., (2006). Global impact: Elucidating plant responses to viral infection. *Mol. Plant-Microbe Interact., 19*(11), 1207–1215.

Winkel-Shirley, B., (2002). Biosynthesis of flavonoids and effects of stress. *Current Opinion in Plant Biology, 5*(3), 218–223.

Wojakowska, A., Muth, D., Narożna, D., et al., (2013). Changes of phenolic secondary metabolite profiles in the reaction of narrow-leaf lupin (*Lupinus angustifolius*) plants to infections with *Colletotrichum lupini* fungus or treatment with its toxin. *Metabolomics, 9*, 575–589 (2013).

Xu, Y., Charles, M. T., Luo, Z., Mimee, B., Veronneau, P. Y., Rolland, D., & Roussel, D., (2017). Preharvest ultraviolet C irradiation increased the level of polyphenol accumulation and flavonoid pathway gene expression in strawberry fruit. *J. Agric. Food Chem., 65*, 9970–9979.

Yang, J., Sun, X. Q., Yan, S. Y., Pan, W. J., Zhang, M. X., & Cai, Q. N., (2017). Interaction of ferulic acid with glutathione S-transferase and carboxylesterase genes in the brown planthopper, *Nilaparvata lugens. J. Chem. Ecol., 43*, 693–702.

Ying, Z., Nanyu, T., Lijin, H., Yongjuan, Z., Xiaoqing, T., & Kangcai, W., (2018). Effects of salt stress on plant growth, antioxidant capacity, glandular trichome density, and volatile exudates of *Schizonepeta tenuifolia* briq. *Int J Mol Sci., 19*(1), 252.

Yokoi, S., Quintero, F. J., Cubero, B., Ruiz, M. T., Bressan, R. A., Hasegawa, P. M., & Pardo, J. M., (2002). Differential expression and function of *Arabidopsis thaliana* NHX Na+/H + antiporters in the salt stress response. *Plant J., 30*, 529–539.

Zeeshan, S., Khalid, P. A., Amjad, H., Nighat, S., & Imran-Ul-Haqaand, S. A. K., (2014). Biochemical alterations in leaves of resistant and susceptible cotton genotypes infected systemically by cotton leaf curl Burewala virus. *Journal of Plant Interactions, 9*(1), 702–711.

Zengin, F., (2015). Effects of exogenous salicylic acid on growth characteristics and biochemical content of wheat seeds under arsenic stress. *Journal of Environmental Biology, 36*, 249–254.

Zhang, F. Q., Wang, Y. S., Lou, Z. P., & Dong, J. D., (2007). Effect of heavy metal stress on antioxidative enzymes and lipid peroxidation in leaves and roots of two mangrove plant seedlings (*Kandelia candel* and *Bruguiera gymnorrhiza*). *Chemosphere, 67*(1), 44–50.

Zhou, H., Finkemeier, I., Guan, W., Tossounian, M. A., Wei, B., Young, D., et al. (2018). Oxidative stress-triggered interactions between the succinyl- and acetylproteomes of rice leaves. Plant Cell Environ. 41, 1139–1153.

Zhao, S., & Li, Y., (2021). Current understanding of the interplays between host hormones and plant viral infections. *PLoS Pathog., 17*(2).

Zrig, A., Tounekti, T., Vadel, A. M., Mohamed, H. B., Valero, D., Serrano, M., Chtara, C., & Khemira, H., (2011). Possible involvement of polyphenols and polyamines in salt tolerance of almond rootstocks. *Plant Physiol. Biochem., 49*,1313–1322.

Zidorn, C. Schubert, B. Stuppner, H. 2005. Altitudinal differences in the contents of phenolics in flowering heads of three members of the tribe Lactuceae (Asteraceae) occurring as introduced species in New Zealand. Biochem. Syst. Ecol. 33, 855–872.

CHAPTER 13

Nitrogen and Organic Acid Metabolism in *Portulaca oleracea* L. Under Drought Stress

VISHAL V. NAIK,[1] AMOL V. PATIL,[2] and BABURAO A. KARADGE[3]

[1]*Department of Botany, Bharati Vidyapeeth's M. B. S. K. K. M., Kadegaon, Sangli, Maharashtra, India, E-mail: purslanevvn@gmail.com*

[2]*Department of Botany, Dr. Bapuji Salunkhe College, Miraj, Maharashtra, India*

[3]*Department of Botany, Shivaji University, Kolhapur, Maharashtra, India*

ABSTRACT

Portulaca oleracea L. is a C_4 succulent belonging to the family Portulacaceae. It is a nitrogen-loving plant with high oxalic acid content. The main objective of the present study was to determine the relationship between nitrogen and organic acid metabolism and the drought-tolerant nature of this plant. In this context, one-month-old seedlings of *P. oleracea* raised in acid-free silica sand with ½ strength Hoagland's culture medium were water stressed by watering them at an interval of 4, 8, and 15 days separately in separate pots, while control plants were watered every alternate day. Results indicated that there is an increase in total nitrogen, nitrate, and nitrite nitrogen in the roots, stem, and leaves of *P. oleracea* imposed by water stress as compared to well-watered control. The investigation into organic acid metabolism revealed that under water stress conditions, the accumulation of oxalic acid content in the root, stem, and leaves of *P. oleracea* was followed by increased activity of the enzyme glycolate oxidase. The role of nitrogen and oxalic

Plant Metabolites under Environmental Stress: Mechanisms, Responses, and Adaptation Strategies.
Nivas M. Desai, Manasi Patil, Umesh R. Pawar (Eds.)

acid metabolism in *P. oleracea* in the water stress tolerance mechanism is discussed.

13.1 INTRODUCTION

Water stress (drought) is regarded as one of the common threats to the agricultural systems throughout the world. Today, agriculture accounts for 70% of total water use. As shown in Figure 13.1, most part of the world is facing water unavailability for agricultural purpose. In India, about 51% of the total area is under cultivation. While, water withdrawal for agricultural purposes is about 90% of total water resources (FAO, 2015). Hence, water unavailability is a major constraint for agricultural production.

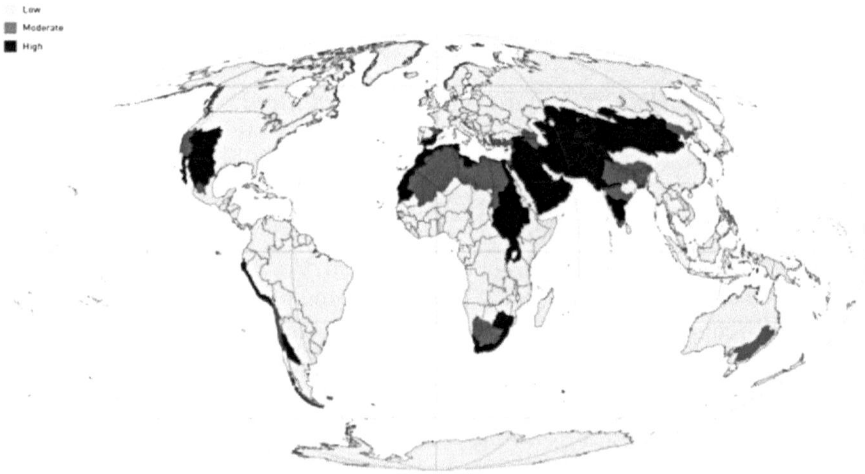

FIGURE 13.1 Global freshwater scenario most of the parts of India under water scarcity.
Source: www.fao.org/nr/solaw.

Water is an essential factor limiting plant's growth and development. The fresh weight of most of the herbaceous plants accounts for 80–90% while, for woody plants it is over 50% (Kramer and Boyer, 1995). The processes like photosynthesis and mineral uptake require water. Low water availability tends towards decreased growth and productivity of plants. Water stress is often associated with high temperature, cold or frozen soils, which can reduce water uptake (Schutzki and Cregg, 2007). The tolerance limits for metabolic activity in vacuolated cell, tissues require turgor pressure of about 0.05–0.5

MPa against a negative water potentials commonly ranging from –0.5 to –15.0 MPa. There are mainly two strategies of plants to cope with water stress, i.e., stress avoidance and tolerance. Water stress avoidance mechanisms include stomatal closure, leaf movement and leaf shading (Osmond et al., 1987). Stress tolerant plant faces adverse conditions by altering morphological, anatomical, and physiological characters (adaptations). In recent years, research on the development of stress-tolerant genotypes has attracted many researchers. However, in the environment, there are some species that show adaptations towards water stress conditions, e.g., xerophytes. While, some plants other than xerophytes show alterations in their normal metabolism to cope with water unavailability.

Portulacaceae is one of the 19 families of terrestrial plants in which species having C_4 photosynthesis have been found. The representative species from the genus *Portulaca* such as *P. oleracea* is a C_4 plant-based on carbon isotope values, Kranz anatomy and expression of key C_4 enzymes (Voznesenskaya et al., 2010). *P. oleracea* is one of the water stress tolerant plant. The leaves and stem of *P. oleracea* are succulent. It is one of the plants with high water use efficiency. However, several workers have reported that *P. oleracea* shows weak CAM type of photosynthesis when exposed to water stress (Kennedy, 1977; Koch and Kennedy, 1980; Karadge, 1981; Lara, Drincovich, and Andreo, 2004; D'Andrea et al., 2015). Furthermore, this plant is nitrophilous, i.e., love to grow under high nitrogen environment and possess higher concentration of organic acids, predominantly, oxalic acid (Karadge, 1981). It is possible that nitrophilous nature and higher oxalic acid concentration play a role in water stress tolerance of *P. oleracea*. The main objective of the present investigation was to find relationship between water stress tolerance of this plant and nitrophilous nature with higher oxalic acid concentration.

13.2 PLANT MATERIAL AND METHODOLOGY

13.2.1 PLANT MATERIAL

Seeds of *P. oleracea* were collected from field in the months of February to April. It was germinated in silica sand in plastic pots (32 cm diameter × 12 cm height), each pot containing 6 kg acid free silica sand. Plants were grown for 1 month with the supply of ½ strength Hoagland solution (Table 13.1) (1:1) alternating with water under natural conditions (Light intensity: 30,000 lux; Temperature: 30 ± 2°C) in botanical garden of Shivaji University, Kolhapur

(MS). After one month of their growth, plants were watered at an interval of 4, 8, and 15 days separately in separate pots while; control plants were watered every alternate day. Fresh plants from each plot were randomly sampled and washed first with tap water and then with distilled water and blotted dry. Root, stem, and leaves were cut into pieces, weighed accurately and immediately used for analysis. All operations were done at room temperature (28 ± 2°C). Salts of microelements were as shown in Table 13.1 and dissolved in distilled water and diluted to 1 liter with distilled water. About 1 ml/0.5 ml from this stock solution was taken for preparation of the culture medium.

TABLE 13.1 Preparation of Hoagland Culture Medium (Hoagland and Arnon, 1950)

		Reagents	ml Taken to Prepare 1 Liter Full Strength Medium	ml Taken to Prepare 1 Liter Half Strength Medium
A.			Macro Elements	
	1.	1 M KH_2PO_4	1	0.5
	2.	1 M KNO_3	5	2.5
	3.	1 M $CaNO_3$	5	2.5
	4.	1 M $MgSO_4$	2	1
B.			Microelements	
	1.	H_3BO_3 (2.86 g)	1	0.5
	2.	$MnCl_2.4H_2O$ (1.81 g)		
	3.	$ZnSO_4.7H_2O$ (0.22 g)		
	4.	$CuSO_4.5H_2O$ (0.08 g)		
	5.	H_2MoO_4 (0.02 g)		
C.		0.5% Fe-EDTA	1	0.5

13.2.2 METHODOLOGY

13.2.2.1 NITROGEN METABOLISM

13.2.2.1.1 Estimation of Total Nitrogen

Total nitrogen was estimated following the method given by Hawk et al. (1948). Oven-dried powdered plant material (0.5 g) was taken in Kjeldahl's flask with a pinch of microsalt (200 g K_2SO_4 + 5 g $CuSO_4$, dehydrated) and to it 5 ml H_2SO_4 (1:1) were added. Few glass beads were added to avoid bumping and the material was digested on low flame. After complete digestion, a faint yellow solution was obtained which was cooled to room temperature, transferred to a volumetric flask, and diluted to 100 ml with distilled water. Then it was filtered through Whatman No. 1 filter paper and used for the estimation of nitrogen.

In a set of Nessler's tubes, 2 ml of plant extract and different concentrations of standard ammonium sulfate solution (0.236 g of oven dried ammonium sulfate dissolved in distilled water and few drops of conc. H_2SO_4 were added. The volume was made to 1,000 ml with distilled water. This solution contains 0.05 mg of nitrogen per ml) were taken. To each of these tubes, one drop of 8% $KHSO_4$ was added and volume was made to 35 ml with distilled water. To this 15 ml of freshly prepared Nessler's reagent were added (Reagent A: 7 g KI + 10 g HgI_2 in 40 ml distilled water, Reagent B: 10 g NaOH in 50 ml distilled water. A and B were mixed in proportion of 4:5 at the time of estimation).

The reaction between the sample and the reagent gives the product $NH_4Hg_2I_3$ which has orange-brown color. The intensity of this color was measured after 15 minutes at 520 nm on double beam UV-VIS spectrophotometer (Shimadzu UV-1800, Japan).

13.2.2.1.2 Enzymes of Nitrogen Metabolism

1. **Nitrate Reductase Assay (NR; EC. 1.6.6.2):** *In vivo* activity of this enzyme was determined following the method by Jaworsky (1971). Plant material (fresh roots, stems, and leaves) was cut into small pieces of about 0.5 cm^2 and incubated in 10 ml incubation medium containing 1 ml, 1 M KNO_3, 2 ml 5% n-propanol, 5 ml 0.2 M phosphate buffer, pH 7.5 and 2 ml, 0.5% Triton-X-100, for 1 h in dark. After incubation, 1 ml of the reaction medium was taken out for determination of nitrite and was mixed with 1 ml each of 1% sulfanilamide in 1 M HCl and 0.02% NEEDA. The absorbance was read at 540 nm on a double beam UV-VIS spectrophotometer (Shimadzu UV-1800, Japan) using a reagent blank. The standard curve was prepared with 0.03 mM KNO_2 (0.0026 mg NO_2 ml^{-1}) in the range of concentrations from 0.3 mM to 1.5 mM KNO_2 against a mixture of 1 ml incubation medium, 1 ml sulfanilamide and 1 ml NEEDA as a blank. The soluble proteins in the enzyme extract were determined according to the method of Lowry et al. (1951). The activity of the enzyme is expressed as μg NO^{2-} liberated h^{-1} mg^{-1} protein.

2. **Nitrite Reductase Assay (NiR; EC. 1.6.6.4):** The activity of NiR was determined following the method by Jaworsky (1971) already described for NR except that KNO_3 was replaced by 0.3 mM KNO_2 in the incubation medium and the incubation was done in light. The amount of KNO_2 remained in the incubation medium after the enzymatic activity for 1 h was determined by recording the absorbance of the reaction mixture containing 1 ml incubation medium, 1 ml sulfanilamide and 1 ml NEEDA. The difference between the two readings, one at 0 min and the other after 1 h enzymatic reaction, gave the amount of KNO_2 utilized by the enzyme. The standard curve of KNO_2 was prepared as described earlier. The soluble proteins in the enzyme extract were determined according to the method of Lowry et al. (1951). The activity of the enzyme is expressed as μg NO^{2-} reduced h^{-1} mg^{-1} protein.

13.2.2.1.3 Estimation of NO_3^- and NO_2^-

After determination of nitrate reductase (NR) and nitrite reductase (NiR) activity, the plant material from the incubation medium was removed carefully and washed thoroughly first with distilled water and then with 0.2 M phosphate buffer, pH 7.5. It was then homogenized and extracted in 10 ml 0.2 M phosphate buffer, pH 7.5. The extract was filtered through Whatman No. 1 filter paper. The filtrate was used for the estimation of nitrate (NO^{3-}) and nitrite (NO^{2-}), following the methods as described below:

1. **Estimation of Nitrate (NO^{3-}):** The method used is based on the reaction between Diphenylamine-H_2SO_4 reagent and nitrate (Kolhoff and Noponen, 1933). The reaction mixture contained 1 ml plant extract, 1 ml distilled water and 1.8 ml diphenylamine-H_2SO_4 reagent (1 g diphenylamine 100 ml^{-1} H_2SO_4). After vigorous shaking, the blue color was allowed to develop for 10 min. The absorbance was recorded at 590 nm on a double beam UV-VIS spectrophotometer (Shimadzu UV-1800, Japan). Instead of extract, distilled water was used to prepare blank. The standard curve was prepared from 0.1 M KNO_3 and from this, nitrate content was calculated. Values of nitrate content are expressed as g 100^{-1} g fresh tissue.
2. **Estimation of Nitrite (NO^{2-}):** Nitrite content was determined following the method by Jaworsky (1971). The reaction mixture contained 5 ml plant extract to which 1 ml sulfanilamide (1% in 1

M HCl) and 1 ml NEEDA (0.02% N-1-Naphthylethylene diamine dihydrochloride) were added. After 15 min the absorbance was read at 540 nm on double beam UV-VIS spectrophotometer (Shimadzu UV-1800, Japan). By using 0.03 mM KNO_2, the standard curve for NO_2 was prepared. The blank was prepared with distilled water in place of nitrite. Nitrite content is expressed as mg 100^{-1} g fresh tissue.

13.2.2.2 OXALIC ACID METABOLISM

13.2.2.2.1 Oxalic Acid

Oven-dried root, stem, and leaves of *Portulaca oleracea* L. were used to determine oxalic acid content as per the method described by Naik et al. (2014). About 500 milligram leaves were powdered and transferred to 50 ml capacity volumetric flask. To which, 30 ml 0.25 N HCl was added and kept in boiling water bath for about 15 min, cooled to room temperature and volume was made with 0.25 N HCl. This solution was used as an extract for determination of oxalic acid.

Assay mixtures contained 1 ml plant extract, 5 ml of 2 N H_2SO_4 and 2 ml of 0.003 M $KMnO_4$. This mixture was incubated for 10 min. at room temperature (27 ± 2°C). After 10 min. absorbance was recorded at 528 nm on Shimadzu, UV 190 spectrophotometer. Reagent blank was prepared with distilled water. Absorbance for blank solution was recorded as Ab and for sample it was As. Oxalic acid content in plant parts was calculated from the calibration curve of std. oxalic acid ranging from 0.1 to 1 mg/ml. ΔA is obtained from Ab-As and used for calculation. The amount of oxalic acid was expressed as g 100^{-1} g dry tissue.

13.2.2.2.2 Determination of Activity of Glycolate Oxidase

Glycolate oxidase activity was estimated according to the method of Hess and Tolbert (1967). About 500 milligram of plant material homogenized in 10 ml cold 0.1 M Tris-HCl buffer (pH 7.5) containing 1 M KCl, 0.01 M EDTA and 0.4 ml 0.2 M β-mercaptoethanol in chilled mortar with pestle. The resultant suspension was filtered through 4 layered cheese cloths and filtrate was centrifuged at 10,000 g for 20 min at 0°C. The supernatant served as crude enzyme source.

The assay mixture contained 2.4 ml 0.1 M phosphate buffer (pH 8.3), 0.1 ml phenylhydrazine hydrochloride (0.1 M, pH 6.6), 0.1 ml 0.1 M cysteine hydrochloride, 0.3 ml sodium glycolate (0.2 M) and 0.1 ml crude enzyme. The enzymatic reaction was initiated by the addition of sodium glycolate. Activity of the enzyme was measured at 340 nm on Shimadzu UV-1800, Japan, spectrophotometer in which absorbance was increased due to formation of phenylhydrazone. Enzyme activity is expressed as μ moles glycolate oxidized min^{-1} mg^{-1} protein.

13.3 RESULTS AND DISCUSSION

13.3.1 EFFECT OF WATER STRESS ON NITROGEN METABOLISM

13.3.1.1 ACCUMULATION OF TOTAL NITROGEN

Total nitrogen content of root, stem, and leaves of *P. oleracea* as influenced by water stress is recorded in Table 13.2. It is clear from the data that total nitrogen content was increased with increase in water stress. The highest total nitrogen was recorded in the root, stem, and leaves (12.36 ± 0.70, 7.36 ± 0.39, and 11.90 ± 0.50 g 100^{-1} g dry tissue, respectively) of plants exposed to 15 DWS. From these observations, it seems that nitrogen was accumulated in response to water stress in this nitrophilous plant, *P. oleracea.*

P. oleracea is a nitrophilous plant (Karadge, 1981; Bharucha and Dubash, 1951; Bharucha and Joshi, 1957). Bharucha and Dubash (1951); and Kaskar et al. (2009) estimated 155 ppm of nitrate concentration in cell-sap of *P. oleracea*. Karadge (1981) noticed that the total nitrogen content of *P. oleracea* ranges from 6–8 g 100^{-1} g dry tissue. D'Andrea et al. (2014) studied metabolic changes in *P. oleracea* in response to water stress and re-watering. They found increased level of nitrogenous compounds like free amino acids and mostly urea in the leaves of *P. oleracea*. According to them, accumulation of these nitrogenous compounds in *P. oleracea* during water stress enhances its adaptation to drought.

13.3.1.2 ACCUMULATION OF NO^{3-} INDUCES OSMOTIC ADJUSTMENT

P. oleracea exposed to water stress showed an accumulation of NO^{3-} in the root, stem, and leaves of *P. oleracea*. It is observed in most of the plants that water stress increases nitrate concentration in plant parts (Erskine et

TABLE 13.2 Effect of Water Stress on Total Nitrogen, Nitrate, and Nitrite Contents of Root, Stem, and Leaves of *P. oleracea**

Plant Parts	Treatments (DWS: Day of Water Stress)	Total Nitrogen ($g\ 100^{-1}$ g Dry Tissue)	Nitrate Content ($g\ 100^{-1}$ g Fresh Tissue)	Nitrite Content ($mg\ 100^{-1}$ g Fresh Tissue)
Root	0 (control)	6.42 ± 0.49[d]	0.340 ± 0.038[d]	0.797 ± 0.024[c]
	4	8.37 ± 0.62[c]	0.476 ± 0.054[c]	0.968 ± 0.023[b]
		(+30.35)	(+39.93)	(+21.42)
	8	9.56 ± 0.52[b]	0.516 ± 0.040[b]	1.273 ± 0.143[ab]
		(+48.92)	(+51.75)	(+42.85)
	15	12.36 ± 0.70[a]	0.703 ± 0.049[a]	1.632 ± 0.187[a]
		(+92.5)	(+106.70)	(+81.63)
Stem	0 (control)	3.99 ± 0.09[d]	0.236 ± 0.035[c]	0.668 ± 0.024[c]
	4	4.86 ± 0.32[c]	0.254 ± 0.017[c]	0.720 ± 0.011[b]
		(+21.83)	(+7.33)	(+6.09)
	8	6.07 ± 0.52[b]	0.291 ± 0.030[b]	0.772 ± 0.017[b]
		(+52.29)	(+22.93)	(+14.63)
	15	7.36 ± 0.39[a]	0.372 ± 0.030[a]	0.952 ± 0.030[a]
		(+84.48)	(+57.33)	(+42.68)
Leaves	0 (control)	6.81 ± 0.76[d]	0.476 ± 0.034[c]	0.860 ± 0.021[c]
	4	7.17 ± 0.87[c]	0.541 ± 0.036[c]	1.569 ± 0.262[b]
		(+5.38)	(+13.69)	(+57.28)
	8	9.56 ± 0.12[b]	0.602 ± 0.034[b]	2.418 ± 0.219[a]
		(+40.40)	(+26.48)	(+160.19)
	15	11.90 ± 0.50[a]	0.741 ± 0.048[a]	2.940 ± 0.276[a]
		(+74.74)	(+55.70)	(+217.47)

* Values in parenthesis indicate percent increase (+) or decrease (–) over the control. Values in a column followed by the same letter are not significantly different at the 0.05 level, as determined by Duncan's Multiple Range Test. Each value is the mean of three different readings. ±SD.

al., 1996). The drought-induced changes in osmotic adjustment in expanded leaves of *Brassica napus* and *B. juncea* were mainly due to the accumulation of nitrate (Ma et al., 2004). It is possible that under water-deficit conditions purslane accumulates more NO_3^- to adjust osmotic pressure within the cell.

It is reported that NO^{3-} also acts as an inducer of NR in *Spirulina platensis* (Jha et al., 2007) and NiR in barley seedlings (Aslam and Huffaker, 1989). Hence, accumulation of NO_3^- in *P. oleracea* may be an adaptive measure to combat with water scarcity.

13.3.1.3 NITRITE (NO^{2-}) CONTENT

NO^{2-} content of the root, stem, and leaves of *P. oleracea* grown under water stress was increased under water stress conditions. NO^{2-} is never allowed to accumulate in the cells. It forms diazo compounds with amino groups of nucleobases ($R\text{-}NH_2$), which are converted into alcohols with the release of nitrogen (Heldt and Piechulla, 2004). NO^{2-} was reported to acidify the stroma, which may inhibit the enzymatic reactions in the Calvin cycle (Purczeld et al., 1978). Higher activity of NiR (Figure 13.2) may eliminate the toxic effects of NO^{2-}. It is also possible that an increased concentration of NO_2^- in the plant parts due to water stress stimulates NiR activity and detoxification occurs.

13.3.1.4 ENHANCEMENT IN ACTIVITY OF ENZYME NITRATE REDUCTASE (NR)

The activity of enzyme NR was increased significantly in the root, stem, and leaves of *P. oleracea*, water-stressed for 8 days in comparison to the control (Figure 13.1). However, the activity was lowered in all plant parts after 15 DWS. NO^{3-} concentration in the cell sap is an important factor which induces NR activity. NO^{3-} modulates and stabilizes NR transcripts (Galangau et al., 1988). It is observed that during water stress the decrease in the NR activity was accompanied by a rapid decline in NR transcript levels (Foyer et al., 1998).

13.3.1.5 INCREASED ACTIVITY OF NITRITE REDUCTASE (NIR) DETOXIFIES NO^{2-}

The activity of enzyme NiR was increased significantly in the leaves of *P. oleracea* for 8 DWS over that in control leaves (Figure 13.2). But, due to 15

DWS, its activity in the leaves was decreased significantly. It is observed that NiR is more stable than NR under the adverse environmental conditions like drought. This stability of NiR may be associated with its chloroplastic location (Gupta and Beevers, 1983). To eliminate toxicity of higher NO^{2-} concentrations, the activity of NiR was increased.

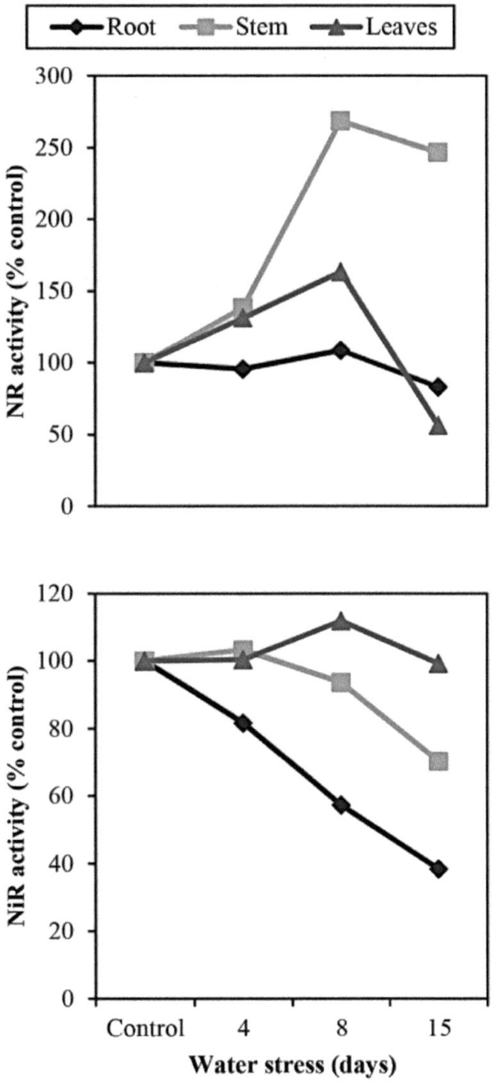

FIGURE 13.2 Effect water stress on activity of enzyme nitrate reductase (NR) and nitrite reductase (NiR) in *P. oleracea*.

13.3.2 EFFECT OF WATER STRESS ON OXALIC ACID METABOLISM IN *P. OLERACEA*

13.3.2.1 ACCUMULATION OF OXALIC ACID

Water stress caused a significant increase in the total oxalic acid content of root, stem, and leaves of *P. oleracea*. The highest amount of oxalic acid was found in the leaves of 15 DWS plants, i.e., 8.32±0.089 g 100^{-1} g dry tissue (Table 13.3). Oxalic acid in plants is a primary chelator of calcium and can complex with calcium to form highly insoluble calcium oxalate crystals (Francheschi and Nakata, 2005). The calcium regulation, plant protection and metal detoxification are some of the highlighted functions of oxalic acid in plants (Francheschi and Nakata, 2005). Furthermore, oxalic acid is one of the natural antioxidant present in plants (Kayashima and Katayama, 2002). Guo et al. (2013) found that increase in oxalic acid content in root and leaves of wheat (*Triticum aestivum*) due to 15% PEG-6000 induced water stress. According to them oxalic acid is involved in maintaining intracellular osmotic balance under water stress.

TABLE 13.3 Effect of Water Stress on Concentration of Total Oxalic Acid in Root, Stem, and Leaves of *P. oleracea* Under Water Stress*

Treatment (DWS)	Total Oxalic Acid Content (g 100^{-1} g Dry Tissue)		
	Root	**Stem**	**Leaves**
0 (control)	2.61 ± 0.098^{c}	3.48 ± 0.063^{c}	6.35 ± 0.092^{c}
4	3.14 ± 0.121^{b} (+20.30)	4.25 ± 0.080^{b} (+22.12)	6.91 ± 0.109^{b} (+8.81)
8	3.67 ± 0.080^{b} (+40.61)	4.87 ± 0.063^{ab} (+39.94)	7.24 ± 0.075^{b} (+14.01)
15	4.14 ± 0.092^{a} (+58.62)	5.14 ± 0.086^{a} (+47.70)	8.32 ± 0.089^{a} (+31.02)

Values in parenthesis indicate percent increase (+) or decrease (–) over the control. Values in a column followed by the same letter are not significantly different at the 0.05 level, as determined by Duncan's Multiple Range Test. Each value is the mean of five replications of three different batches of plants. ± SD.

13.3.2.2 ACTIVITY OF ENZYME GLYCOLATE OXIDASE

Influence of water stress on the activity of enzyme glycolate oxidase in *P. oleracea* is recorded in Table 13.4 and represented in Figure 13.3. From the results, it is clear that the activity of enzyme glycolate oxidase was increased in the root and leaves and it was decreased in the stem. Glycolate oxidase is considered as a key enzyme in oxalate biosynthesis (Xu et al., 2006). It is observed that glycolate oxidase activity was increased in response to drought stress (Rizhsky, Liang, and Mittler, 2002). Karadge (1981) has noticed that high glycolate oxidase activity in the leaves of *P. oleracea* is related to the oxalic acid production. He also noticed that external supply of oxalic acid stimulates the activity of this enzyme. Hence, increased activity of glycolate oxidase in the root and leaves of *P. oleracea* under water stress may be responsible for the production of oxalic acid for adaptation to water stress.

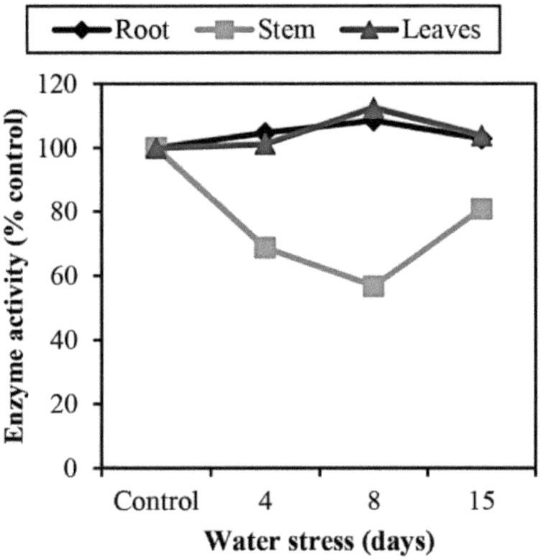

FIGURE 13.3 Effect of water stress on the activity of enzyme glycolate oxidase in *P. oleracea*.

13.4 CONCLUSIONS

From these results, it concluded that nitrogen metabolism has definitely has a role in stress tolerance in *Portulaca oleracea* under water stress conditions. An accumulation of nitrogen and its derivatives possibly helps in improvement of physiological processes to fight against water stress. It is possible that nitrophilous nature of *P. oleracea* is one of the reasons for drought-tolerant nature. Oxalic acid is considered as anti-nutritional factor but, in this plant, it may be acts as an osmoregulator. Significantly higher concentrations of oxalic acid in all vegetative parts of *P. oleracea* under water stress depicts its possible role in stress tolerance mechanism of this plant.

KEYWORDS

- **nitrogen**
- **oxalic acid**
- ***P. oleracea***
- ***Triticum aestivum***
- **water stress**

REFERENCES

Aslam, M., & Huffaker, R. C., (1989). Role of nitrate and nitrite in the induction of nitrite reductase in leaves of barley seedlings. *Plant Physiology, 91*(3), 1152–1156.

Bharucha, F. R., & Dubash, P. J., (1951). The problem of nitrophily. *Vegetatio, 3*(3), 183–194.

Bharucha, F. R., & Joshi, G. V., (1957). Studies in organic acid metabolism in *Portulaca oleracea* Linn. *Naturwissenschaften, 44*(8), 263–263.

D'Andrea, R. M., Andreo, C. S., & Lara, M. V., (2014). Deciphering the mechanisms involved in *Portulaca oleracea* (C_4) response to drought: Metabolic changes including crassulacean acid-like metabolism induction and reversal upon re-watering. *Physiologia Plantarum, 152*(3), 414–430.

D'Andrea, R. M., Triassi, A., Casas, M. I., Andreo, C. S., & Lara, M. V., (2015). Identification of genes involved in the drought adaptation and recovery in *Portulaca oleracea* by differential display. *Plant Physiology and Biochemistry, 90*, 38–49.

Erskine, P. D., Stewart, G. R., Schmidt, S., Turnbull, M. H., Unkovich, M., & Pate, J. S., (1996). Water availability–a physiological constraint on nitrate utilization in plants of Australian semi-arid muiga woodlands. *Plant, Cell and Environment, 19*(10), 1149–1159.

FAO, (2015). *FAO Statistical Pocketbook: World Food and Agriculture*. Food and Agriculture Organization of the United Nations, Rome. http://www.fao.org/3/a-i4691e.pdf (accessed on 2 June 2022).

Foyer, C. H., Valadier, M. H., Migge, A., & Becker, T. W., (1998). Drought-induced effects on nitrate reductase activity and mRNA and on the coordination of nitrogen and carbon metabolism in maize leaves. *Plant Physiology, 117*(1), 283–292.

Francheschi, V. R., & Nakata, P. A., (2005). Calcium oxalate in plants: Formation and function. *Annual Review of Plant Biology, 56*, 41–71.

Galangau, F., Daniel-Vedele, F., Moureaux, T., Dorbe, M. F., Leydecker, M. T., & Caboche, M., (1988). Expression of leaf nitrate reductase genes from tomato and tobacco in relation to light-dark regimes and nitrate supply. *Plant Physiology, 88*(2), 383–388.

Guo, R., Hao, W. P., Gong, D. Z., Zhong, X. L., & Gu, F. X., (2013). Effects of water stress on germination and growth of wheat, photosynthetic efficiency and accumulation of metabolites. In: In: Maria C. H. S., (ed.), *Soil Processes and Current Trends in Quality Assessment*. ISBN: 978-953-51-1029-3, InTech, doi: 10.5772/51205.

Gupta, S. C., & Beevers, L., (1983). Environmental influences on nitrite reductase activity in *Pisum sativum* L. seedlings. *Journal of Experimental Botany, 34*(11), 1455–1462.
Hawk, P. B., Oser, B. L., & Summerson, W. H., (1948). *Practical Physiological Chemistry* (p. 1323). (Publ.) The Blockiston Co. USA.
Heldt, H. W., & Piechulla, B., (2004). *Plant Biochemistry* (pp. 279–280). Academic Press.
Hess, J. L., & Tolbert, N. E., (1967). Glycolate pathway in algae. *Plant Physiology, 42*(3), 371–379.
Hoagland, C. R., & Arnon, D. I., (1950). The solution culture method for growing plants without soil. *California Agric. Exp. Cir.*, 347. Berkeley, California.
Jaworsky, E. G., (1971). Nitrate reductase assay in intact plant tissue. *Biochemical and Biophysical Research Communications, 43*, 1274–1279.
Jha, P., Ali, A., & Raghuram, N., (2007). Nitrate-induction of nitrate reductase and its inhibition by nitrite and ammonium ions in *Spirulina platensis. Physiology and Molecular Biology of Plants, 13*(2), 163–167.
Karadge, B. A., (1981). *Physiological Studies in Succulents*. A Ph. D. Thesis submitted to Shivaji University Kolhapur.
Kaskar, C., Fernandez, J. A., Ochoa, J., Ninirola, D., Conesa, E., & Tuzel, Y., (2009). Agronomic behaviour and oxalate and nitrate content of different purslane cultivars (*Portulaca oleracea*) grown in a hydroponic floating system. *Acta Horticulturae, 807*, 521–526.
Kayashima, T., & Katayama, T., (2002). Oxalic acid is available as a natural antioxidant in some systems. *Biochimica et Biophysica Acta (BBA)-General Subjects, 1573*(1), 1–3.
Kennedy, R. A., (1977). The effects of NaCl, polyethyleneglycol and naturally-induced water stress on photosynthetic products, photosynthetic rates, and CO_2 compensation points in C_4 plants. *Zeitschrift für Pflanzenphysiologie, 83*(1), 11–24.
Koch, K., & Kennedy, R. A., (1980). Characteristics of crassulacean acid metabolism in the succulent C_4 dicot, *Portulaca oleracea* L. *Plant Physiology, 65*(2), 193–197.
Kolhoff, I. M., & Noponen, G. E., (1933). Diphenylamine sulfonic acid as a reagent for the colorimetric determination of nitrate. *Journal of American Chemical Society, 55*, 1148–1153.
Kramer, P. J., & Boyer, J. S., (1995). *Water Relations of Plants and Soils*. Academic Press, San Diego.
Lara, M. V., Drincovich, M. F., & Andreo, C. S., (2004). Induction of a crassulacean acid-like metabolism in the C4 succulent plant, *Portulaca oleracea* L.: Study of enzymes involved in carbon fixation and carbohydrate metabolism. *Plant and Cell Physiology, 45*(5), 618–626.
Lowry, O. H., Rosenbrough, N. J., Furr, A. L., & Randall, R. J., (1951). Protein measurement with folin phenol reagent. *Journal of Biological Chemistry, 193*, 262–263.
Ma, Q., Turner, D. W., Levy, D., & Cowling, W. A., (2004). Solute accumulation and osmotic adjustment in leaves of Brassica oilseeds in response to soil water deficit. *Crop and Pasture Science, 55*(9), 939–945.
Naik, V. V., Patil, N. S., Aparadh, V. T., & Karadge, B. A., (2014). Methodology in determination of oxalic acid in plant tissue: A comparative approach. *Journal of Global Trends in Pharmaceutical Sciences, 5*(2), 1662–1672.
Osmond, C. B., Austin, M. P., Bilings, J. A., Boyer, J. S., Dacey, W. H., Noberl, P. S., Smith, S. D., & Winner, W. E., (1987). Stress physiology and the distribution of plants. *Bioscience 37*(1), 38–48.
Purczeld, P., Chon, C. J., Portis, A. R., Heldt, H. W., & Heber, U., (1978). The mechanism of the control of carbon fixation by the pH in the chloroplast stroma. Studies with nitrite-mediated proton transfer across the envelope. *Biochimica et Biophysica Acta (BBA)-Bioenergetics, 501*(3), 488–498.

Rizhsky, L., Liang, H. J., & Mittler, R., (2002). The combined effect of drought stress and heat shock on gene expression in tobacco. *Plant Physiology, 130*, 1143–1151.
Schutzki, R. E., & Cregg, B., (2007). Abiotic plant disorders symptoms, signs and solutions: A diagnostic guide to problem solving. *Extension Bulletin, E-2996*, 1–16.
Voznesenskaya, E. V., Koteyeva, N. K., Edwards, G. E., & Ocampo, G., (2010). Revealing diversity in structural and biochemical forms of C_4 photosynthesis and a C_3-C_4 intermediate in genus *Portulaca* L. (*Portulacaceae*). *Journal of Experimental Botany, 61*(13), 3647–3662.
Xu, H. W., Ji, X. M., He, Z. H., Shi, W. P., Zhu, G. H., Niu, J. K., Li, B. S., & Peng, X. X., (2006). Oxalate accumulation and regulation is independent of glycolate oxidase in rice leaves. *Journal of Experimental Botany, 57*(9), 1899–1908.

Index

β

β-aminobutyric acid, 53
β-carotene, 223
β-carotenoid, 138
β-oxidation, 64, 223, 225, 226, 234

γ

γ-amino butyric acid, 62
γ-lactone dehydrogenase, 225

A

Abelmoschus esculentus L., 12
Anthocephalus indicus, 100, 102
Azadirachta indica, 100, 117, 121, 123–125, 127–132, 134, 136–139, 141–143, 145
Abiotic stress, 1–3, 17, 19, 26, 27, 34, 56, 61, 63, 150, 153, 164, 203, 204, 206, 211, 215, 226–228, 237, 250, 251, 254, 267
  drought, 3
  heat, 8
  metals, 14
  salt, 12
Abitulon indicum, 102
Abscisic acid (ABA), 7–9, 12, 13, 16, 19, 26, 36, 64, 67, 68, 151, 152, 162, 163, 169, 170, 226, 228, 233
  independent genes, 162
  responsive element-binding proteins (AREB), 162
Acacia
  arabica, 93, 101
  decurrens, 15
Acclimatization, 50, 51
Accumulation of,
  pollutants, 91, 110
  polyphenols, 252
  sludge, 77
  stress metabolite, 169
  vital solutes, 168
Achnatherum sibiricum, 5
Acidic pollutants, 141
Acidosis, 62
Active
  defense genes, 263
  phytomonitor, 106
Acyrthosiphon pisum, 266
Adaptation process, 51
Adverse environmental conditions, 204, 236, 287
Aegiceros corniculatum, 38
Aerenchyma, 62
Aerobic
  bacterial degraders, 74
  conditions, 77
  organisms, 27, 234
Agricultural
  production, 278
  water consumption, 150
Agrochemical markets, 215
Air
  conditioners, 108
  pollutants, 86, 87, 92, 117, 142
  pollution, 97, 108
    tolerance index (APTI), 88, 92, 99, 100, 102, 104, 106, 110, 118, 119, 122, 137, 140–143, 145
  quality
    deterioration, 143
    index (AQI), 93–95, 110
    management, 144
    monitoring stations, 110
Albumin-globulins, 4
Alfalfa, 101
  mosaic virus (AMV), 264
Alkaloids, 1, 17, 27, 28, 165, 204, 205
Alkoxy radical (RO•), 62, 63, 220
Allium cepa, 79
Alpine, 2
Alternaria, 5, 15, 19
  alternata, 17
Alternative oxidase (AOX), 225

Aluminum salts, 77
Alvespimycin, 19
Amaranthus, 140
Ambient air quality monitoring stations (AAQMS), 94, 95
Amendment (organic manures), 155
Amino acid degradation, 120
Ammonium transporters, 54
Ammopiptanthus mongolicus, 231
Amphan, 185, 186, 193, 196, 197
Anabaena
  subcylindrical, 80
  variabilis, 78
Anaerobic
  azo-reductase, 79
  environment, 77
  process, 79
Anatomical parameters, 132
Andrographis paniculate, 157
Angiosperms, 2, 213
Anoxia, 61, 62
Anthocyanin, 205, 210, 211, 252, 253, 256, 257, 264
Anthraquinone, 74
  dyes, 75
Anthropogenic, 45, 55, 56, 61, 91, 97, 191, 196
  stresses, 45, 55, 91
  threats, 56
Anthropological disturbances, 193
Antiarthritic, 1, 17
Antibiotic activity, 3
Antidiabetic, 1, 17
Antifungal compounds, 212
Antiherbivore activity, 210
Anti-inflammatory
  activity, 17
  agents, 1
Anti-microbial, 210, 214
  compounds, 251
Anti-nutritional
  enzymes, 266
  factor, 289
Antioxidant, 34, 49, 53, 67, 119, 139, 169, 207, 219–222, 233, 234, 237, 250, 251, 263, 267
  capacity, 253
  enzymes, 13, 16, 28, 53, 54, 220, 232, 237, 251, 255
  potential, phenolics, 249
Antioxidative enzyme, 27, 152, 169
Anti-proliferative activity, 3
Antispasmodic, 26
Aphids, 3
Aphrodisiac, 26
Apocynaceae, 103, 106
Apoplast, 226
  enzymes, 65
Aquaporin, 162, 164
  proteins, 163
Aquatic
  bodies, 75, 76, 80
  pollution, 75, 76
Arabidopsis, 7, 8, 211, 227, 256, 257
  plants, 211
  thaliana, 7, 8, 257
Arabinose, 5
Arabis hirsuta, 15
Arachis hypogaea, 158, 213
Arenaria serpyllifolia, 15
Arizona fescue (AF), 5, 19
Aromatic amines, 77, 79
Artemisia absinthium, 18
Ascorbate
  glutathione (ASC-GSH), 6, 235
  peroxidase (APX), 3, 6, 8, 13, 53, 67, 221, 225, 232, 234–236, 255
Ascorbic acid (AA), 8, 9, 67, 88, 92, 99, 117–122, 126, 132, 137, 139–142, 145, 207, 236
Aspartic acid, 168
Aspergillus, 8, 9, 13, 16, 19
  aculeatus, 13
  japonicus EuR-26, 9
  violaceofuscus, 8
Asteraceae, 103, 106
Atmospheric, 38, 65, 87, 95, 98, 110, 120, 140, 164, 190, 191
  deposition, 87, 95, 110
  pollutants, 87, 98
  pollution, 87
  resistance, 120
Aurobassium pullunts, 6
Auto-exhaust pollution, 92
Autogenic succession, 188
Automobile exhaust, 117, 138
Autotrophs, 91

Auxin
biosynthetic genes, 256
related cytochrome P-450, 7
Avicennia sp., 15
Azo
dyes, 73–75, 77, 79, 80
reductases, 80

B

Bacillus licheniformis, 264
Bruguiera gymnorrhiza, 189, 255
Bacteria pathogens, 232
Barley yellow dwarf virus-PAV (BYDV-PAV), 265
Barringtonia asiatica, 105
Bauhinia purpurea L., 104
Bean yellow mosaic virus infection, 232
Beneficial microbes, 56
Benthic fauna, 190
Benzene, 144, 249, 251
Bermuda grass, 13, 27, 34
Beta-amino butyric acid, 261
Betaine, 3, 168
Bioaccumulation, 14, 77
Bioaccumulators, 91, 92
Biochemical
alterations, 152
changes, 62, 118, 141
characters, 92
defense mechanisms, 250, 265
enzymes, 265
markers, 120
parameters, 92, 118, 121, 122, 132, 135–137, 140–142
process, 99
Biochemistry, 149, 150
Biodegradation, 78
Bio-diversity, 188
Sundarbans, 195
Bio-filters, 95
Biogeochemical cycle, 91
Bio-geographical zones, 186
Bioindicators, 91, 121
Biological
diversity, 55
productivity, 190
Biomass productivity, 192
Biomonitoring, 141
Bioremediation, 81
Biosensors, 91
Biosorption, 14, 74, 77–81
Biosynthesis, 7, 27, 52, 153, 211, 215, 234, 235, 253, 256–259, 289
phenolic compounds, 256
Biotic stress, 12, 203, 220, 227, 231, 232, 237, 250, 267
phenolics, 259
microbial pathogens, 260
pests phenolics, 265
Black biopolymer melanin, 14
Bleaching process, 75
Blue
carbon storage, 197
green algae, 74
Boerhaavia diffusa, 140
Boreal forest, 2
Botrytis cinerea, 18, 213
Bougainvillea
species, 106
spectabilis, 97, 104–106
Willd., 104–106
Brassica
campestris L., 16
crops, 254
napus, 230, 286
Brassinolide, 50
Brassinosteroid, 47, 165
Bryophyllum pinnatum, 103, 105, 106
Bryophytes, 92
Buffering, 119

C

Cassia
angustifolia, 35
fistula, 100, 102, 104, 139
siamia, 139
Ceratocystis polonica, 214
Clethra barbinervis, 14
Colletotrichum capsici, 260, 261
Cadmium, 212, 230, 255
toxicity, 230
Caftaric acid, 205, 210
Cajanus cajan, 67, 231
Cakile maritima, 228
Calcineurin B-like protein 1, 7
Calcium
mobilization, 233
oxalate, 228, 288

Calluna vulgaris, 53
Calmodulin-binding activator 2, 8
Calotropis
  gigantea, 97, 102
  gigantean, 98
  gigantia, 104
Calvin cycle, 161, 224, 286
Canna indica L., 104
Canopy architecture, 158
Capsicum annuum, 4, 17
Carbohydrate
  metabolism, 157
    proteins, 52
  production, 54
Carbon, 3, 14, 33, 74, 77, 79, 80, 87, 95, 108, 109, 120, 142, 152, 155, 162, 163, 185, 186, 191–194, 196, 197, 210, 214, 220, 228, 279
  dioxide fixation, 74, 142
  sequestration, 191
Carbonate, 15, 63
Carboxylation efficiency, 229
Carcinogenic, 16, 74, 76
Carminative, 26
Carotenoid, 13, 28, 117, 120, 121, 126, 137–139, 141, 207, 221, 223
  content, 6, 13, 126
  pigment, 138, 142
Caryophyllaceae, 6
Catalase (CAT), 5–9, 12, 13, 16, 25, 28, 30, 34, 36–40, 67, 92, 120, 152, 207, 221, 232, 234, 235, 253, 255, 264, 265
  activity, 6, 12, 16, 30, 38, 92
Caterpillar salivary glands, 232
Catharanthus, 32, 34
  roseaus, 259
  roseus, 154
Cation-anion balance, 27, 34
Ceiba pentandra, 104
Cell
  elongation, 153
  membrane, 162
    elasticity, 151
  mortality, 207
  multiplication, 207
  organelles, 63, 65, 232
  permeability, 119
  specific manner, 233
  wall, 64, 227, 235, 260
    elasticity, 167
    linked polyamine oxidase, 226
Cellobiose dehydrogenase, 78, 80
Cellulases, 213
Cephalandra, 140
Cessation (enzyme activity), 76
C-glycosylflavone, 259
Chaetomium, 4, 5, 19
  globosum, 4, 5
Chalcone, 205, 206, 210
  isomerase, 206, 211
Chamomile
  cultivation, 255
  plants, 255
Change detection analysis, 194
Chemical
  essential organic compounds, 74
  interaction, 265
  stability, 74
  tissue analysis, 91
  toughness, 213
Chenopodium
  album L., 16
  quinoa, 6, 231
Chickling pea plant, 156
Chili seedlings induced pathogenesis-related protein, 234
Chlorine, 75
Chlorogenic acid, 13, 211, 212, 254, 257, 261, 264, 266
Chlorophyll, 3, 6, 9, 12–16, 18, 28, 49, 50, 53, 63, 92, 117–122, 125, 132, 137–139, 141, 142, 145, 149, 152, 153, 157, 160, 161, 223, 255, 256, 258
  fluorescence, 6, 50, 161
    kinetic parameters, 50
  pigments, 119, 142, 161
  protein, 49
Chloroplast, 34, 38, 50, 54, 99, 119, 137, 138, 140, 141, 160, 161, 164, 205, 207, 220, 223–225, 230, 232, 234–236
  location, 287
  MDHAR expression, 236
Chlorosis, 12, 91, 93
  dwarf genes, 7
Chromophore groups, 75
Chromosomal abnormalities, 76

Chrysomyxa rhododendri, 261
Cinnamic acid, 206, 210, 215, 255
Cinnamyl alcohol dehydrogenase, 255
Cistus clusii, 257
Citrus, 93
  limon, 104
  reticulata, 6
  species, 101
Cladosporium, 5, 15
  cladosporioides, 5
Climate
  change, 149, 150, 153, 194, 196, 197
  smart crop production practices, 2
  threats, 196
Chlorine (Cl), 26, 75
Coastal communities, 193
Cochliobolus lunatus, 12, 13
Cocoa swollen shoot virus disease (CSSVD), 265
Codiaeum variegatum, 104
Coffea arabica, 229
Cold
  acclimated, 52
    leaves, 52
  regulated proteins, 51
  stress, 26, 56, 231
Colletotrichum
  circinans, 212
  tracutatum, 260
Colobanthus quitensis, 6
Colonization, 3, 6, 8, 14, 18
  induced ubiquitin-mediated protein degradation, 14
Commelina benghalensis L., 93
Compatible osmolytes, 168
Complex
  aromatic molecular structures, 76
  unsaturated organic compounds, 74
Condensed tannins, 212
Conifers oaks, 93
Conservation of,
  natural resources, 196
  nutrients, 190
Conventional
  methods, 73, 77
  modes, 73, 77
  physicochemical monitors, 118, 143
Cooling diuretic, 26
Copper, 53, 95, 208, 209, 211, 212, 231, 234, 255
Cost-effective alternative method, 74
Cotton leaf curl Burewala virus (CLCuBuV), 265
Coumaric acid, 210
Coumarins, 205, 210
Crassulaceae, 103, 106
Crataegus, 257
  laevigata, 257
Crop protection, 3
Cross-pollination, 55
Crotalaria retusa, 34
Cryoprotectants, 52
Cucumber
  mosaic virus, 263
  sativus, 231
Curvularia
  affinis, 15
  lunata, 9
  protuberata, 4
Cuticular waxes, 101
Cyanide sensitive respiratory pathway, 27
Cyanobacteria, 73, 74, 78–81
Cytokinin, 7, 163
Cytosol, 6, 34, 54, 205, 226, 234–236, 252
Cytotoxic, 16, 28

D

Dalbergia sissoo, 100, 102
Daldinia eschscholtzii, 9
Delonix regia, 108, 117, 118, 121, 123–125, 127–133, 136, 137, 139, 141–143
Drypetes roxburghii, 100, 102
D-arabino-1,4-lactone oxidase, 162
Decolorization
  process, 77
  textile effluents, 78
Defense
  mechanism, 34, 138, 168, 221, 227, 232, 233, 255, 257, 258, 263, 265
  related enzymes, 233, 234, 259–262
Deforestation, 56, 186, 191
Deglycosylated, 214
Degradation of,
  membrane fluidity, 231
  textile dyes, 77
Dehydrated, 36, 67, 162, 164, 167, 168

avoidance, 167
leaf mesophyll cells, 3
responsive element-binding proteins (DREB), 8, 152, 162
stresses, 67
Dehydrins, 51, 52, 152, 163
Dehydroascorbate reductase (DHAR), 6, 221, 234, 236
Delphinidin, 211
Demethylation (aromatic rings), 214
Denaturation, 137, 164, 229
Dendranthema grandiflora, 266
De-nova synthesis, 120
Depsipeptides, 1, 17
Desaturation, 52
Deschampsia antarctica, 6, 66
Desert plant-associated endophytic fungi, 8
Desiccation, 152
Detoxification, 119, 168, 207, 219, 220, 228, 232, 234, 250, 286, 288
Dibenzylbutane, 213
Dicumarol, 210
Didymella applanata, 261
Differentially expressed genes (DEGs), 49
Diffusivity, 120
Digitalis lanata, 212
Dihydrokaempferol, 211
Dihydroquercetin, 211
Diketopiperazines, 19
Dimeric compounds, 214
Diospyros, 93
species, 101
Disease-causing microbes, 212
Diurnal tidal, 190
Diversification (carbon stock), 192
Diversity (microbial pathogens), 259
Dodonaea viscosa, 32
Dose-dependent manner, 254
Dothideomycetes, 5, 6
Double-layered cheese-cloth, 29
Drought, 4, 8, 49, 149–169, 228–230, 256, 257, 277
adaptation processes, 168
avoidance, 165–169
escapers, 165
heat stress, 56
induced ABA synthesis, 233
related symptoms, 151
resistance, 149, 165, 166
resistance mechanisms, 169
resistance response, 165
drought avoidance, 166
drought escape, 165
drought tolerance, 167
responsive genes, 162
stress, 3–8, 17, 18, 47, 49, 149–170, 211, 229, 233, 256, 257, 289
condition, 154, 165
tolerance, 162, 167, 168
nature, 277, 289
varieties, 256
Dry biomass, 155, 156
Duncan multiple range test, 288
Duranta plumieri Jacq., 105
Dust
absorbing force, 119
capturing capacities, 97, 99, 103, 104
fall index (DFI), 103–106
interception
accumulation capacity, 101
capacity, 96
pollution, 96, 98
retention indices (DRI), 102–106
D-viniferin, 214
Dye, 73–81
concentration, 75, 78
containing textile effluents, 74
Dysmenorrheal, 26
Dyspepsia, 26
Dysphania ambrosioides, 15
Dysuria, 26

E

Ecological
availabilities, 104
indicators, 91
Economic ornamental woody plants, 55
Eco-sensitive zones, 193
Eczema, 26
Edaphic factors, 45
Efficient traffic management measures, 143
Electrochemistry, 77
Electrolyte leakage, 8, 13, 50, 229–231
Electron
transfer processes, 36
transport
chain (ETC), 34, 64, 223, 225, 226, 258
mechanisms, 149
withdrawing groups, 79

Electrostatic affinity, 78
Elymus dahuricus, 17
Embolism, 167
Embryogenesis, 52, 56, 164, 170
Encroachment, 100, 194
Endophyte, 9, 10, 13, 15, 53, 54
  colonized plants, 12
  fungi, 2, 17, 18
  microfungi, 2
  microorganisms, 45
  streptomyces, 56
Endoplasmic reticulum (ER), 205, 207, 223, 224, 227
Endosymbiont, 3
Engine suspension systems, 95
Entomopathogenic activity, 3
Environmental
  adversities, 203
  fluctuations, 203
  hazards, 73
  protection, 100
  services, 187
  stress, 36, 45, 47, 120, 219–222, 224, 227, 230, 233–237, 249–252, 256
    condition, 234, 236, 249
Enzymatic, 28, 30, 50, 52, 67, 74, 77, 78, 80, 81, 207, 219, 224, 225, 234, 250, 258, 282, 284, 286
  antioxidant production, 234
  approach, 74, 78, 80, 81
  defense systems, 28
  inhibition, 207, 221
  methodology, 77
Ephemeral local perennial plants, 166
Epichloë
  coenophiala, 6
  sinica, 6
  typhina, 8
Epicuticular
  lignin, 101
  wax (plants), 140
Epidemiological studies, 144
Ericaceae, 46
Ericoid mycorrhizal (ERM), 53, 56
Erwinia amylovora, 232
Erysiphe necator, 213
Erythrina indica, 108
Estimation of
  nitrate, 282
  nitrite, 282
Estuarine
  plants, 2
  water, 189, 190
Ethylene, 8, 26, 62, 168, 235
  related Feronia, 7
Eukaryotes, 77
Euphorbiaceae, 103, 106, 189
Evaporation processes, 150
Evapotranspiration, 108
Evergreen perennials, 2
Excitation energy, 36
Excited carbonyl (RO), 63
Exogenous H_2O_2 induced gibberellic acid signaling, 233
Exophiala sp., 15
Experimental methods-materials, 28
  enzyme
    proteins, 30
    related to defense mechanism, 29
  proline, 28
  titratable acid number (TAN), 29
Expressed sequence tags (ESTs), 52
Exserohilum sp, 9
External environmental changes, 226
Extracellular
  chelation, 14
  matrix, 140

F

Ficus
  benghalensis, 93, 100, 102, 105, 108
  benjamina, 96–98, 104–106
    L. var. *nuda*, 103–106
  glomerata, 100, 102, 104
    Roxb., 102, 104
  hispida L.f., 104
  religiosa, 93, 100, 102, 108
Fenton reaction, 224
Fermentative metabolism, 62
Fertilizers, 254
Ferulic acid, 205, 254, 261, 266
Filtration, 77
Flavanols, 205, 261
Flavanones, 205, 210
Flavonoids, 8, 9, 12, 13, 28, 165, 205–208, 210–212, 214, 251–253, 255–259, 262, 266
Flavonols, 205, 210, 257, 259
Flouin-containing oxidase, 36

Fluoride accumulation, 91, 93
FMN-linked oxidoreductases, 8
Foliar
dust deposition, 97, 104
tissues, 92
Food webs, 186
Foreign endophyte colonization, 18
Forest management, 47
Formaldehyde, 144
Formamide, 163
Fossil fuel, 87
combustion, 191
Free oxyradical scavenging system, 119
Freshwater
habitats, 78
nutrients, 192
Fungal
endophyte, 1, 6, 14, 15, 18
research, 19
Fungicidal activity, 18
Furandiones, 1, 17
Fusarium, 4, 13, 14, 19, 262
culmorum, 4
oxysporum, 262
verticillioides, 13

G

Gaeumannomyces cylindrosporous, 18
Glycine max, 8, 9, 12, 13
Galactopyranose, 5
Gas-liquid diffusion, 120
Gastropods, 190
Geldanamycin, 19
Gene expression, 152, 169, 170, 232, 233
Genetic
diversity, 55
mechanism of pathogen, 260
Genomics, 170, 237
Genotoxicity, 16, 229
Gentisic acid, 205
Gibberellic acid, 6, 17, 233
Gliocladium cibotii, 9
Global
climate change, 150, 153
environmental monitoring systems, 144
warming, 150, 190, 191, 193, 195, 197
Glucosidases, 213
Glutamate synthetase, 54
Glutamine synthetase, 28, 54
Glutathione, 3, 6, 7, 9, 13, 16, 28, 53, 67, 120, 207, 221, 235, 254
reductase (GR), 6, 9, 13, 53, 67, 207, 221, 234–236
activity, 5
Gluthathion, 139
Glyceraldehyde 3-phosphate dehydrogenase, 7
Glycerol, 163
Glycolate oxidase, 8, 64, 66, 225, 226, 277, 283, 288, 289
activity, 283
reaction, 225
Glycosides, 1, 17, 140
Glyoxysomes, 28, 34
Gmelina arborea, 102
Gossypium barbadense, 158
Grain-filling stage, 158
Green
belt, 100, 102
development plan, 100, 102
house gasses, 191
Guaiacol peroxidase (GPX), 221, 234, 235
Gymnosperms, 2, 213

H

Helianthus annuus, 8, 9, 158, 210, 215, 257
Heart diseases, 88
Heat
shock proteins (HSPs), 51, 163, 164, 170
tolerance, 50, 53
Heavy metal, 254
toxicity, 53, 250, 254
Heme-containing enzyme, 38, 234, 235
Hemorrhage, 76
Hepatics, 2
Herbaceous
foliage, 107
plants, 2, 278
Herbicide treatment, 27
Herbivores, 203, 205, 210, 212–214
Heritiera fomes, 187, 189, 195
Heterogeneity
groups, 205
lignin, 213
mangrove vegetation, 191
Hibiscus cannabinus L., 264

Hoagland culture medium, 277
Hordeum, 35, 40, 66
vulgare, 66
Hormonal
metabolism, 7
regulations, 55, 56
signaling, 50
Horticulture plants, 45
Hydraulic conductivity, 167
Hydrocarbons, 87, 144
Hydrogen
peroxide (H2O2), 6, 8, 27, 28, 30, 34, 36–38, 41, 50, 53, 62, 64–67, 160, 207, 211, 220, 222–235, 259, 260
sulfide, 88
Hydrolysable tannins, 212
Hydroperoxides, 63
Hydroperoxyl radical, 62, 220
Hydroxybenzoic acids, 210
Hydroxycinnamic, 205, 210, 254, 261
acid, 205, 210, 254, 261
Hydroxyl (OH), 27, 34, 63, 64, 66, 205, 207, 220, 222–224, 227, 230, 232, 234, 249, 256
groups, 207, 249, 251, 256
phosphate amino, 78
radical, 63, 64, 66, 207, 220, 222
Hydroxylation, 210, 214
Hylocomium splendens, 93
Hypericum
brasiliense, 256
perforatum L., 261
pruinatum, 211, 254
triquetrifolium Turra, 261
Hyperosmotic stress, 230
Hypersensitive reaction (HR), 220, 221, 232, 250
Hypnum cupressiforme, 93
Hypobromous acid (HOBr), 63, 68
Hypochlorous acid, 63
Hypoiodous acid (HOI), 63
Hypolipidemic, 1, 17
Hypoxia, 61
Hyptis suaveolens, 104, 105

I

Immobilized cyanobacterial strain, 79
Immune-suppressive, 1, 17
Impaired mitosis cell division, 153
Improved photosynthetic efficiency, 13
In vitro analysis, 52
In vivo activity, 281
Incertae sedis, 5
Incubation medium, 281, 282
Indian Meteorological Departments (IMD), 94
Indigo dyes, 74
Indole
3-pyruvic acid, 54
3-acetic acid, 12, 19, 54
3-butyric acid response 5, 7
acetic acid (IAA), 8, 12, 13, 15, 17, 54, 235, 256, 267
Industrial sources, 88
Industrialization, 2, 53, 87
Infrastructural establishments, 186
Inoculated grass seedlings, 16
Inorganic phosphates, 79
In-silico interaction studies, 19
Integrated
marine-coastal management, 194
urban environmental quality monitoring, 144
Internal ionic media, 120, 138
Intracellular
compartmentalized, 14
osmotic balance, 288
Ion uptake translocation, 157
Ionic strength, 78
Irregular irrigation, 61
Isocoumarines, 1, 17
Isoflavones, 13, 205
Isoflavonoid, 214, 259
phytoalexins, 214
Isoflavonols, 210
Isoquercetine, 211
Ixora coccinea, 97, 98, 105

J

Jasmonic acid, 12, 13, 16, 55, 163

K

Kandelia candel, 15, 189, 255
Kaempferol, 211, 256, 257, 259
Keto acids, 32
Kjeldahl flask, 280

L

Lactuca sativa, 6, 13
Landscape erosion, 91
Lantana camara, 105
Laser beam monitoring, 145
Late embryogenesis abundant (LEA), 52, 152, 164, 170
Leaf
  abscission, 3
  metabolism, 120
  proline concentration, 6
  scorching, 156
  surface temperature, 156
  transpiration, 3
  water potential, 6, 152, 157–159, 169
Lecanora conizaeoides, 92
Leguminous plants, 165
Lentil genotype, 39
Leptosphaera coniothyrium, 261
Lichens, 92
Light-harvesting complex, 258
Lignification, 213
  process, 213
Lignin, 78, 80, 210, 212–214, 227, 234, 249, 251, 252, 260, 262, 265
Lillium spp, 93, 102
Lipid peroxidation (LPO), 3, 6, 13, 28, 66, 119, 161, 207, 223, 224, 229–231, 255
Lipoxygenase (LOX), 28, 227, 230, 231
Long-distance water transportation, 213
Lycopersicon esculentum, 213
Lycopersicum esculantum, 34
Lyngbya majuscule, 78

M

Mangifera indica, 93, 100, 102, 105
Millingtonia hortensis, 100, 102
Magnesium sulfate, 211
Maharashtra Pollution Control Board (MPCB), 86, 94–96, 110
Malachra capitata Linn., 104
Malondialdehyde (MDA), 7–9, 12–14, 16, 49, 50, 53, 54, 66, 67, 120, 229–231, 255, 256
  production, 54, 230, 231
Malonyl co-A, 206
Malus X domestica, 7
Malvidin, 211
Mangrove, 15, 185–190, 195
  ecosystem, 189–192, 196, 197
  swamp forestland, 185
  tree-derived seedlings, 188
  vegetation, 185–187, 192, 194, 195
Mannitol, 5, 168
Marsh vegetation, 190
Matric effect, 153
Matricaria chamomilla, 255
Mean net photosynthetic rate, 54
Mechanical
  injury, 211, 228, 263
  stability, 79
  stress, 6, 220, 227
Medicago sativa L., 91
Medicinal plants, 25–28
Mehler reaction, 64, 224, 235
Melanin-bound heavy metals, 14
Membrane
  networks, 163
  peroxidation, 230
Mentha
  arvensis, 230
  piperita, 262
Mesophyll resistance, 120
Mesophytic bioinvasive plants, 187
Mesorhizobium sp., 261
Meta-analysis, 7, 17
Metabolic
  activity, 278
  impairment, 161
  mechanisms, 169
  pathways, 47, 62, 66
  processes, 62, 63, 65, 66, 205, 226
Metabolism inhibition, 266
Metabolomic, 237
  approaches, 1
Metal transformation, 14
Meteorological conditions, 98, 101
Methyltransferase 1, 8
Meyerozyma caribbean, 12
Microarthropods, 190
Microbial
  biodegradation, 77
  infection, 211, 260
  pathogen, 259, 260
    attack, 250
Microclimatic conditions, 97, 108

Microflora, 61
Micromorphological, 99
  parameters, 132
Micropropagation, 55
Mineral
  deficiencies, 27
  distribution, 254
  nutrition association, 158
Mineralization, 77
Mirabilis jalapa, 13
Miscanthus, 158
Mitochondria, 27, 34, 38, 119, 164, 207, 220, 223–225, 232, 234–236
  electron
    transfer, 230
    transport chain (mtETC), 225
Mitogenactivated protein, 162
Moderate drought-challenged condition, 17
Modern-day adulteration, 75
Molecular
  chaperones, 164, 169
  mechanisms, 151, 170, 215, 256
Mollusks, 190
Monodehydroascorbate reductase (MDHAR), 6, 221, 234, 236
Monolignols, 210, 212
Morpho-anatomical changes, 3
Morphological, 3, 9, 62, 92, 93, 99, 106, 121, 149, 151–154, 156, 165, 279
  parameters, 93, 121
  drought stress, 154
    leaf, 156
    root system, 154
    shoot system, 155
Morpho-physiological, 168
Mosses, 93
Mucous membranes, 76
Multifaceted regulatory system, 151
Multifarious soil-borne plant pathogenic fungi, 210
Multiple
  progeny plants, 55
  signal transduction pathway, 151
Mung bean plants, 229
Municipal Corporation, 95
  of Greater Mumbai (MCGM), 86, 96
Muntingi acalabura L., 105
Muntingia calabura L., 105
Muscodor, 19
Mutagenicity test, 79
Mutualistic, 2
Mycorrhizal fungi, 54

N

N,N-dimethylphenylenediamine (DMPD), 76
N-1-naphthylethylene diamine dihydrochloride, 283
NADH
  dehydrogenase, 27
  dependent reductases, 80
NADPH-dependent oxidases, 226
Nannong
  jingyan, 158
  xuefeng, 158
National planning, 101
Natural
  disasters, 195
  disturbances, 196
  ecosystems, 151, 192
  hazards, 196
N-demethylase, 78, 80
Necrosis
  development, 233
  on leaves, 159
Nematodes, 190, 214, 259, 263
Neocosmospora, 16
  genus, 16
Neosartorya
  hiratsukae, 15
  stramenia, 15
Neotyphodium, 5
  coenophialum, 5
Nerium
  indicum, 140
  odorum, 96–98, 103–106
    Aiton., 96, 103–106
  odorumm, 97
  species, 106
Nesslers reagent, 281
Net photosynthetic rate (PN), 18, 50, 53, 152
Nickel (Ni), 15, 16, 211, 230, 254, 255
Nicotiana
  benthamiana, 5
  tabacum, 91, 92, 232, 236
Nigrospora oryzae, 9
Nilaparvata lugens, 266

Nitrate reductase (NR), 161, 281, 282, 286, 287
Nitrite reductase (NiR), 282, 286, 287
  assay, 282
Nitrogen, 5, 6, 13, 17, 54, 79, 95, 120, 137–139, 203, 255, 277, 279–281, 284, 286, 289, 290
  compounds, 52, 284
  dioxide, 120
  metabolism, 26
  oxides, 120
Nitrophilous nature, 279, 289
Non-acclimated
  rhododendrons, 51
  state, 52
Non-biological stresses, 203
Non-colonized plants, 6
Non-cyclic electron transport, 159, 223
Non-endophytic plants, 5
Non-enzymatic, 28, 207, 250, 258
  antioxidants, 221, 250
Non-flavonoids, 205
Nonfunctional phaeophytin molecules, 141
Noninoculated plants, 17
Non-mycorrhizal conditions, 53
Non-stomatal inhibition, 160
Non-structural carbohydrates, 168
Normal
  carbohydrate metabolism, 62
  metamorphosis process, 213
Nostoc muscorum, 78
Nucleobases, 286
Nutraceutical, 215
Nutrient
  abundance, 188
  metabolism, 152, 157
  mineralization, 249
  transportation, 167
Nutritional deficiency, 229

O

Odor detection threshold, 88
Oligosaccharide chains, 38
Operculina turpethum, 104
Organic
  acids, 5, 12, 14, 27, 33, 34, 168, 279
  carbon dynamics, 191
  chemical coloring compounds, 73
Ornamental plant, 45, 54, 95
  species, 96, 108
Orobanche cernua, 210, 215
Orthrophosphoric acid, 29
Oryza sativa, 13, 14, 230
Oscillatoria, 79
  brevis, 79
  curviceps, 80
Osmolytes, 163, 170
Osmoprotectants, 167
Osmoregulator, 289
Osmotic
  adjustment, 31, 32, 40, 152, 167–169, 286
  modification mechanism, 163
  potential, 151
  pressure, 26, 286
Over-expression (antioxidant components), 237
Oxalate oxidase, 36, 226
Oxalic acid, 29, 226, 277, 279, 283, 288–290
  metabolism, 283, 288
Oxidative
  damage, 120, 211, 221, 237, 265
  metabolisms, 225, 230
  metabolites, 67
  stress, 17, 28, 35, 36, 47, 120, 168, 208, 209, 213, 219–223, 228–231, 234, 236, 237, 249–252, 254, 256, 258, 267
Oxidoreductase flavoprotein, 235
Oxygen
  evolving mechanisms, 76
  hampered mitochondrial, 66
Oxygenates, 87
Ozone, 36, 63, 92, 138, 144
  exposure, 36

P

Paecilomyces
  formosus, 12, 15
    LHL 10, 15
    LWL1, 12
Passiflora edulis, 264
Peltophorum
  inerme, 108, 117, 118, 121, 123–125, 127–133, 136–140, 142, 143
  pterocarpum, 108
Penicillium, 4, 6, 12, 13, 15, 17–19
  brevicompactum, 6, 13
  chrysogenum, 6, 13

citrinum, 6
janthinellum, 15
minioluteum, 6, 12
resedanum, 4, 17
Phialocephala fortinii, 14
Polyalthia longifolia, 93, 100, 102, 117, 118, 121–133, 136–139, 142, 143, 145
Portulaca, 277, 279, 283
oleracea, 277–279, 283–290
Pseudomonas aeruginosa, 261, 263
Psidium
guajava L., 100, 102
spicigera, 100
Palisade tissues, 157
Particulate matter (PM), 87, 94–96
Pathogenic, 262
infection, 226
organisms, 213, 214
Pathogens, 149, 205, 208, 213, 214, 250, 259, 260, 263
P-coumaroyl Co-A, 206
Pedalium, 25, 26, 28, 31–41
murex, 25, 26, 28, 31–40
Pedilanthes
tithymaloides, 96–98
Pedilanthus
tithimyloides, 97
tithymaloides, 103–106
Poit. *variegatus*, 103, 106
Poit., 103–106
Pepper leaf curl virus (PepLCV), 265
Peroxidases (POXs), 6–9, 12, 13, 16, 17, 25, 28, 30, 34, 36, 38–40, 50, 53, 64, 67, 78, 80, 92, 120, 139, 152, 207, 221, 226–228, 232, 235, 253, 255, 260–265, 267
activities, 255, 267
Peroxidation, 76, 119, 207, 221
lipids, 76, 221
membrane lipids, 207
Peroxisomal, 28, 34, 64, 66, 207, 220, 224, 225, 234, 236
matrix, 225
membrane, 64
Peroxy radical (ROO•), 62, 63
Peroxyacetyl nitrate, 91
Perylene derivatives, 1, 17
Pesailotia disease, 56
Pestalotiopsis sydowiana, 56
Pesticides, 254
Pests attack, 265
Petroselinum crispum L., 157
Peyronellaea, 15
Phaseolus
plant, 154
valgaris, 230
vulgaris, 38
Phenol derivatives, 207
Phenolic, 8, 9, 12–14, 28, 140, 204, 205, 214, 215, 221, 249–254, 256, 258, 260–262, 264–267
accumulation, 251, 252, 255, 256, 258, 260–265
acids, 205, 207, 212, 251, 252, 254–257, 259, 261, 266
components, 265
compounds, 3, 14, 53, 67, 119, 140, 141, 203–209, 211, 213, 214, 249–251, 255, 256, 258, 259, 261, 262, 267
enriched cultivars, 267
metabolism, 255, 261
molecules, 257
rich cultivar, 267
secondary metabolites, 260
Phenology, 101
Phenoloxidases, 80
Phenolphthalein, 29
Phenylalanine
ammonia-lyase (PAL), 206, 208, 215, 255, 261–266
lyase, 260
Phenylhydrazone, 284
Phenylpropanoid, 19, 206, 207, 210–212, 214, 250, 251, 256, 260, 262, 264
biosynthetic pathway, 250
pathway, 206, 207, 210–212, 251, 256, 260, 262
Phoma, 15
Phormidium, 79
autumnale, 79
Phosphoglycolate, 66
Phospholipid signaling enzymes, 163
Phosphoric acid, 31
Photo oxidation, 161
Photochemical efficiency, 5, 50, 152, 169
Photoinhibition, 52
photosynthesis, 160

Photo-oxidation, 142
Photo-protectant, 205
Photorespiration, 66, 160, 166, 223, 225, 226, 229
  oxidation, 234
  pathway, 160
  rate, 66
Photosynthesis, 3, 14, 28, 36, 49, 50, 62, 66, 76, 118, 120, 138, 140, 141, 145, 149, 151, 152, 157–162, 166, 169, 194, 223, 226, 231, 254, 258, 278, 279
  activities, 3, 4, 6, 50, 137, 142, 161, 249
  apparatus, 161
  carbon assimilation pathway, 33
  factors, 152
  machinery, 159, 169
  mechanism, 159, 258
  pigments, 92, 161
  products, 162
  rates, 169
Photosystem I (PSI), 5, 50, 66, 223, 224, 258
Phragmites
  australis, 16
  karka, 229
Phyllotaxy, 96, 99, 101, 106
Physical interaction, 265
Physicochemical
  methods, 73
  properties, 119
Physiological, 3, 28, 62, 88, 91, 117–119, 121, 138, 141, 149, 151–154, 156–159, 165–167, 169, 211, 213, 219, 223, 250, 255, 256, 260, 265, 279, 289
  defense mechanisms, 141
  estimations, 91
Phytoalexins, 214
Phytoestrogens, 213
Phytohormones, 7, 12, 36, 50, 55, 56, 152, 163, 264
Phytomonitoring, 91–93, 95–97, 106
Phytomonitors, 85, 96, 105
Phytopathogenic bacteria, 214
Phytophthora cinnamommi, 54
Phytoplankton, 190, 194
Phytoremediation, 3, 96, 104–106
Picea, 213
Pinocembrin, 13
Pinus, 93, 213
  nigra, 140
Piriformospora indica, 6, 7, 13
Pirimiforma indica, 8
Pisonia alba Span., 105
Pisum sativum, 32, 230, 256, 262
Plant
  adaptability, 25, 40
  breeding tactics, 170
  cell membrane, 226
  colonization, 1
  defense, 203, 207, 210, 213, 214, 232, 234, 237, 265, 267
    system, 203, 213, 214
  growth, 92
    promoting bacteria, 260
    promoting rhizobacteria (PGPR), 234, 261–264
    regulator, 256
    survival, 150
  hormones, 62
  immunity, 228
  metabolism, 149, 153, 220, 251
  pathogen interaction, 226, 233, 260
  pest interaction, 232
  production, 119
  stress physiology, 204
  water relations, 163
Plasma membrane, 99, 164, 207, 226, 232, 236
Plasmalemma, 65
Plasmodesmata, 264
Plasmopara viticola, 213
Plectosphaerella, 15
Pleurozium schreberi, 93
Poaceae, 6
Pollutant accumulation, 91
Pollution-exposed plants, 139
Polyamines, 120, 227, 259
Polyathia lotigifolia, 102
Polycentric strategy, 196
Polychaetes, 190
Polyethylene glycol (PEG), 6, 230, 236, 288
Polygalacturonases, 213
Polyketides, 1, 17
Polymeric protein fraction, 4
Polyphenol, 29, 205, 258
  oxidase, 3, 16, 80, 255, 260
Polyunsaturated fatty acids (PUFA), 227
Populus species, 155

Porostereum spadiceum, 12
Portulacaceae, 277, 279
Post
anthesis development, 166
flooding stress, 66
Potassium transport 3, 7
Precautionary activities, 119
Preservative industries, 254
Preussia sp., 15
Primary metabolism, 140
Proanthocyanidins, 257
Productive ecological alliance, 196
Programmed cell death (PCD), 37, 66, 221, 223
Prokaryotes, 77
Proline, 27, 28, 31, 32, 40, 163
glycine, 3
Prosopis juliflora, 102
Protein
phosphatase, 162
phosphorylation, 151, 233
cascades, 151
synthesizing enzyme, 120, 138
Protocatechuic acid, 205
Protonation reaction, 223
Protoplasmic water, 152
PR-protein accumulation, 259
Pseuderanthemum carruthersii, 105
Pseudococcus maritimus, 266
PSII mechanism, 62
Public awareness, 194
Puccinia graminis, 213
Purpureocillium lilacinum A5, 15
Pyrocatechinic, 212

Q

Quaternary ammonium compounds, 168
Quercetine, 211
Quercus petracea, 138
Quinones oxidation, 65

R

Rabbosia eriocalyx, 15
Rhododendron, 45–50, 52–56
apradae, 46
arboreum, 46, 54
catawbiense, 52
cultivars, 50, 53
fortunei, 54
grande, 46, 47
hainanense, 50
hyperythrum, 50
ponticum, 53
pulchrum, 49
species, 52
Raised epidermal cell margins, 101
Reactive oxygen
intermediates (ROI), 63, 164, 170
species (ROS), 3, 5, 6, 8, 17, 25, 27, 28, 34–36, 40, 41, 49, 61–68, 151, 152, 160, 163–165, 168–170, 207, 219–234, 236, 237, 249–252, 254–260, 263, 267
Reaumuria soongorica, 154
Redox
mediator, 79
sensitive proteins, 233
Reductases, 79
Reoxygenation, 67
Repercussions, 195
Reproduction
growth, 155, 165, 169
rate, 47
Resistant-tolerant cultivar, 250
Respiratory
burst oxidase homolog D (RBOHD), 66
enzymes, 26
Reverse transcription-polymerase chain reaction (RT-PCR), 262, 264
Rhamnose, 5
Rhizodermea veluwensis, 14
Rhizopus oryzae, 9
Rhizoscyphus sp., 14, 15
Rhododendron, 53, 56
Ribulose bisphosphate carboxylase (RuBPC), 66
Ricinus communis L., 104
Roegneria kamoji, 6
Role of,
cyanobacteria in remediation of textile dyes and effluent, 78
biosorption, 78
enzymatic method, 79
phenols in stress management, 208
coumarins, 210
flavonoids, 210
lignin-lignans, 212

    - phenylpropanoids, 208
    - phytoalexins, 214
    - stilbenes, 213
    - tannins, 212
- Romaine lettuce, 253
- Root
  - architecture, 162, 167, 256
  - diameter, 6, 154
  - fungal endophytes, 14
- *Rosa longicuspis*, 15
- Rough pubescent sunflower leaves, 99
- RuBP oxygenation, 160

S

- *Samanea saman*, 104, 108, 117, 118, 121, 123–125, 127–132, 134, 136–139, 142, 143
- *Solanum*
  - *lycopersicum*, 7, 13, 67, 230, 231
  - *tuberosum*, 7, 208, 263, 264
- *Swietenia. mahogany*, 117, 118, 121, 123–125, 127–132, 134, 136–143
- Salicylic acid, 7, 13, 50, 210, 254–256
- Salinity, 2, 12–14, 17, 25–28, 31–40, 53, 150, 188, 191, 192, 194–196, 211, 220, 228, 229, 236, 250–254
  - plant metabolism, 27
  - regimes, 192
  - stress, 53, 228
- Salt
  - resistant species, 195
  - stressed okra plants, 13
  - tolerant, 34, 40
    - ability, 54
    - mangrove forests, 187
    - plants, 27
    - species, 26, 34, 40
- *Salvia mirzayanii*, 253
- Sandstorm, 87
- Saskatchewan microbial collection database (SMCD), 3, 5, 19
- Scavenging, 19, 65–67, 119, 163, 165, 221, 257
- Scientific community, 249
- *Sclerotium rolfisii*, 261
- Sea spray flagging, 107
- Secondary
  - messenger, 222, 233
  - metabolites, 8, 47, 56, 170, 203–205, 210, 215, 219, 249–252, 257, 260, 262, 264
  - plant metabolites, 253
  - pollutants, 86
- Sediment
  - accumulation, 194
  - deposition, 189
- Seed germination, 6, 16, 26, 54
- Selective sensitive species, 109
- Self-medication, 267
- Semiquinone ($SQ^{\bullet-}$), 63, 64
- *Serendipita indica*, 16
- *Sesamum indicum* L., 66
- *Sesbania*, 35, 38, 40
- *Setaria italica*, 32
- Shikimic acid pathway, 204, 206
- Short-term flooding, 61
- Siderophores, 14, 19
- Signal transduction, 151, 162, 204, 227, 233, 249, 266
- Simple
  - inorganic compounds, 73, 77
  - organic compounds, 79
- Singlet
  - chlorophyll (1Chl), 64
  - oxygen (O–), 28, 62, 66, 207, 220, 222, 232
- Sites of production
  - ROS, 63
    - apoplast, 64
    - cell wall, 65
    - chemical nature (ROS), 65
    - chloroplast, 63
    - mitochondria, 64
    - peroxisome, 64
    - plasma membrane, 65
- *Sitobiaon avenae* F., 266
- Skin ulceration, 76
- Sleepy grass (SG), 5, 19
- Small interference RNAs (siRNAs), 264
- Sodium, 31, 62, 77, 211, 229, 284
  - glycolate, 284
- Soil
  - conservation practices, 101
  - contamination, 254
  - fertility, 54
  - microbes, 54
  - moisture, 150, 153, 154
  - salinity, 26, 186, 194, 228
- *Solanaceae*, 214
- Solar radiation, 108, 118, 121, 156, 258

Soybean plant stem, 156
Specific
  biochemical activity, 266
  metabolic enzymes, 141
  metabolites, 118
Spermatorrhea, 26
Spirulina platensis, 286
Stachytarpheta indica Vahl, 105
Standard
  deviation, 117, 122
  traditional methods, 77
Stem circumference, 156
Stevia rabaudiana, 157
Stilbenes, 213
Stilbenoids, 251
Stomach cancer, 88
Stomata, 3, 50, 52, 53, 62, 64, 92, 120, 151, 152, 159–161, 163, 167–169, 223, 228, 229, 233, 279
  closure, 151, 159, 163, 228
  conductance, 3, 152, 159–161, 163, 169
Streptomyces, 56
  cellulosae, 264
Stress
  adaptability, 188, 233
    plants, 62
  initiated
    alterations, 169
    modifications, 158
  management, 17, 45, 150, 267
    mechanism, 45
    strategies, 150
  related genes, 162
  relieving metabolites, 1, 17
  responsive gene, 162
  tolerance, 3, 12, 18, 19, 49, 50, 53, 54, 56, 163, 164, 169, 222, 234–237, 251, 257, 264, 278, 279, 289
    genotypes, 279
    mechanism, 278, 289
    transgenic plants, 235
Structural genes, 51
Substantial
  complex food web, 190
  mangrove management, 195
Sucrose
  non-fermenting 1-related protein kinase 2, 7
  phosphate synthase, 161
Sugar alcohols, 5
Sulfanilamide, 281, 282
Sulfur, 93, 95, 137
  dioxide pollution, 93
Sulphonate groups, 80
Sulphosalicylic acid, 29
Sundarbans, 185, 187–189, 191–197
  ecosystem, 194, 196
Superoxide, 3, 6, 8, 25, 27, 28, 34–36, 40, 41, 50, 53, 62, 63, 66, 67, 120, 138, 152, 207, 212, 220–222, 253
  anion, 63, 67, 207, 220
  dismutase (SOD), 3, 6–9, 13, 14, 25, 28, 30, 34–36, 40, 50, 53, 67, 120, 152, 207, 221, 223–226, 232, 234, 253, 255, 264, 265
  radicals, 27, 28
Suspended particulate matter (SPM), 87, 94, 97, 110, 132, 136, 143
Sustainable
  biological indicator, 144
  cultivation, 267
Symplocos paniculata, 15
Symptomless colonizers, 2
Synchronized traffic signals, 143
System (air quality), 94
Systemic
  augmentation, 210
  resistance, 262, 263
Syzygium cumini, 139

T

Tabernaemontana coronaria, 140
Talaromyces omanensis, 6
Tamarindus indica, 93, 102
Terminalia
  arjuna, 100, 102
  bellerica, 103
  catappa L., 104, 108
Trichoderma
  atroviride, 6
  harzianium, 5
  harzianum, 5, 17, 261
Tannins, 205, 212
Tartrazine, 80
Tectona grandis, 102, 159
Telosma mosaic virus (TeMV), 264
Tetranychus urticae, 266

Textile
dyeing industries, 74
dyes, 73, 74, 76–78, 80, 81
effluent, 73–75, 77, 79, 81
fibers, 75
industries, 73–75, 77, 78, 80
wastewater, 77, 80
Theobroma cacao, 213
Thermal
aggregation of proteins, 164
power plants, 87
Thespesia
populnea, 103, 108
populnia, 93
Thicker palisade tissue, 157
Thiobarbituric acid reactive substances (TBARS), 6, 67, 68, 209, 229, 231
Thuja occidentallis, 93, 102
Thylakoid membrane, 50, 137, 138, 161
Thymine dimerization, 258
Tissue
culture, 55
damage, 165
dehydration, 167
Tithonia
diverisfolia, 96
rotundifolia, 103, 105, 106
Titratable acid number (TAN), 25, 27–29, 32–34, 40, 41
Tobacco mosaic virus, 263, 264
Toluene chromatophore, 29
Tomato cultivar, 257, 262, 264
Total dry weight (TDW), 17, 19
Toxic oxygen radicals, 120
Toxification, 254
Trachyspermum ammi sprague, 105
Traditional treatments, 77
Traffic
management, 143
movement, 143
Trametes hirsuta, 16
Transcription factors (TFs), 8, 50, 151, 162, 164, 170, 233
Transgenic tobacco, 14
Translocation processes, 155
Transmembrane transportation, 62
Transpiration physiognomies, 153
Treatment (chili seedlings), 233
Trema orientalis, 105
Tribulus, 25, 26, 28, 31–41
terrestris, 25, 26, 28, 31–41
Trichome, 99, 101, 104–106, 157, 168
density, 92, 104
Tridax procumbens, 256
Trigonella plant, 139
Triplet chlorophyll (^{3}Chl), 64
Triticum
aestivum, 7, 16, 17, 66, 288, 290
turgidum L., 5
Tuberculosis, 88
Turgor pressure, 156–159, 161, 163, 278
Tyrosinase, 78, 80
Tyrosine ammonia-lyase enzyme, 210

U

Ubiquinone-cytochrome b region, 27
Ubiquitin-mediated degradation, 14
Ultraviolet radiations, 250
United States Environmental Protection Agency (USEPA), 88
Urban
air quality management, 118, 143
ecosystem, 107
environment, 96, 118
quality, 144
industrial areas, 88, 92
regeneration, 100
vegetation, 95, 99, 196
Urbanization, 87, 186
Urea, 163, 284
Ureidoglycine aminohydrolase, 8
Urinary discharges, 26
Urtica urens L., 91
Urticaceae, 103, 106
UV radiation, 36, 47, 258
stress, 258
UV-B radiation, 258, 259
UV-tolerant rice cultivar, 259
UV-VIS spectrophotometer, 281–283

V

Vaccinium
corymbosum, 163
macrocarpon, 53, 213
Vachellia farnesiana, 16

Vanillic acid, 205, 210
Vascular bundles, 157
Vegetation
  growth, 157, 169
  sink for dust, 98
    dust pollution studies, 98
    green belt, 99
    objectives, 100
    studies on plants, 102
Vehicle
  emissions, 95
  pollution load, 143
  prohibition, 143
Venation, 101, 106
Venturia inaequalis, 261
Verticillium, 15
Vesicle calculi, 26
Vicia faba, 137–139, 232
Vigna
  mungo, 40
  radiata, 53, 230
  radiate, 32
  unguiculata, 32
Violaxanthin, 138
Virus-plant interactions, 264
Vitis vinifera, 7, 213
Volatile organic compounds (VOC), 87, 110
Volcanic emissions, 87

W

Wastewater treatment plants, 75
Water
  consumption, 153, 156
  deficit condition, 159
  deprivation-related gene upregulation, 7
  excess availability, 3
  holding capacity, 188, 229
  logging, 61
    situation, 61
    stress adaptive characteristics, 62
    stress, 61–63, 65–67
    susceptible plants, 67
    treatment, 66
  retention, 157
  scarcity, 3, 150, 156, 162, 165, 167, 168, 170, 203, 211, 278, 286
  stress, 5, 6, 31, 47, 151–160, 162, 164, 165, 170, 277–279, 284, 286–290
  transparency, 74
Waxy deposition, 101, 106
Weather forecasting, 94
Westiellopsis prolifica, 79, 81
Wildlife sanctuaries, 187
Wind
  driven processes, 87
  erosion, 101
Wound-induced plant cells, 228

X

Xanthine oxidase (XOD), 64, 68, 225
Xanthomonas campestris pv. *malvacearum*, 227
Xanthones, 1, 17
Xanthophyll, 13
Xanthoria parietina, 92
Xeric climate, 2
Xeromorphic
  characteristics, 167
  characters, 156
Xerophytes, 279
Xylella fastidiosa, 263
Xylitol, 5

Y

Yarrowia lipolytica, 13
Yellowing leaves, 156

Z

Zea mays, 6, 7, 18, 66, 211, 266
Zinc-finger proteins, 162
Ziziphus
  jujuba Mill., 104, 105
  jujube, 93, 97, 98
Zizypus jujuba, 102